METAL PROMOTED SELECTIVITY IN ORGANIC SYNTHESIS

Catalysis by Metal Complexes

VOLUME 12

Editors:

R. UGO, *University of Milan, Milan, Italy*
B. R. JAMES, *The University of British Columbia, Vancouver, Canada*

Advisory Board:

The titles published in this series are listed at the end of this volume.

METAL PROMOTED SELECTIVITY IN ORGANIC SYNTHESIS

Edited by

A.F. NOELS, M. GRAZIANI*, AND A.J. HUBERT

Laboratoire de Synthèse Organique et de Catalyse
Université de Liège
Belgium

*Dipartimento di Scienze Chimiche
Università degli Studi di Trieste
Italy

KLUWER ACADEMIC PUBLISHERS
DORDRECHT / BOSTON / LONDON

Library of Congress Cataloging-in-Publication Data

```
Metal promoted selectivity in organic synthesis / edited by A.F.
  Noels, M. Graziani, and A.J. Hubert.
      p.   cm. -- (Catalysis by metal complexes ; v. 12)
    Includes bibliographical references and index.
    ISBN 0-7923-1184-1 (HB : printed on acid free paper)
    1. Catalysis. 2. Organic compounds--Synthesis.
  3. Stereochemistry.   I. Noels, A. F.   II. Graziani, M.
  III. Hubert, A. J.   IV. Series.
  QD505.M455   1991
  547.2--dc20                                        91-11398
```

ISBN 0–7923–1184–1

Published by Kluwer Academic Publishers,
P.O. Box 17, 3300 AA Dordrecht, The Netherlands.

Kluwer Academic Publishers incorporates
the publishing programmes of
Martinus Nijhoff, Dr W. Junk, D. Reidel and MTP Press.

Sold and distributed in the U.S.A. and Canada
by Kluwer Academic Publishers,
101 Philip Drive, Norwell, MA 02061, U.S.A.

In all other countries, sold and distributed
by Kluwer Academic Publishers Group,
P.O. Box 322, 3300 AH Dordrecht, The Netherlands.

Printed on acid-free paper

Printed in The Netherlands

TABLE OF CONTENTS

PREFACE

The demand for selective organic reactions is growing more acute everyday. Indeed, greater product selectivity has an important impact on energy and resource utilization, in terms of reduced process energy requirements for product separation and purification, in terms of low-value by-products, and in terms of environmental acceptance and compatibility. Moreover, more and more chemicals, especially pharmaceuticals, have to be sold in an optically active form.

The search for selectivity constitutes a tremendous challenge for the chemists. In the last two decades, homogeneous transition metal based catalysis has emerged as one of the most promising tools for obtaining selectivity.

In connection with developments in this area, this book contains updated and expanded versions of most of the lectures presented at a Comett course held in Trieste (Italy) in 1989 and sponsored by the European Community.

A primary aim is to cultivate a deeper understanding of the parameters that govern the selectivities and stimulate a wider utilization of transition metal based catalysis in organic synthesis. All aspects of selectivity, chemo-, regio-, stereo- and enantioselectivity are considered and illustrated by applications in various fields or organic synthesis. The impact of catalysis in oxydation, reduction, carbonylation reactions, carbene chemistry, in Ni and Pd promoted dimerizations, oligomerizations as well as fonctionalisations is stressed, quite often with special emphasis laid on reaction mechanisms. In this aspect, the last chapter examplifies the interest of high pressure NMR and IR when investigating the nature of reaction intermediates in homogeneous reactions.

The approach of several papers is a very didactic one, stressing the basic principles of homogeneous catalysis and the most significant aspects affecting the course of the reaction. Therefore, the book should be of interest not only to the chemists engaged in organic synthesis and catalysis, but also as an initiation for advanced students.

A. Noels et al. (Eds.), Metal promoted selectivity in organic synthesis, xiii.
© 1991 Kluwer Academic Publishers. Printed in the Netherlands.

W. KEIM

INDUSTRIAL ASPECTS OF SELECTIVITY APPLYING HOMOGENEOUS CATALYSIS

The value of an industrial product is generally determined by: a) cost to manufacture it, b) safety of process and/or product, c) environmental acceptance and compatibility. Regarding the chemical industry, the selectivity of a reaction is often the key feature in obtaining the above targets. This implies that search for selectivity provides a great challenge to the chemical industry necessitating substantial research and development efforts. In aiming at selectivity, catalysis plays an important role. In the last decades, homogeneous transition metal based catalysts have emerged as especially promising in obtaining selectivity.

In this contribution, the following outline will be used to describe specific impacts of homogeneous catalysis on selectivity:
  1. Contributions of homogeneous catalysis to selectivity
  2. Selectivity and homogeneous, industrial processes

1. Contributions of homogeneous catalysis to selectivity

In describing selectivity various classifications are used whereby the most common one differentiates: chemoselectivity, regioselectivity, stereoselectivity, shapeselectivity.

1.1. CHEMOSELECTIVITY

When two or more reaction sites of similar chemical structure exist separately in the substrate, the selection to choose one particular site is called chemoselective. Chemoselectivity allows a control of the different intrinsic reactivity in various bond types. Also differentiation between different functional groups can be obtained. Equations (1)–(3) demonstrate this principle based on hydrogenation.

1

A. F. Noels et al. (Eds.), Metal promoted selectivity in organic synthesis, 1–15.

$$\text{Ph-CH=CH-NO}_2 \xrightarrow[\text{H}_2]{\text{(Ph}_3\text{P)}_3\text{RhCl}} \text{Ph-CH}_2\text{-CH}_2\text{-NO}_2 \tag{1}$$

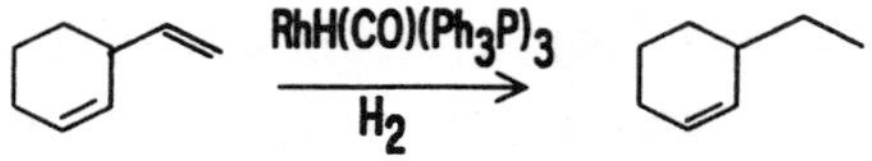

$$\xrightarrow[\text{H}_2]{\text{RhH(CO)(Ph}_3\text{P)}_3} \tag{2}$$

$$\xrightarrow[\text{H}_2]{[\text{Co(CO)}_3(\text{PBu}_3)_2][\text{Co(CO)}_4]} \tag{3}$$

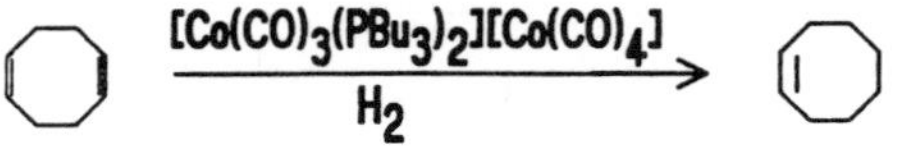

The carbon-carbon double bond can be chemoselectively hydrogenated with the Wilkinson catalyst as shown in Equation (1). Neither the aromatic ring nor the nitro group react. In Equation (2) the vinyl-group in vinylcyclohexene is preferentially hydrogenated [1]. In Equation (3) cyclooctadiene is hydrogenated selectively to cyclooctene.

An other example demonstrating chemoselectivity is shown in Equation (4).

$$\text{cis/trans-octatriene-1,3,7} \xrightarrow{\langle \text{Pd} \rangle} \text{C}_{16}\text{-dimer} + \text{cis-octatriene-1,3,7} \tag{4}$$

Only the trans isomer of octatriene-1,3,7 is dimerized [2]. In this way it is possible to isolate pure cis octatriene-1,3,7.

It can be anticipated that in the future, chemoselectivity will grow in importance. The potential to select specific bond types is inherent to organometallic complexes, which form the backbone of homogeneous catalysis. The better our understanding in this area develops – and it has dramatically in the last years – the greater will be its utilization.

1.2. REGIOSELECTIVITY

Regioselective reactions belong to the most important areas of applications of homogeneous catalysis.

In Equation (5) the regioselective carbon-carbon coupling of propene are presented.

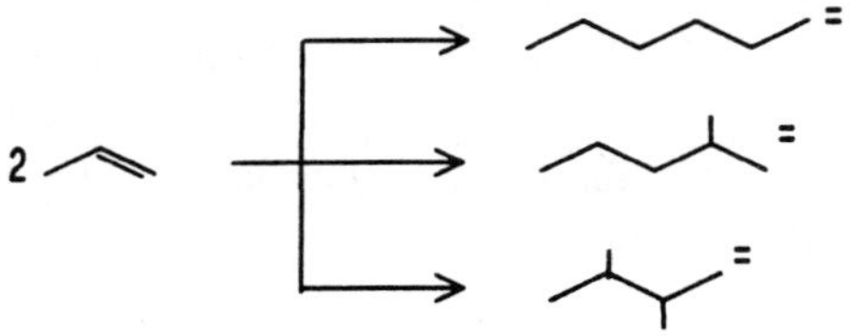

$$\tag{5}$$

Propene can be dimerized in a linear and in a branched fashion. The branched dimers are useful in improving the octane number in gasoline. The linear dimerization of monoolefins is important for platicisers. To meet the requirements of both markets it is important to learn to control the regioselectivity.

One of the most impressive examples of control of regioselectivity (Table I) was first reported by Bogdanovic and Wilke [3].

TABLE I

Dimerization of propene with $C_3H_5NiCl/AlClEt_2$.
Influence of the ligand R_3P on the selectivity. Conditions: 25°C, 10 bar

R_3P	Hexenes [%]	2-Methylpentenes [%]	2,3-Dimethylbutenes [%]
Ph_3P	21,6	73,9	4,5
Me_3P	9,9	80,3	9,8
iPr_2tBuP	0,1	19,0	80,9
$iPrtBu_2P$	1,0	81,5	17,5

With trialkyl- or arylphosphane modified nickel systems, hexenes, 2-methylpentenes, or 2,3-dimethylbutenes can be produced depending upon the substituent on the phosphorus (see Table I). The catalytic activity is exceptionally high, and even at low temperatures good turnover numbers can be obtained. It should be noted, however, that the selectivity leading to 2,3-dimethylbutene in the case of tert-butyldiisopropylphosphane can be increased to 96.4%, at −70°C.

This chemistry is used by Institut Français du Pétrole (IFP). Since the first plant was established in the USA in 1977, IFP has given about 45 licences, which include the following variants: in the Dimersol G process, propene, produced by catalytic cracking, is dimerized to yield C_6-olefins useful in octane-number improvements (about 20 plants). The Dimersol X variant has found application in the dimerization of butene or codimerization of propylene/butene for the preparation of plasticizers. In the Dimersol E process, high-octane gazoline is prepared from FCC-based olefins (FCC = fluid catalytic cracking process for the processing of crude oil).

The complex **1** is highly regioselective in dimerizing α-olefins to predominantly (> 80%) dimers [4].

(Complex 1)

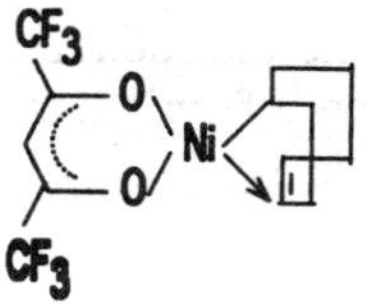

1

Unfortunately, until now, the turn over numbers are too low for an industrial application.

An other very important industrial reaction where regioselectivity provides the key for its application is the hydroformylation of olefins exhibited in Equation (6).

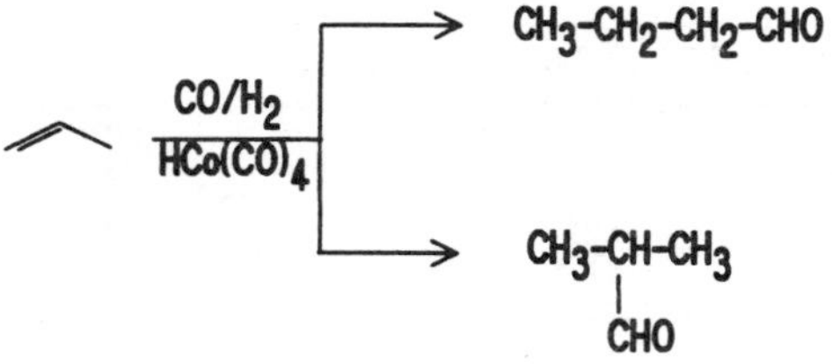

$$CH_3-CH_2-CH_2-CHO$$

$$\underset{\underset{CHO}{|}}{CH_3-CH-CH_3}$$

(6)

Linear and branched products are obtained. Sometimes the linear and sometimes the branched products are wanted. Control can be obtained by phosphine modification. The phosphine modified Shell catalyst gives a higher linear: branched product ratio by exerting mechanistically more control over the regiochemistry of the cobalt-hydride to olefin addition step. Control can also be obtained by the metal applied. This is demonstrated for rhodium in Equation (7), a process carried out in Japan for the synthesis of C_4-oxygenated products.

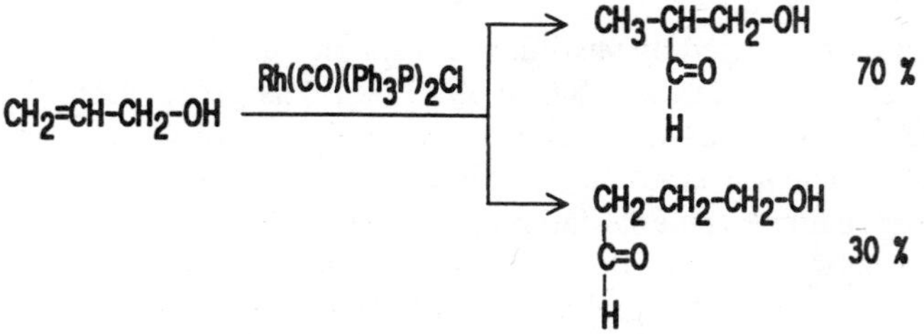

$$CH_2=CH-CH_2-OH \xrightarrow{Rh(CO)(Ph_3P)_2Cl}$$

$$\underset{\underset{H}{|}}{\overset{\overset{CH_3-CH-CH_2-OH}{|}}{C=O}} \qquad 70\ \%$$

$$\underset{\underset{H}{|}}{\overset{\overset{CH_2-CH_2-CH_2-OH}{|}}{C=O}} \qquad 30\ \%$$

(7)

Mechanistically these C–C-linkages discussed involve metal-alkyl additions to olefins, which can occur in a Markovnikov (path a in Equation (8)) or anti-Markovnikov (path b in Equation (8)) addition.

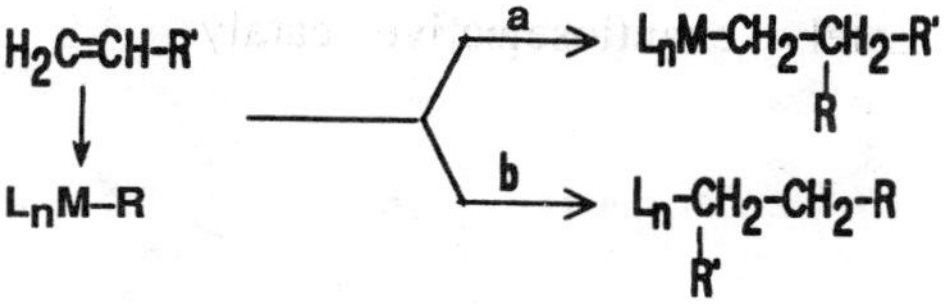

$$H_2C=CH-R' \quad \xrightarrow{\quad\quad} \quad \begin{cases} \xrightarrow{a} & L_nM-CH_2-CH_2-R' \\ & \quad\quad\quad\quad\; R \\[4pt] \xrightarrow{b} & L_n-CH_2-CH_2-R \\ & \quad\quad\quad\quad\; R' \end{cases} \tag{8}$$

$$\downarrow$$
$$L_nM-R$$

The factors which control the direction of addition are very complex; both steric and electronic effects are discussed. In practise it is very difficult to predict the control based on the influence of «ligand tailoring» [5].

Control in regioselectivity is also important in the double hydrocyanation of dienes (Equation (9)) and monoenes practized, for instance, by Du Pont [6] in the adipic acid synthesis.

$$\xrightarrow[\text{Ni}^0\text{-Phosphite}]{\text{HCN}} \quad \underset{CN}{CH_2-CH_2-CH_2-CH_2} \underset{CN}{} \tag{9}$$

This reaction is thought to occur in three stages, resulting in the desired regiochemistry in the first reaction (Equation (10)), both 1,4- and 1,2-addition of HCN occur.

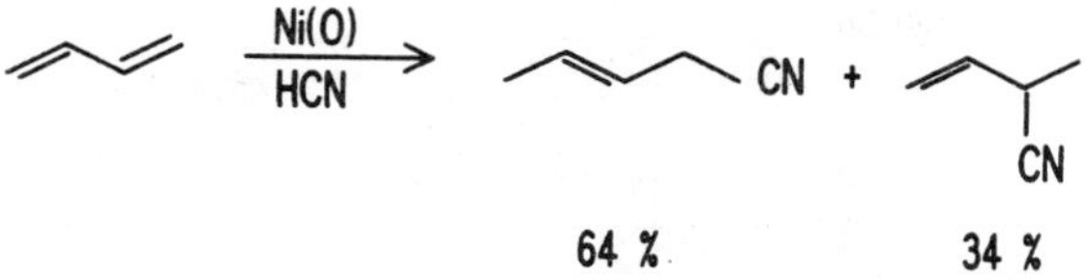

$$\xrightarrow[\text{HCN}]{\text{Ni(0)}} \quad \text{\big(} CN \text{\big)} + \text{\big(} CN \text{\big)} \tag{10}$$

64 %   34 %

Isomerization and additional hydrocyanation of the terminal double bond yields the adiponitrile.

A variety of other addition reactions occurring regioselectively are know: hydroboration, hydroalumination, hydrosilylation, hydrozirconation.

1.3. STEREOSELECTIVITY

Stereoselective syntheses provide challenging targets in industry. Here homogeneous catalysis is of special interest. Figure 1 elucidates the potential based on butadiene oligomerization.

W. Keim

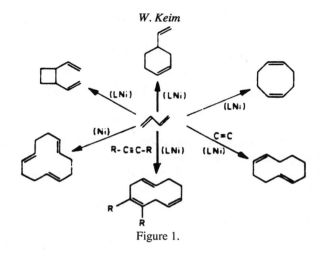

Figure 1.

Various ringsystems possessing different cis and trans double bonds are obtainable in high selectivity. Here, a prudent choice of the ligand or metal (Equation (11)) is essential.

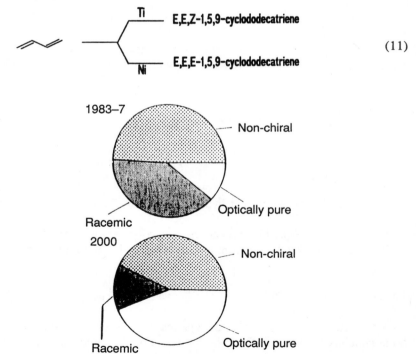

$$\text{(11)}$$

Figure 2.

Lately enantioselective catalysis has attracted substantial industrial interest. Although the number of practised processes is rather small the potential is high. Figure 2 exemplifies the future market for optically active drugs underlining the importance of chirality in this field.

Other important future markets include: agriculturals, polymers, fine chemicals and many others more.

A particular advantage of using metals in enantioselective complex catalysis is that the reaction can be «tuned» by ligand variations: for instance, chiral ligands can be applied. A number of asymmetric reaction catalyzed by transition metals are known [7]: hydrogenation, cyclopropanation, codimerization, allylic alkylation, epoxidation, hydroformylation, hydroesterification, hydrosilylation. In this context the polymerization of propylene must be mentioned [8].

In our laboratory, we are interested in the enantioselective telomerization [9]. Telomerizations of butadiene/formaldehyde (Equation (12)), butadiene/β-dicarbonyl compounds (Equation (13)), butadiene/nitropropane (Equation (14)) could be carried out obtaining ee values up to 50%. In all these reactions chiral ligands were used.

$$(12)$$

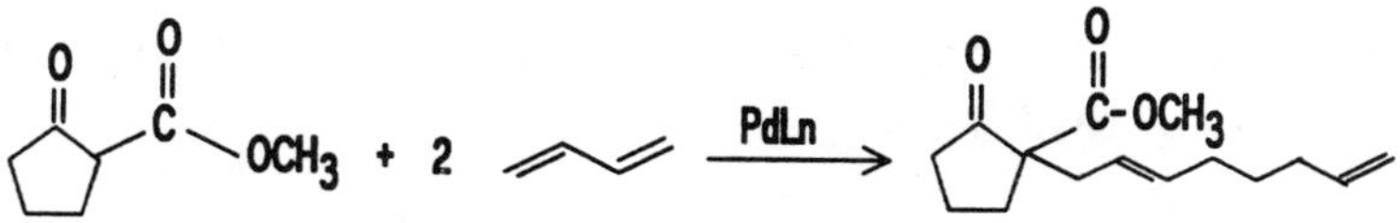

$$(13)$$

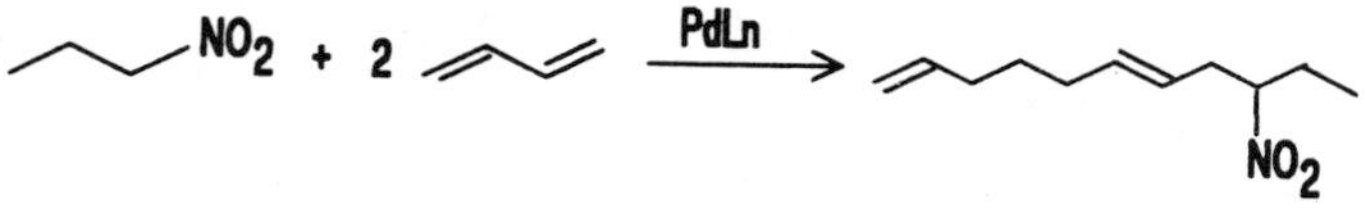

$$(14)$$

It is possible to conduct the telomerization with monoketones or aldehydes via enamines as intermediates as shown in Figure 3 for cyclohexanone. Up to 90% ee values are obtainable.

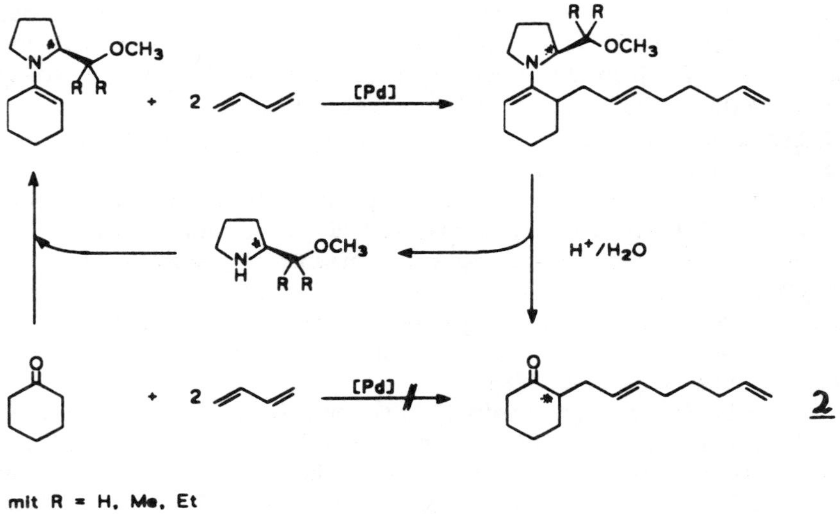

mit R = H, Me, Et

Figure 3.

In the first step, cyclohexanone is converted into the chiral pyrrolidine enamine. Telomerization with butadiene followed by hydrolysis yields the alkyl chain substituted cyclohexanone derivative **2**.

It is noteworthy that in this telomerization optical activity is introduced via the chiral auxiliary (S)-(−)-2-(methoxymethyl) pyrrolidine, which has been very successfully used by Enders [10] and is commercially available [11].

Oxidation of the chiral **2** according to Equation (15) leads to an useful optically active aldehyde [12].

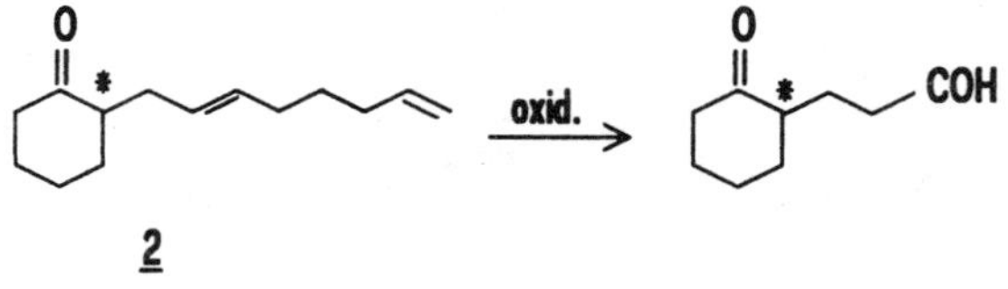

$$(15)$$

2

1.4. SHAPESELECTIVITY

In nature shapeselectivity is often utilized to control chemical syntheses (biocatalysis). Carbonic anhydrase schematically shown in Figure 4 is one well known example.

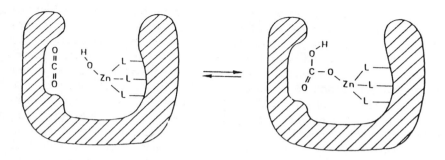

Figure 4.

Carbonic anhydrase is a metalloprotein, whose activity is based on a central zinc atom which is bound to the protein chain by three imidazole rings L. At the fourth free coordination site of the zinc, water is split into hydrogen and hydroxyl ions. Carbon dioxide is stored in a hydrophobic cavity of the protein and inserts into the zinc hydroxy bond.

Shape selectivity is also exerted by chemists when using zeolites. Figure 5 exemplifies three types of shapeselectivity often practiced in the petroleum industry (catalytic cracking, isomerization, alkylation, dewaxing, methanol to gazoline/Mobil MTG).

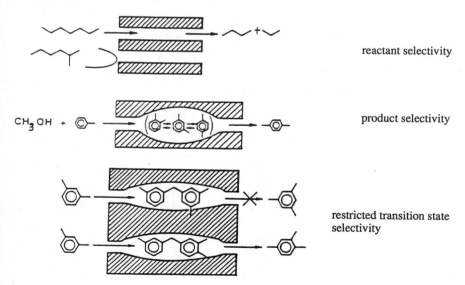

reactant selectivity

product selectivity

restricted transition state selectivity

Figure 5.

In homogeneous, transition metal based chemistry, shapeselectivity by proper chemical manipulation of the reactive site microenvironment is in its infancy. An intriguing example of this type of microengineering was described by Grove and coworkers [13]. A porphyrin-incorporated bilayer membrane was constructed synthetically. Hydroxylation of saturated hydrocarbons and cholesterol was demonstrated. With cholesterol, hydroxylation occurred selectively at the C_{25} tertiary carbon. This selective hydroxylation, in preference to attack at the more reactive Δ^5 double bond, was attributed to the imposed proximity of C_{25} to the active site. It can be speculated that this type of microenvironment design approach could lead into a new area of catalysis bridging biocatalysis and homogeneous catalysis. Biocatalysis applying principles of synthetic organometallic chemistry can be seen on the horizon.

In our laboratory we follow an other approach outlined in Figure 6. An organometallic complex **3** is synthesized within the cavities of a zeolite (ship-in-the-bottle approach).

The complex **3** – if big enough – should not be able to leave the zeolite. Catalysis with **3** offers a possibility to exclude certain molecules from entering. So far, we believe that we could obtain this goal with complex **1**.

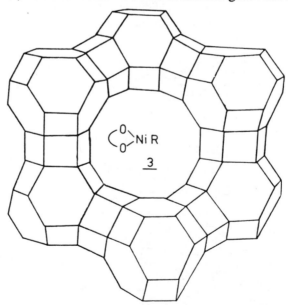

Figure 6.

Complex **1**, to our knownledge, is the only catalyst-precursor-complex which dimerizes α-olefines into highly linear dimers. Indeed, using 1-butene the zeolite/**1** catalyst yielded dimers which are linear up to 70% [15].

Besides the classification of selectivity used above sometimes selectivity is referred to as: substrate selectivity, product selectivity.

Substrate selectivity deals with selective reaction of specific starting materials and is broken up in identity selection, type selection, stereoselection.

Identity selection is often found in enzymatic reactions: only one chemical species is reacted in the presence of hundreds of other similar species.

2. Selectivity and homogeneous, industrial processes

The above examples elucidate the importance homogeneous catalysis plays in industry. In Table II advantages and disadvantages of homogeneous and heterogeneous catalysis are compared.

TABLE II
Advantages and disadvantages of homogeneous catalysis

	homogeneous	heterogeneous
New products	in both possible	
Selectivity	often better, high purity compounds attainable	
Activity	metal atoms are used	only surface atoms are used
Reaction conditions	mild (20–200°C)	severe (> 250°C)
Diffusion	practically unknown	given
Reproducibility	given	often difficult (know-how)
Understanding	in situ spectroscopic methods (NMR, IR, ESR, etc.) can be applied	difficult, often impossible under reaction conditions
Catalyst recycle	full of problems	easy

Much of the selectivity observed with soluble catalysts originates in the process control that is attainable in the liquid phase. Not only are temperature and mixing better controlled that in heterogeneous systems, but also the nature of the active catalytic species is regulated more effectively. Control of catalyst and ligand concentration is better than that attainable on the surface of a solid.

The potential advantages of a homogeneous catalysts can be outlined as follows:

The existence of (at least in theory!) only one type of active centre, implying that every site takes part in the catalytic reaction resulting in higher activity per metal atom.

Most importantly this uniqueness of homogeneity of the active centres can result in very high selectivities.

Due to the relative ease of study of homogeneous catalyst systems and the wealth of analytical techniques available, it is often possible to fully characterize the catalyst (precursor) and to study the reaction mechanism in detail.

Having characterized the system as just described, it is often possible to modify or «tailor» the catalyst to improve activity and selectivity or even to alter the reaction course to afford different products.

The greatest virtue of homogeneous catalysis is selectivity. This is the major reason that has led to the widerspread adaptation of homogeneous catalysis by industry. In this context two pillars of homogeneous catalysis will be focused upon: optimal utilization of starting materials, avoidance of by-products.

2.1. OPTIMAL UTILIZATION OF STARTING MATERIALS

This point is quite obvious. If a reaction proceeds with high selectivity, the loss of starting materials is minimized. For instance, the process to manufacture ethyleneoxide from ethene and oxygen proceeds with an overall-selectivity of ca. 80%. About 20% of the ethene is burned to undesired by-products.

To demonstrate the selectivity advantages of homogeneous catalysis elucidated to above on one process, a commercial process, Shell's «SHOP»-process, is chosen [16].

A block scheme of this process is shown in Figure 7.

The first step comprises the oligomerization of ethene to α-olefins. The α-olefins formed follow a geometric distribution, which can be controlled to a certain degree. The various α-olefins produced in the geometric distribution

SHOP - Process

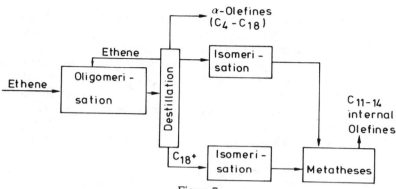

Figure 7.

are difficult to balance on the market. Therefore only the α-olefins which can be sold – mainly the C_8–C_{20} range – are distilled off. The olefins outside of the market range $< C_{10}$ and $> C_{20}$ are isomerized in a second step. The mixture of internal olefins enters metathesis in a third step.

Metathesis converts $< C_{10}$- and $> C_{20}$-olefins into products in the C_{10}–C_{14}-detergent range, which can be taken out via distillation. The combination of isomerization and metathesis paired with distillation and recycle offers an unique technology to tailor the carbon number distribution. This process scheme gives a high degree of flexibility to provide linear olefins of any carbon number and amount desired.

The key to development was the discovery of a selective, novel homogeneous oligomerization catalyst. Olefins of $> 95\%$ α-olefin content and $> 99\%$ linearity are formed. The selectivity to linear olefins is essential for carrying out metathesis. Branched product olefins react very slowly under metathesis conditions leading to a build-up of non-reactive and branched olefins, which have a low market value.

Here the selective formation of linear products demonstrates impressively that selectivity obtained by homogeneous catalysis was the key to develop a new process, which, with approximately a one million ton capacity in 1990, is one of the largest applications of homogeneous catalysis by a transition metal.

The high selectivity to linear α-olefins obtained in the SHOP-process ensures also optimal utilization of the starting material namely ethylene. Practically all products manufactured can be marketed and the by-products are practically zero.

2.2. AVOIDANCE OF BY-PRODUCTS

Processes operating with high selectivities are essential for the future. If a process produces more than one product, there is always the problem of marketing the by-products. A two and more product process is not liked by industry. For instance, the phenol process producing phenol and acetone in a fixed ratio, faces difficulties in selling the two products. Often the by-products possess only heating value. Furthermore, often the by-products must be disposed as waste. It is obvious that selective processes will be favoured by environmental constraints, especially when producing disposable by-products. Therefore, by-products must be minimized in the future.

There is also an increasing demand for purity of chemicals. Products must be pure from a manufacturer point of view. Ethylene for polymerization to polyethylene, for instance, must have a purity of > 99%. Or the high purity of chemicals necessary in the electronic industry should be mentioned here. Purity of products is also a necessity in pharmaceuticals and many other fine-chemicals. Purity can be obtained by separation processes, which, of course, increase the cost and sometimes are even impossible.

In all these examples homogeneous catalysts with their built in selectivity potential can contribute to technical solutions of the chemical industry.

Finally, a self explanatory point in benefits of selectivity is the minimized cost to operate a plant.

*Institut für Technische Chemie und Petrolchemie der
Rheinisch-Westfälischen Technischen
Hochschule Aachen, Worringer Weg, 1,
D-5100 Aachen (Germany)*

References

1. O'Connor, C. and Wilkinson, G. *J. Chem. Soc.* **1968**, *A.*, 2668.
2. Keim, W.; Chung, H. *J. Org. Chem.* **1972**, *37*, 947.
3. Bogdanovic, B. *Adv. Organomet. Chem.* **1979**, *17*, 105–137.
4. Keim, W.; Behr, A.; Kraus, G. *J. Organomet. Chem.* **1983**, *251*, 377.
5. Keim, W. *Angew. Chem. Int. Ed. Engl.* **1990**, *29*, 235–244.
6. Parshall, G.W. *Homogeneous Catalysis*, Wiley-Interscience, New York 1980, 235–244.
7. Bosnich, B. *Topics in Inorganic and Organometallic Stereochemistry*, Vol. 12, Geoffroy, G.L. Ed., John Wiley 1986, p 119.

8. Kaminsky, W. *Catalytic Polymerization of Olefins*, Keii, T.; Soga, K. Ed., Elsevier Amsterdam, 1986, p 293.
9. Keim, W.; Koehnes, A.; Roethel, T.; Enders, D. *J. Org. Chem.* **1990**, *382*, 295.
10. Enders, D.; Kipphardt, H. *Nachr. Chem. Tec. Lab.* **1985**, *33*, Nr 10-992.
11. Merck Schuchard, Frankfurter Str. 250, D-6100 Darmstadt (FRG).
12. Keim, W.; Koehnes, A.; Roethel, T. *J. Organic Chem.* (in preparation).
13. Groves, J.T.; Neumann, R. *J. Am. Chem. Soc.* **1989**, *111*, 2900.
14. Keim, W. *J. Molec. Cat.* **1989**, *52*, 19.
15. Mehlborn, A. unpublished (on going thesis) RWTH Aachen.
16. (a) Keim, W. *Chem.-Ing.-Tech.* **1984**, *56*, 850; (b) Freitas, E.R.; Gum, C.R. *Chem. Eng. Prog.* **1979**, *75*, 73; (c) Behr, A.; Keim, W. *Arabian J. Sci. Eng.* **1984**, *10*, 377; (d) Keim, W.; Kowaldt, F.H.; Goddard, R.; Krüger, C. *Angew. Chem.* **1978**, *90*, 493; (e) *Angew. Chem. Int. Ed. Engl.* **1978**, *17*, 466; (f) Keim, W.; Behr, A.; Gruber, B.; Hoffmann, B.; Kowaldt, F.H.; Kürschner, U.; Limbäcker, B.; Sistig, F.P. *Organometallics* **1986**, *5*, 2356.

I. TKATCHENKO

TO WHICH EXTENT DO PHOSPHANES INDUCE SELECTIVITY IN C–C BOND FORMATION?

1. Introduction

Phosphorus compounds have been prepared in many coordination numbers, *i.e.* from P(V) to P(I). Among them the generic compounds $PR_n(OR')_{3-n}$ are the most numerous and offer a large variety of tuning the properties of organometallic complexes. This variety is even greater when are considered the association of several such moieties connected through a hydrocarbon residue, leading to multidentate ligands. They are the most commonly encountered ancillary ligands associated with organometallic compounds. The number of transition metal complexes containing P(I) to P(V) ligands is extremely large. Virtually all the metals of the periodic table stabilized in various oxidation states by hydride, carbonyl, nitrosyl, alkyl, alkene, alkynes, arenes ... ligands are capable of accomodating phosphoranes, **1**, phosphines, **2**, phosphites, **3**, phosphides, **4**, phosphinidenes, **5**, etc.

$M(PR_3)_4$
2 M =Ni, Pd, Pt

$PhP=W(CO)_5$
5

$Ni[P(OEt)_3]_4$
3

$Mo(PCy_2)_4$
4

1

$(OC)_3Fe \overset{\displaystyle \overset{Ph_2}{P}}{\underset{\displaystyle \underset{Ph_2}{P}}{\longrightarrow}} Fe(CO)_3$

6

This topic has been extensively reviewed [1–6] and also comprises bidentate and polydentate phosphines, bridging ligands (such as μ_2-phosphido, **6**). Moreover, chiral phosphines are known and exhibit chirality either at P (*e.g.* **7**) or in the carbon framework (*e.g.* **8, 9**).

17

A. F. Noels et al. (Eds.), Metal promoted selectivity in organic synthesis, 17–45.
© 1991 Kluwer Academic Publishers. Printed in the Netherlands.

7   8

9aP = PPh$_2$ 9bP = P

(DIPHOL)

The use of P(III) compounds as ancillary ligands on transition metal species active in homogeneous catalysis often provides dramatic or subtle modifications of their activity and/or selectivity in the conversion of unsaturated substrates [7, 8]. Such complexes are mainly used in hydrogenation, oligomerisation, telomerisation, polymerisation, and carbonylation reactions. However, their application to oxidation is restricted to some cases where the phosphanes are not readily oxidized. Moreover, it has been found that under the hydrogenation and hydroformylation conditions; the P–C bond in tertiary phosphines degrades through an «oxidative addition» process [5]. Although it has been pointed out that «this problem seriously decreases the usefulness of phosphines for technological applications of homogeneous catalysis» [9], we will show that phosphines remain so far the best compromise for catalyst tailoring in homogeneous catalysis.

Since different applications of regio-, stereo- and enantioselective reactions involving phosphorus ligands are presented in this book, the scope of this contribution is limited to oligomerisation and hydroformylation reactions. The general properties of phosphorus(III) compounds will be presented first.

2. Structural, physical and bonding properties of phosphanes

Changing substituents on phosphorus ligands can cause marked changes in the behaviour of the free ligands and their transition metal complexes. As noted ty Tolman [2], prior to 1970, this behaviour was practically rationalized in terms of electronic effects. However, the dissociation in solution of NiL$_4$ and PdL$_4$:

$$ML_4 \rightleftharpoons ML_3 + L \rightleftharpoons ML_2 + 2L$$

with the trends:

$$M = Ni, L = P(OEt)_3 < PMe_3 < P(O\text{-}p\text{-}Tol)_3 < P(O\text{-}i\text{-}Pr)_3 < PEt_3 <$$
$$P(O\text{-}o\text{-}Tol)_3 < PMePh_2 \ll PPh_3 \quad [10]$$

$$M = Pd, L = PMe_3 \approx PMe_2Ph \approx PMePh_2 < PEt_3 \approx PBu_3 \approx PPh_3$$
$$< PBz_3 < P(i\text{-}Pr)_3 < PCy_3 < PPh(t\text{-}Bu)_2 \quad [11]$$

show the steric effects of the substituants on the rates of L dissociation.

As sketched in Scheme 1, Tolman defines:

(i) electronic effect changes in molecular properties as the result of transmission along chemical bonds (*i.e.* changing $P(p\text{-}C_6H_4OMe)_3$ to $P(p\text{-}C_6H_4Cl)_3$,

(ii) steric effect changes in molecular properties as the result of (non-bonding) forces between parts of a molecule (*i.e.* changing from $P(p\text{-}C_6H_4Me)_3$ to $P(o\text{-}C_6H_4Me)_3$.

Scheme 1

Electronic and steric effects are intimately related and difficult to separate. Actually, steric effects can have important consequences and *vice versa*. For example, an increase of the angles between the substituents on a P(III) atom will result in a decrease of the percentage of *s* character in the phosphorus lone pair. Also, changing the electronegativity of substituents can affect the geometrical parameters.

2.1. THE STERIC PARAMETER θ: THE CONE ANGLE

Upon bonding a P(III) ligand to a metal centre, the coordination number increases to 4, with tetrahedrally distributed bonds. Since the rotation of a tetrahedron around an apical axis describes a cone, the cone angle was introduced by Tolman to define the bulk of P(III) ligands in quantified terms. The concept was based on a bond length of 2.28 Å. The simplest case of a symmetrical PR_3 ligand was further extended to phosphines of the type $PRR'R''$ (Figure 1).

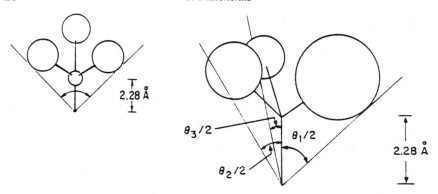

Figure 1. The Cone angle of a symmetrical (left) and non-symmetrical phosphane (right)
(Reproduced with the permission of the author from ref. 2)

2.2. THE ELECTRONIC PARAMETERS

The electronic effects of tertiary phosphanes depend markedly on the nature of the substituents on phosphorus. Several authors have shown that CO stretching frequencies are good indicators of variations in an electronic series of phosphanes. Tolman defined the electronic parameter ν as the frequency of the A_1 carbonyl mode of $Ni(CO)_3L$ in dichloromethane. The additivity of the contributions of substituants, χ_i, shown in equation 1 makes it possible to estimate ν for a variety of ligands $PX_1X_2X_3$ for which it has not been measured.

$$\nu = 2056.1 + \sum_{i=1}^{3} \chi_i \qquad (1)$$

TABLE I
Selected values of $\nu(cm^{-1})^2$

L	$\Delta\nu$	$\Delta\nu$
P(p-Tol)$_3$	2066.7	0.1
P(o-Tol)$_3$	2066.6	
PMe$_3$	2064.1	2.4
PEt$_3$	2061.7	2.5
P(i-Pr)$_3$	2059.2	3.1
P(t-Bu)$_3$	2056.1	

That v is indeed a measure of electronic effects – unaffected by the crowding of $Ni(CO)_3$ by the substituents on phosphorus – is suggested by the similar values observed for $P(p\text{-Tol})_3$ and $P(o\text{-Tol})_3$ (Table I).

2.3. BONDING OF PHOSPHORUS(III) LIGANDS

P(III)-transition metal bonds are essentially covalently coordinated, the P(III) atom providing the electrons. However, this simple picture does not account for all the structural data now available. Three factors are involved in the ligand contribution to the bond: (i) σ-bonding, (ii) π-bonding, and (iii) steric factors which have been discussed above.

2.3.1. Basicity of phosphines and σ-bonding

In providing electrons to the metal centre, phosphanes behave as Lewis bases. The usual measure of phosphane basicity, pK_a (H_2O), is a measure of Brønsted basicity (*i.e.* proton affinity) which is not easily transposed to Lewis basicity. Usually, electron-releasing substituents will increase the electron availability on phosphorus, resulting in a higher pK_a. The increase in the bulkiness of a substituent group on phosphorus will increase the R-P-R angle, which in turn will increase the p character of the lone pair. The strength of the metal-phosphorus σ-bond decreases in the series: $P(t\text{-Bu})_3 >$ $P(OR)_3 \approx PPh_3 > PH_3 > PF_3 > P(OPh)_3$.

2.3.2. π-Bonding

The possibility that π-back-bonding, *i.e.* the interaction of filled metal d-orbitals with vacant d-orbitals on phosphorus, was important in transition-metal complexes has been examined with several physical chemical techniques (IR and NMR spectroscopy, X-ray crystallography, cyclic voltammetry). There are few instances where $d\pi$–$d\pi$ backbonding is required to explain the observed physical properties. Recent theoretical studies show that $d\pi$–$d\pi$ bonding is not important between fluorophosphines and transition metal centres because the phosphorus d-orbitals are too high in energy. The back-bonding is attributed to an interaction between the metal d-orbitals and the σ* orbitals of the phosphine [12].

From various compilations of experimental data, an empirical π-acceptor series can be drawn, and prominent among them are various PX_3 compounds:

$$NO > CO \approx RNC \approx PF_3 > PCl_3 > PCl_2(OR) > PCl_2R > PBr_2R >$$
$$PCl(OR)_2 > PClR_2 > P(OR)_3 > PR_3 \approx SR_2 > RCN > RNH_2 \approx H_2O.$$

2.3.3. Separation of σ- and π-bonding components

Mathematical relationships between reactions parameters (log k, log K) and electronic (pK_a, χ_i values) and steric (cone angles) factors have given poor results, except for some pK_a correlations. These poor correlations may be associated with the failure to separate the steric and electronic components of phosphorus-metal bonding and to separate the electronic contributions into its σ- and π-components. Giering and his co-workers have evaluated quantitatively the relative σ-donicities, π-acidities and π-basicities of phosphorus(III) ligands on the basis of the correlations of the $E_L°$ values of oxidation of η-MeCp(CO)$_2$MnL with the pK_a values of R$_3$PH$^+$ compounds and the symmetric v_{CO} stretching frequencies of Ni(CO)$_3$L as reported by Tolman [13].

On the basis of the correlation of the $E_L°$ with the pK_a values of R$_3$PH$^+$, the ligands can be divided into three classes:

- class I where the ligands are both σ- and π-donors (*i.e.* PR$_3$, with R = Et, Bu, Cy) which suggests an easier oxidation of the complex than is predicted by its pK_a value;
- class II where the ligands are pure σ-donors (*i.e.* PMe$_3$, PR$_{3-x}$Ph$_x$, P(*p*-XC$_6$H$_4$)$_3$, where X = H, Me, OMe) which corresponds to the linear part of the plot of $E_L°$ with pK_a;
- class III where the ligands are both σ-donors and π-acceptors (*i.e.* P(*p*-ClC$_6$H$_4$)$_3$, P(OR)$_3$, with R = Ph, Me, Et, *i*-Pr, R$_3$ = PPh$_2$(OMe)) which indicates a better stabilization of the M(I) state. However, the π-acceptor abilities of class III ligands are small in comparison to those of the strong and π-acids such as NO, CO and *t*-BuNC.

The steric and electronic ligand properties are related to the reaction parameters by the equation:

$$\log k = f_1(\theta) + f_2(pK_a) + f_3(E_{\pi a}) + f_4(E_{\pi b}).$$

The contribution of the last two terms is generally negligible and a graphical method can separate quantitatively the σ-electronic and steric ligand effects. [13] Therefore the ligand θ/pK_a map (Figure 2) is a useful tool for choosing among the phosphanes showing similar properties. It should be also pointed out that the boundaries of the three classes of phosphorus ligands will be dependent on the π-basicity of the metal centres.

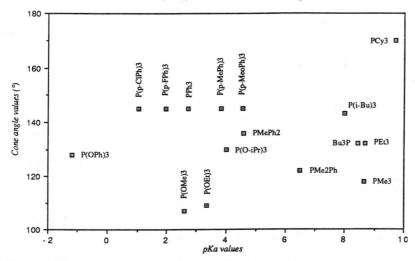

Figure 2. The map cone angle *vs* basicity (pK_a) of selected phosphanes.

3. Oligomerisation of alkenes

The oligomerisation of simple monoenes involves three basic steps: (i) initiation, (ii) propagation, and (iii) termination (Equation (2)).

$$m_T \xrightarrow{\text{(i)}} m_T \begin{array}{c} H \\ R \end{array} \xrightarrow{\text{(ii)}} m_T \diagup\!\!\!\diagdown \begin{array}{c} H \\ R \end{array} \xrightarrow{\text{(iii)}} m_T H \qquad (2)$$

product

Depending on the relative rate of propagation and termination, oligomers (even dimers, trimers) or polymers are obtained and one can expect that ligand effects intervene on each of these elementary steps. The oligomerisation reaction has been extensively reviewed [7]. Both Ziegler-type and soluble late-transition-metal compounds can be used. Only the catalysts of groups 8–10 contain phosphane ligands.

3.1. OLIGOMERISATION OF ETHYLENE

The reaction of ethylene at –20°C under 1 bar pressure with the phosphane-free catalyst prepared from bis(η^3-allylnickel chloride) and ethylaluminium dichloride in chlorobenzene results in the rapid formation of a mixture of ethylene dimers – mainly 2-butenes – with lesser amounts of higher oligomers. This catalytic reaction involves a hydridonickel species **10** stabilized by the Lewis acid which acts as a chelating ligand.

S⋯ ―Cl⋯ ⸗Cl
  Ni⸪   Al⸗
H⸝ ⸜Cl⸝ ⸜R

10

$[-\!\!\!\triangleleft\!\!(\, Ni(cod)\,)]PF_6$

11

The selectivity of dimerisation as well as the degree of oligomerisation can be controlled in a wide range by the addition of phosphines [14]. Table II shows that a gradual decrease in the relative rate constants takes place which is accompanied by a decrease in the activity of the catalyst. The order parallels the increasing steric demands rather than any electronic effect related to the basicity of the phosphine.

TABLE II

Influence of phosphines on the dimerisation and oligomerisation of ethylene[a]

PR_3	C_4 (%)	C_6 (%)	C_8 (%)
PMe_3	>98	1	–
PPh_3	90	10	–
PCy_3	70	25	5
$PEt(t\text{-Bu})_2$	65	25	10
$P(i\text{-Pr})(t\text{-Bu})_2$	25	25	50
$P(t\text{-Bu})_3$[b]			Polyethylene

[a] Catalyst: $(C_3H_5NiBr \cdot PR_3\text{-}Et_3Al_2Cl_3)$; –20°C; 1 bar.
[b] Ni : P = 1 : 4.

The use of cationic complexes leads to a large increase in the selectivity for higher oligomers. [15] The addition of phosphines to **11** leads to an increase in the formation of the kinetic products, *i.e.* 1-butene and 3-methylene pentane (Figure 3). Moreover, there is a strong enhancement of the ramification in the trimers cut going from 70% in the unmodified system to 93% for PBu_3 and PCy_3. Again, these trends can be explained on steric grounds [16].

Terminal olefins in the range C_4–C_{18} having high linearity can be prepared by using catalysts obtained from $Ni(cod)_2$ and various chelating phosphines and phosphites (SHOP process). In fact, three electron-donor ligands present for example in complexes **12-14** are necessary and this could be extended to other more conventional ligands like hfacac in Ni(hfacac) $(\eta^3\text{-}C_8H_{13})$, **15** [17].

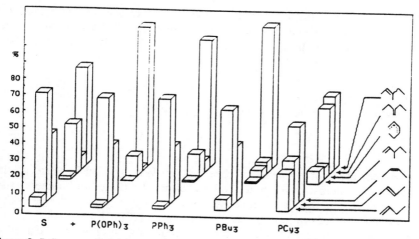

Figure 3. Influence of phosphanes added to **11** on the product distribution in the C_4 and C_6 cuts [16].

12 **13** **14**

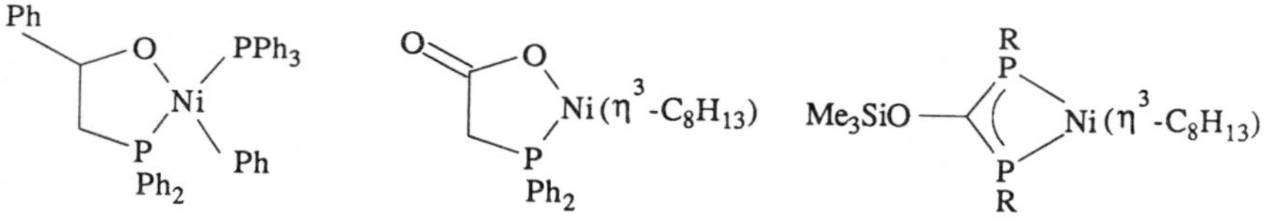

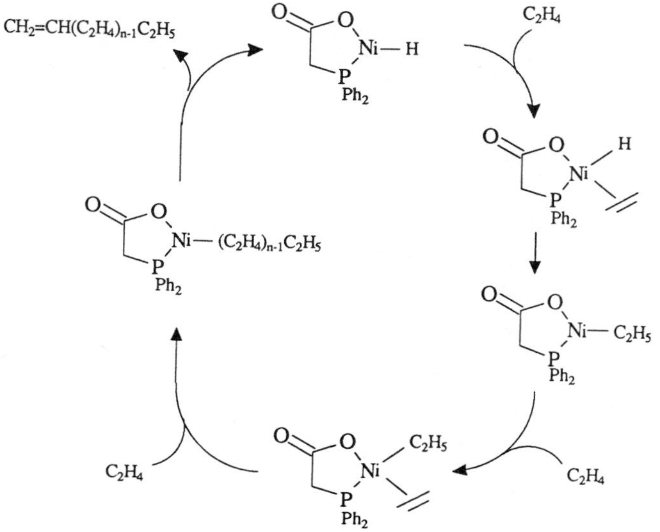

Scheme 2

In all the cases, the catalyst is a hydrido-nickel species. The catalytic reaction follows Michaelis-Menten saturation kinetics. The chain-termination step is the β-hydride elimination (Scheme 2, involving the precursor **13**).

3.2. OLIGOMERISATION OF TERMINAL ALKENES

3.2.1. Propene

The dimerisation of propene has been extensively studied because of the considerable interest of the hydrogenated dimer 2,3-dimethyl butane as fuél additive [18]. The reaction course can be controlled to give methylpentenes, 2,3-dimethylbutenes or hexenes as the main products [14].

The initial hydride migration step can occur to give either a linear alkyl product (Ni $\rightarrow$ C_1) or a branched alkyl product (Ni $\rightarrow$ C_2). Similarly, the alkyl-to-alkene migration, implicit in the next step of the oligomerisation cycle, can also occur in two distinct directions (Scheme 3).

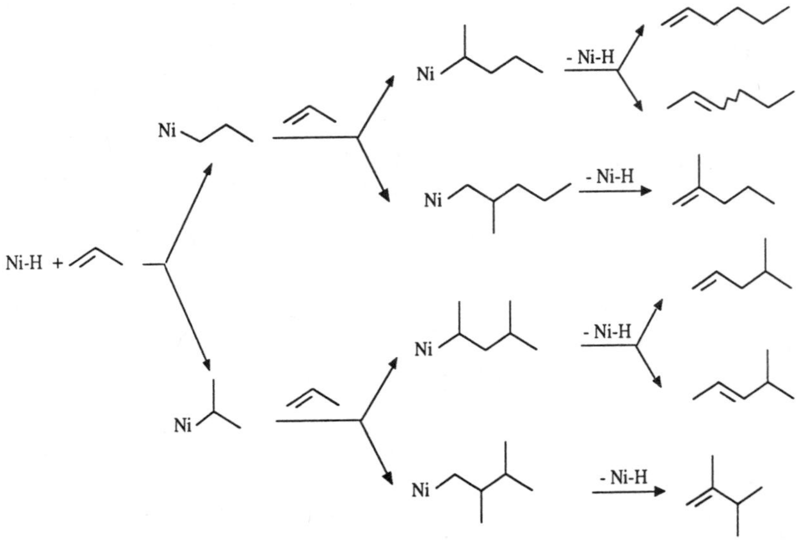

Scheme 3

The use of phosphanes now allows the control of the regioselectivity of the individual reaction steps. Assuming that the reactivity of the nickel-*n*-propyl species towards propene is the same as that of the nickel-*i*-propyl species, and assuming that no isomerisation occurs during the course of the reaction, it is possible to calculate the ratio of Ni → C$_1$ and Ni → C$_2$ addition in both step 1 and step 2 of Scheme 3. The data listed in Table III indicate that, with the exception of the bulky P(t-Bu)$_2$i-Pr, the nature of the tertiary phosphine has little influence on the direction of the first step. This suggests that the polarity of the nickel-hydrogen bond controls the first addition of propene. It is during the second step that the tertiary phosphine ligands exert their steering influence. Increasing the size of L strongly favours formation of the less sterically demanding alkyl (Ni → C$_1$).

Asymmetric catalytic codimerisation of two different alkenes has been observed in several instances [7, 19] and will be presented in another contribution of this book.

TABLE III

Influence of tertiary phosphine ligands on the nickel catalyzed dimerisation of propene

Phosphine	$v(cm^{-1})$	$\theta(°)$	First step Ni → C$_1$:Ni → C$_2$	Second step Ni → C$_1$:Ni → C$_2$
PMe$_3$	2064.1	118	15:85	15:85
PEt$_3$	2061.7	132	17:83	29:71
PBu$_3$	2060.3	132	18:82	34:66
P(i-Pr)$_3$	2059.2	160	20:80	86:14
PCy$_3$	2056.4	170	26:74	83:17
P(t-Bu)$_2$i-Pr	2056.1	175	69:31	98:2

3.2.2. Styrene

Oligomerisation and polymerisation of styrene mediated by transition metal complexes have been reported to be also influenced by added phosphanes [20]. Complex **11** is a good precursor for the oligomerisation of styrene (DP ≈ 13). The characterization in the reaction medium of **16** indicates the intermediacy of a hydrido-nickel species (*i.e.* [HNi(cod)]$^+$), Equation (3), thus affording a good basis for the investigation of the reaction mechanism [21].

$$\left[-\!\!\!\left\langle\!\!\left\langle Ni(cod) \right]PF_6 \ + \ \right. \right. \longrightarrow \ \left[\right. \left\langle Ni(cod) \right]PF_6 \ + \ C_{12}H_{14} \quad (3)$$

16

The 1:1 adducts of **11** with tributyl- and tricyclohexylphosphine afford low DP polymers with a Markov-type distribution [22]. In both cases, ^{13}C NMR spectroscopy clearly indicates the occurrence of Ni $\rightarrow$ C$_2$ coupling. Noteworthy is the high isotactic content (89%) obtained in the case of PCy$_3$ [23]. A terminal C(H)Ph group is observed except for the latter bulky phosphine which leads to a methylene terminal group (Scheme 4).

In that case, the coordination sphere of nickel becomes more crowded, and the increase of interactions with the growing polymer chain determines an inversion of the styrene mode of addition from secondary to primary. Once the primary addition occurs, the η^3-benzylic stabilization is no longer possible and the β-elimination reaction terminates the chain-growth process leaving the 2,3-diphenylbuten-3-yl end group and regenerating the nickel hydride active species. The added effects of an η^3-benzylic coordination of the growing chain and the presence of the basic and bulky tricyclohexylphosphine, giving origin to a very crowded nickel coordination sphere, are believed to be responsible for the high isotactic content and for irregular head-to-head chain transfer through β-hydrogen elimination.

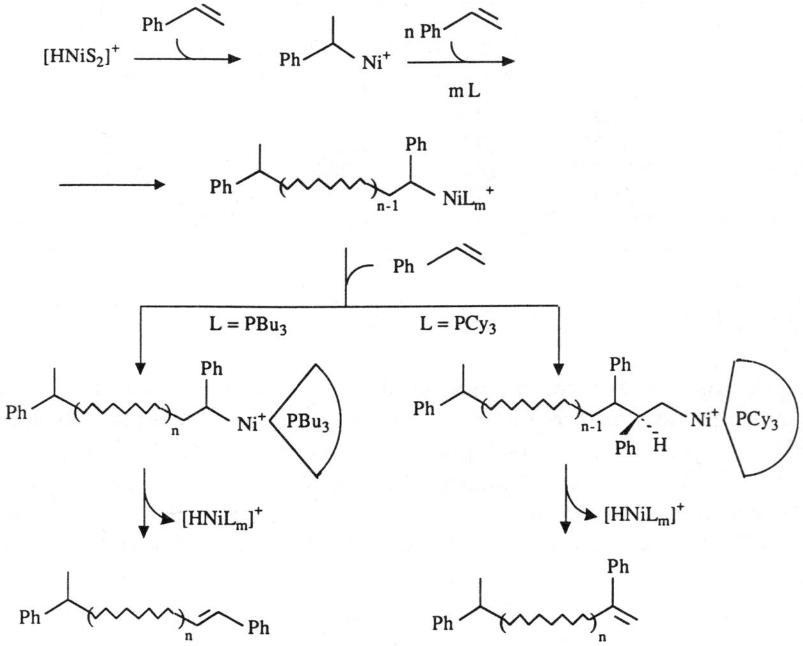

Scheme 4

3.2.3. Dimerisation of acrylates

Dimerisation of functionalized alkenes is worthy mainly in the case where tail-to-tail coupling leading to α,ω compounds could be selectively achieved. This is the case for the dimerisation of alkyl acrylates where only the linear products could be used for the synthesis of methyl adipate through an alternative route avoiding benzene as a starting material (Equation (4)). The introduction of an electron-withdrawing group on the double bond would now reverse the direction of the first addition depicted for propene. This has been achieved again by modifications of the coordination sphere of the nickel centre in **11**. The addition of the basic, non-hindered trimethylphosphine and the replacement of the Lewis acid by a non-coordinating anion such as BF_4^- provides a selective catalyst for the preparation of alkyl hex-2-enoates [19].

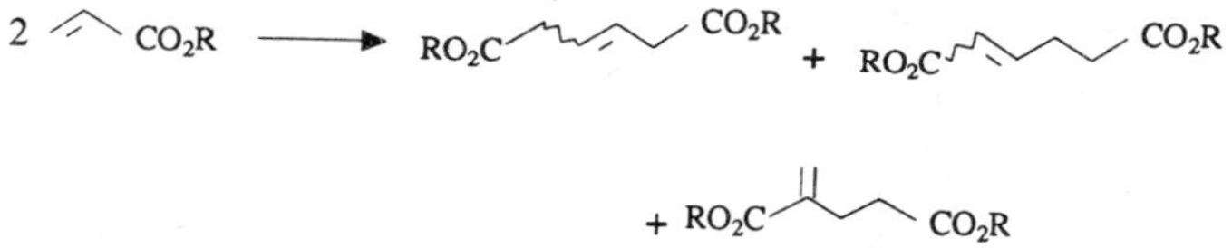

$$2 \quad \diagup\!\!\!\diagdown CO_2R \longrightarrow RO_2C\diagdown\!\!\diagup\!\!\diagdown\!\!\diagup CO_2R + RO_2C\diagdown\!\!\diagup\!\!\diagdown\!\!\diagup CO_2R$$

$$+ RO_2C\diagdown\!\!\diagup\!\!\diagdown CO_2R$$

(4)

Another way for preparing these dimers is to avoid the stepwise mechanism proposed for alkene oligomerisation (Equation 2). A cyclometallation process involving two alkenes has been already suggested in the case of nickel, but requires two vacant sites [24]. This has been achieved in the case of $Pd(acac)_2$. In fact, a catalytic system consisting of $Pd(acac)_2$ and $HBF_4 \cdot OEt_2$ only is selective for the dimerisation of methyl acrylate, but poorly active (1.2 to 2.5 turnovers) [25]. The product resulting from the protonation of $Pd(acac)_2$, $[Pd(acac)(acacH)]BF_4$, **17**, is also active *per se*. However, the activity is still low. The tributylphosphane adduct of complex **18**, $[Pd(acac)(PBu_3)_2]BF_4$, is active for the dimerisation of methyl acrylate, provided that tetrafluoroboric acid is added (turnover frequencies up to $150h^{-1}$, Scheme 5) [26].

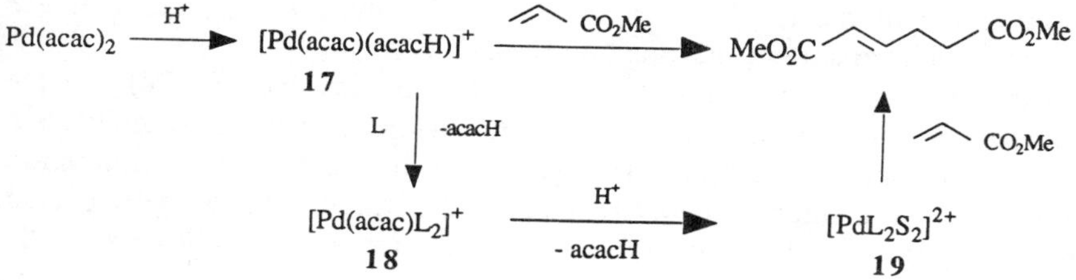

Scheme 5

Reaction of this complex in the presence of nitriles affords dicationic palladium complexes, $[PdS_2(PBu_3)_2]^{2+}$. The succinonitrile complex is fairly active (turnover frequencies up to $20h^{-1}$) and very selective (>95%).

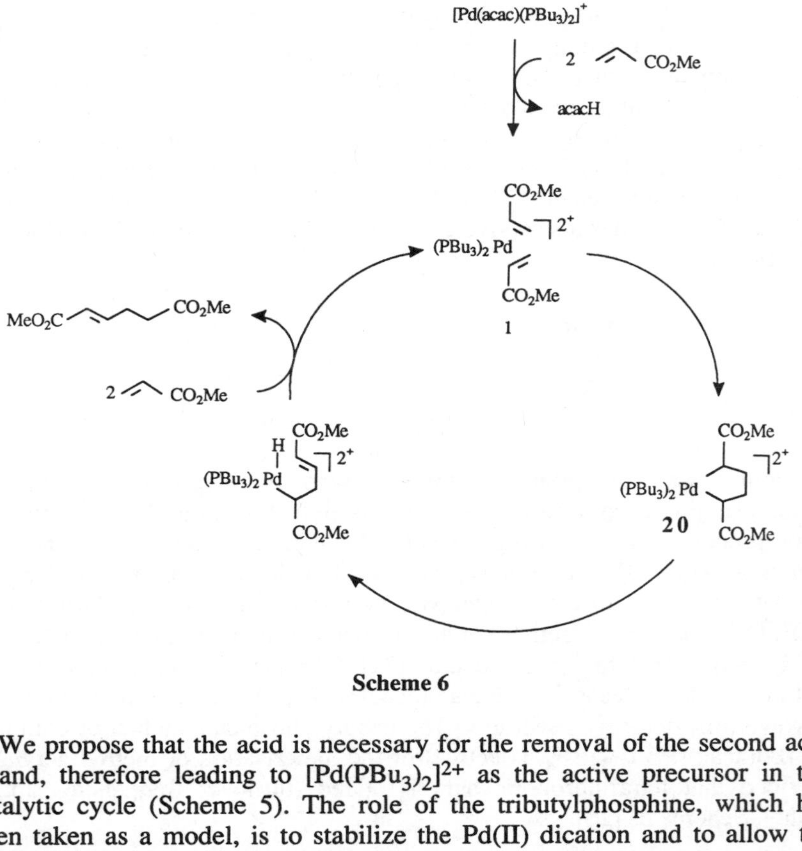

Scheme 6

We propose that the acid is necessary for the removal of the second acac ligand, therefore leading to $[Pd(PBu_3)_2]^{2+}$ as the active precursor in the catalytic cycle (Scheme 5). The role of the tributylphosphine, which has been taken as a model, is to stabilize the Pd(II) dication and to allow the coordination of the acrylate through its double bond by depletion of the electrophilic character of this dication [26].

The catalytic cycle proposed for this reaction involves Pd(IV) species which have little precedent in the literature. Although scheme 6 indicates a possible evolution of a Pd(IV) metallacycle **20**, by a β-hydrogen elimination - reductive elimination tandem process, another route towards the expected products could involve protonolysis of this intermediate followed by β-hydrogen elimination.

It is worth noting that combinations of a palladium complex (*i.e.*, Pd(dba)$_2$, Pd(acac)$_2$), a phosphane and HBF$_4$OEt$_2$ provide efficient catalysts for the selective linear dimerisation of methyl acrylate [27].

4. Oligomerisation of butadiene

The catalytic dimerisation and oligomerisation of 1,3-dienes can lead to a very large number of linear and cyclic products. It provides one of the best examples of the ability of phosphanes to control both rate and product distributions in homogeneous catalysis and shows just how complex catalytic systems can become [7, 28]

4.1. INFLUENCE OF PHOSPHANES ON PRODUCT DISTRIBUTION

In the absence of added phosphorus ligands, nickel complexes converts butadiene into *trans,trans,trans*-1,5,9-cyclododecatriene (*ttt*-CDT), with small amounts (10–20%) of *ttc*-CDT and *tcc*-CDT, 1,5-cyclooctadiene (COD), 4-vinylcyclohexene (VCH) and *cis*-1,2-divinylcyclobutane (DVCB). With the addition of phosphanes, dimers are the main products and the rate of reaction (80°C) can be increased by one magnitude.

Rates and product distributions depend on the L:Ni ratio, as well as ligand type, temperature, and, of course, butadiene conversion. Table IV shows the effect of varying the ligand-to-Ni ratio for PPh$_3$. Optimal formation of COD is observed and is quite general for all the phosphanes examined. Heimbach *et al.* [29] have made extensive studies of cyclooligomer concentrations as a function of L:Ni ratios. By plotting percentages of various products against log(L/Ni), they get «ligand-concentration control maps» (Figure 4). The virtual elimination of CDT between log(L/Ni) = −2 and 0 indicates that the

TABLE IV
Dependence of product distribution on the ratio PPh$_3$/Ni at 80°C

	0:1	1:1	2:1	4:1	8:1
% CDT	87.2	6.0	0.5	0.6	–
% COD		64.0	62.0	56.0	5.0
% VCH	8.2	27.0	36.0	41.0	8.0
% > C$_{12}$	3.6	2.8	1.9	1.5	1.6
Productivity g BD/g Ni	75.0	180.0	185.0	165.0	80.0
Turnover rate hr^{-1}	85.0	200.0	205.0	185.0	90.0

major catalytic trimerization intermediate changes to one which contains one L per Ni. Such intermediates **20, 21** have been characterized when bulky ligands are used [30] The fact that higher L/Ni ratios favour VCH shows that this dimer is formed by an intermediate which should have two ancillary ligands.

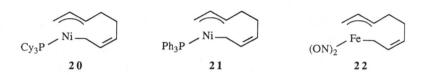

$$Cy_3P \overset{Ni}{\underset{}{}} \qquad Ph_3P \overset{Ni}{\underset{}{}} \qquad (ON)_2 \overset{Fe}{\underset{}{}}$$

      20   **21**   **22**

It is interesting to note that VCH is selectively formed (> 99%) with [Fe(NO)$_2$] as active species, presumably through **22** [31].

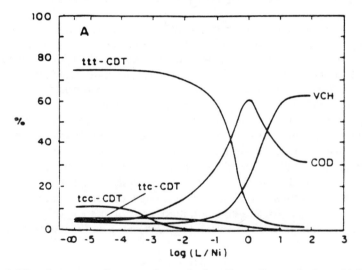

Figure 4. Ligand-concentration control map for butadiene oligomerisation with L = PPh$_3$. (*Taken from ref 29, with the permission of the authors*)

4.2. STRUCTURE – SELECTIVITY RELATIONSHIPS

With bulky, weakly basic ligands such as P(OC$_6$H$_4$-o-Ph)$_3$ (v = 2085 cm^{-1}, θ = 192°), COD is the predominant product of the dimerisation (96%). With a more strongly basic ligand, of roughly comparable size, *e.g.* PCy$_3$

($v = 2056.4$ cm^{-1}, $\theta = 170°$), COD and VCH are produced in about equal amounts (41 and 40%, respectively) and DVCB makes up some 14% of the product mixture. These results have been rationalized on the basis of an equilibrium between the three bis(allyl) species, **23, 26, 27**, which is controlled by electronic and steric properties (Scheme 7). Thus, increasing the basicity, electron-donor properties of the ligand should favour the formation of VCH. This is indeed evidenced by the results obtained with $L = P(OPh)_3$ ($v = 2085.3$ cm^{-1}, 4% yield), $L = PPh_3$ ($v = 2068.9$ cm^{-1}, 27% yield), and $L = PCy_3$ ($v = 2056.4$ cm^{-1}, 40% yield), while the COD yield decreases from 81 to 64, to 41%.

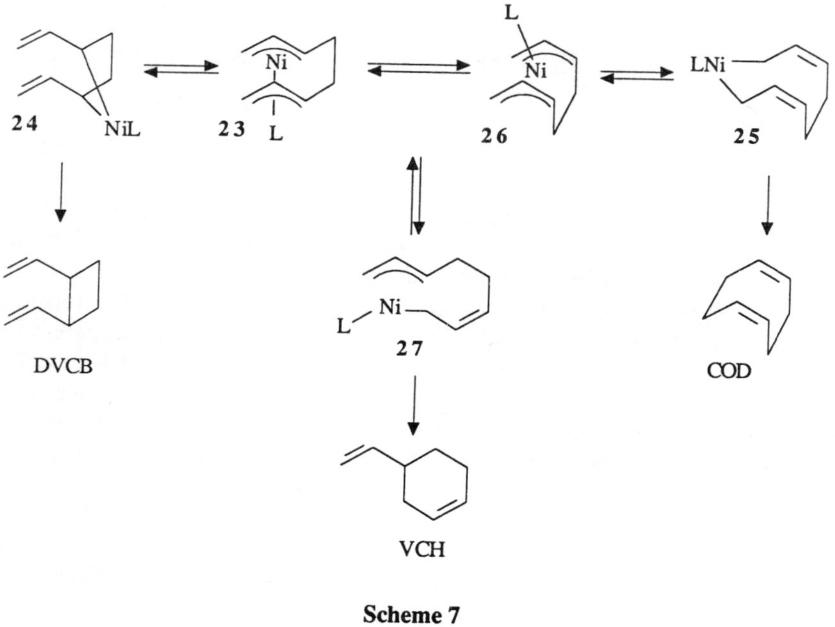

Scheme 7

The steric properties of the ligand also play a role in directing the course of the reactions. With weakly basic ligands, *i.e.* tertiary phosphites, the steric bulk influences the relative concentration of **23, 26** and hence the relative yields of COD and DVCB. Thus, with $P(OC_6H_4\text{-}o\text{-}Ph)_3$ and $P(OPh)_3$ which have similar electronic properties ($v = 2085$ and 2085.3 cm^{-1}, respectively), the relative yields of COD and DVCB are 96%, 81%, and 0.2%, 9.2%, respectively, reflecting the smaller bulk of $P(OPh)_3$ ($\theta = 128°$) with regard to $P(OC_6H_4\text{-}o\text{-}Ph)_3$ ($\theta = 192°$).

4.3. PREPARATION OF HIGHER OLIGOMERS

The formation of linear trimers and tetramers is scarcely reported in the literature [32]. The same is true for the corresponding telomers [33]. Convenient synthesis of the two series of products should provide starting materials for the preparation of C_{12} and C_{16} functional derivatives. In the presence of phosphanes (PBu$_3$, PPh$_3$), complex **28** which was used for the telomerisation of butadiene with alcohols is also operative for the reaction with phenol. In this case, phenoxydodecatrienes are obtained together with phenoxyoctadienes (Scheme 8).

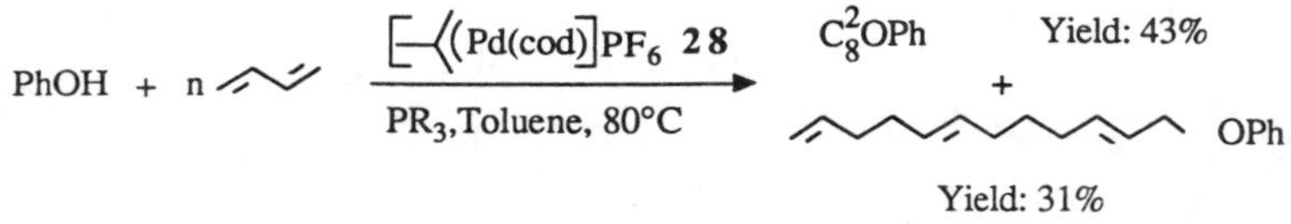

Scheme 8

The C_{12} compounds are in fact obtained by the codimerisation of butadiene and phenoxyoctadienes. Indeed, no co-dimer is formed in the attempted reaction between the half-hydrogenated C_8 telomer and butadiene, therefore suggesting that the C-C coupling reaction involves the terminal double bond of the telomer (Scheme 9) [34]. A cationic hydridopalladium species, stabilized by the phosphanes is proposed as the active species. The crotyl-alkene complex resulting from the reaction of this species with butadiene could react further by an «insertion» process because of the presence of the ancillary phosphane. No reaction occurs in the absence of phosphanes [35].

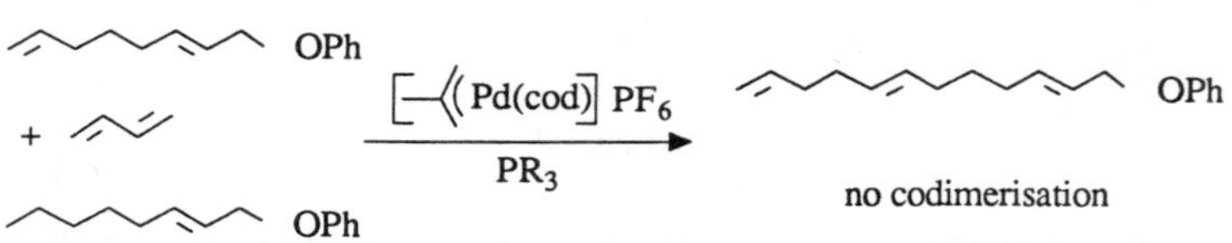

Scheme 9

Another approach to the synthesis of higher oligomers of butadiene is to build up the oligomeric chains on two metallic sites. Reaction of complex **28** with the cyclic aminophosphonites **29** (R = H) provides the cationic complexes **30** which are fairly active in the conversion of butadiene to C_8, C_{12} and C_{16} oligomers. Noteworthy is the formation of large amounts of trimers and tetramers. Interestingly, the C_8 and C_{12} cut consists mainly of octatrienes and hexadecapentaenes whereas the C_{12} fraction also contains branched oligomers.

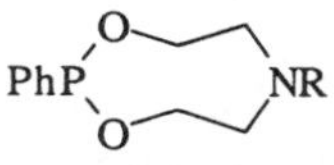

29 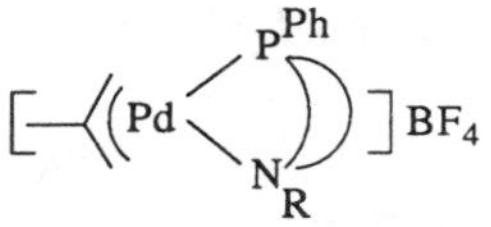   30

This pattern suggests that the C_{16} oligomers derive principally from the C_8 ones through dimerisation and that the C_{12} oligomers are produced by stepwise Markovnikov/anti-Markovnikov «insertions» of butadiene into an (η^3-1-methylallyl-palladium) species. The reported linear trimerisation of butadiene is said to occur on a dinuclear species which contains bridging acetate ligands. It is thus tempting to invoke a bis(allylpalladium) species **31** for the tetramerisation of butadiene. The reaction may arise from the intramolecular coupling of two η^3-1-3-octa-2,7-diene-1-yl ligands through C–1 and C–8′ carbon atoms and further β-hydrogen elimination. As expected, the bis(substituted) aminophosphonite complexes of **29** are much less active and give only C_8 oligomers [36].

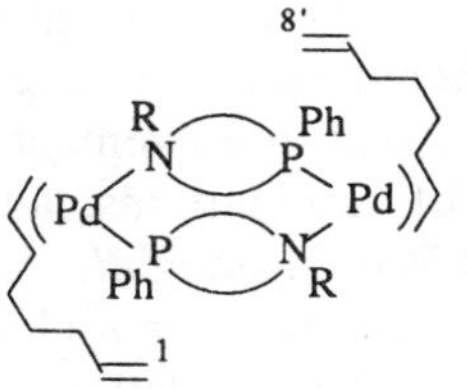

31

5. Hydroformylation of alkenes

Since its discovery by Roelen in 1938, using cobalt catalysts, the hydrofor- mylation reaction, Equation (5), has been thoroughly studied with both cobalt and rhodium catalysts [7, 37].

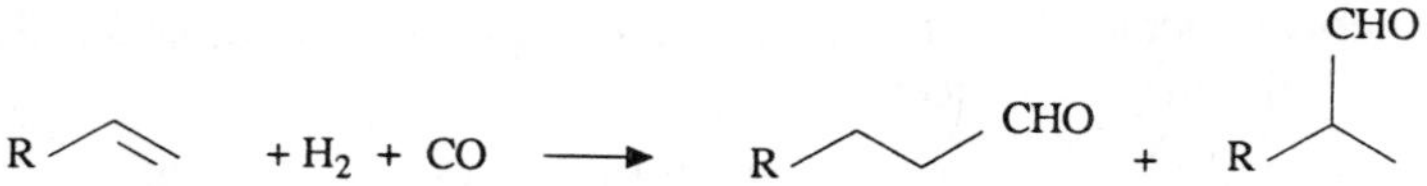

(5)

Three types of homogeneous transition metal complexes are used in industrial hydroformylation processes:

(i) simple cobalt carbonyl, or rather hydrido cobalt carbonyl, complexes,

(ii) hydrido-cobalt carbonyl complexes having tertiary phosphine ligands, and

(iii) tertiary phosphine hydrido-rhodium carbonyl species.

TABLE V
Comparison of the basic hydroformylation processes

Parameter	$CoH(CO)_4$	Catalyst $CoH(CO)_3PBu_3$	$RhH(CO)(PPh_3)_3$
Catalyst regeneration	yes	no	no
Total pressure (bar)	250–300	50–100	30
Temperature (°C)	140–180	180	90–120
n/iso butanals	3	10	> 10
Aldehyde hydrogenation	low	high	low

The need for linear products and the search for reduced reaction pressure and temperature leaves room for improvement of the basic «oxo process». The first important breakthrough in overcoming the drawbacks of this process occurred in the early sixties when Slaugh and Mullineaux [38] discovered that adding tertiary phosphine ligands to cobalt carbonyl led to complexes which did not depend on high CO pressures for their stability and which were able to catalyse the conversion of 1-alkenes to normal alcohols with over 90% selectivity. The second improvement, appeared in the early seventies, was the introduction of rhodium instead of cobalt. The LPO process [39] based on the work of Wilkinson *et al.* offers milder temperature and pressure conditions. High selectivities for *n*-aldehydes is also a main advantage of this process. However, it is limited to lower alkenes, although variants may overcome this limitation. The most interesting one is the use of water-soluble complexes which are the basis of the Ruhrchemie-Rhône-Poulenc process [40].

5.1. TERTIARY-PHOSPHINE MODIFIED COBALT CARBONYL SYSTEMS

Reaction of dicobalt octacarbonyl with L ligands provides complexes **32** according to the following sequence:

$$Co_2(CO)_8 + 2L \; \rightleftharpoons \; Co_2(CO)_6L_2 + 2CO \qquad (6)$$

$$Co_2(CO)_6L_2 + 2H_2 \; \rightleftharpoons \; CoH(CO)_3L \qquad (7)$$

Owing to the increased steric hindrance, resulting from the presence of a bulkier ligand L (θ for CO is $\approx 95°$), the unsaturated 16-electron species, $CoH(CO)_2L$ shows a marked preference for terminal over internal alkenes when it comes to forming π-complexes $CoH(CO)_2L$ (alkene). Similarly, in all the subsequent ligand migration reactions in the catalytic cycle (Scheme 10), transition states leading to the least substituted alkyl or acyl complex are much favoured. In these modified cobalt systems, it is the steric effect of the tertiary phosphine ligands which is mainly responsible for the higher *n-iso* ratios. However, the maximum influence is apparently achieved with a ligand cone angle of *ca.* 130°. The most frequently used phosphine is PBu_3 which has a Cone angle of 132°.

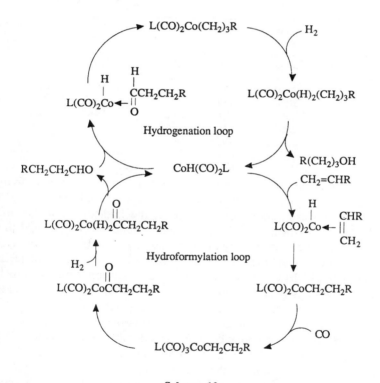

Scheme 10

Thus, in the catalytic hydroformylation of 1-hexene using $CoH(CO)_3L$ as catalyst precursor, the percentage of linear product formed is essentially independent of whether L = P(i–Pr)$_3$ (θ = 160°) or PPr$_3$ (θ = 132°). Table VI shows that while the product linearity is fairly independent of the phosphine cone angle, it is not independent of the ligand basicity or its electronic parameters. The selectivity towards linear products decreases as v increases (pK_a decreases), this effect being most marked for the arylphosphine ligands, PEt$_2$Ph, PEtPh$_2$ and PPh$_3$. Concurrent with this decrease in linear selectivity, there is a decrease in the hydrogenation activity of the system, as reflected in the increasing aldehyde to alcohol ratio, but an increase in the overall hydroformylation activity, as reflected in the increasing k values [41].

TABLE VI

Hydroformylation of 1-hexene using Co$_2$(CO)$_6$L$_2$ as catalyst precursor
(T = 160°C, p = 70 bar, H$_2$/CO = 1.2/1)

L	pK_a	v(cm^{-1})	θ(°)	10^3k (min^{-1})	n %	Aldehyde to alcohol ratio
P(i-Pr)$_3$	9.4	2059.2	160	2.8	85.0	–
PEt$_3$	8.7	2061.7	132	2.7	89.6	0.9
PPr$_3$	8.6	2060.9	132	3.1	89.5	1.0
PBu$_3$	8.4	2060.3	132	3.3	89.6	1.1
PEt$_2$Ph	6.3	2063.7	136	5.5	84.6	2.2
PEtPh$_2$	4.9	2066.7	140	8.8	71.7	4.3
PPh$_3$	2.7	2068.9	145	14.1	62.4	11.7

As far as hydrogenation activity is concerned, the superior activity of cobalt/tertiary-phosphine systems compared to the simple cobalt carbonyl system has been ascribed to an increase in the hydridic character of the cobalt-hydride on substituting CO by L. Tributylphosphine (a class I ligand) will increase the electron density at the metal centre and thus at the hydride hydrogen centre. Enhancing the negative charge at this centre will facilitate the nucleophilic hydride ligand migration to acyl carbon. Similarly, increasing the electron density at the metal centre will favour the oxidative addition of hydrogen. Both effects would assist the hydrogenation cycle.

Finally, as there is more electron density available for the metal-carbonyl bonding orbitals, the complexes are considerably more stable, which prevents decomposition of the catalyst under the reaction conditions.

The increased steric bulk of the tertiary-phosphine ligand also alters the nature of the rate-determining step. Whereas hydrogenolysis of the cobalt-acyl species appears to be rate-determining with unmodified cobalt catalysts, initial alkene coordination or possibly hydride ligand transfer to coordinated alkene is the slowest step in the catalytic cycle with the modified system.

5.2. TERTIARY-PHOSPHINE MODIFIED RHODIUM COMPLEXES

With olefins, CO and H_2, catalytic hydroformylation takes place even at 25°C and subatmospheric pressure in the presence of $RhH(CO)(PPh_3)_3$, **33**. This complex is in equilibrium with several species in solution. At a concentration of *ca.* 10^{-3} mol.l^{-1}, the bis(phosphine) complex $RhH(CO)$ $(PPh_3)_2$, **34**, predominates. At lower rhodium concentrations, further dissociation of phosphine yields the more active species, $RhH(CO)PPh_3$, **35**. The reaction of CO with a solution of **33** involves rapid conversion into $RhH(CO)_2(PPh_3)_2$, **36**, and the dimer $[RhH(CO)_2(PPh_3)_2]_2$, **37**. Under hydroformylation conditions, the equilibrium between these two species is assumed to lie mainly on the side of the hydridorhodium species, **36**.

Two catalytic cycles have been suggested for this hydroformylation reaction [42]. The first cycle (Scheme 11) involves initial dissociation of a triphenylphosphine ligand from **36** to give an unsaturated species which coordinates alkene. This reaction is followed by (i) rapid hydride ligand migration, (ii) triphenylphosphine coordination, (iii) acyl ligand formation followed by (iv) oxidative addition of dihydrogen, (v) reductive elimination of aldehyde, and (vi) coordination of CO giving back **36**. The slow step is thought to be the oxidative addition of dihydrogen.

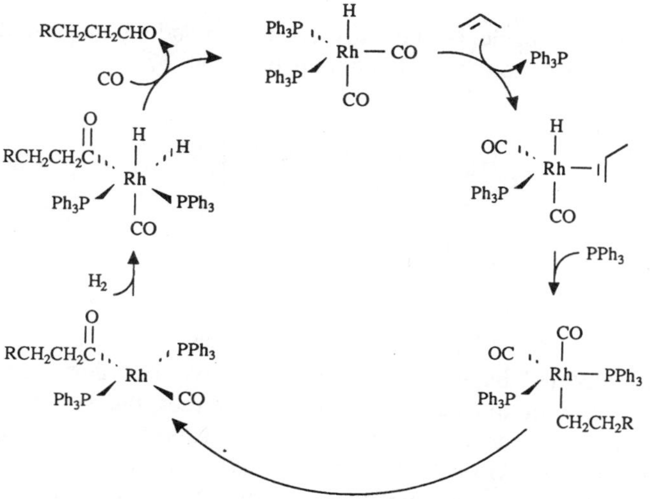

Scheme 11

In the associative mechanism (Scheme 12), the alkene is considered to add directly to the pentacoordinate species, **36**, without prior dissociation of either a phosphine or a carbonyl ligand.

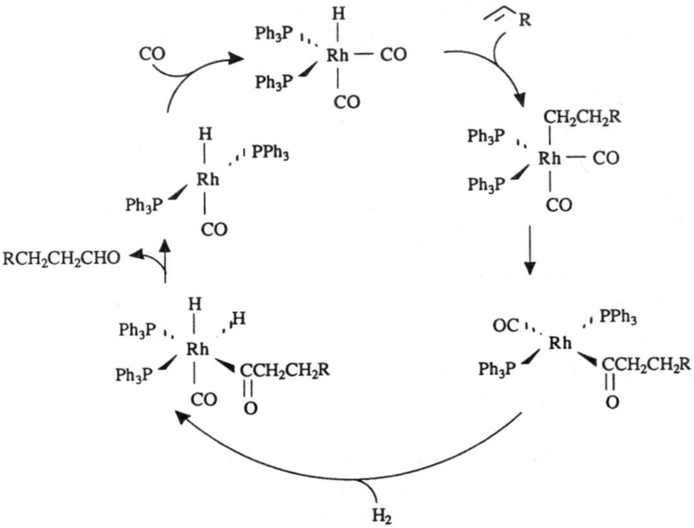

Scheme 12

The associative cycle is sterically more demanding and would favour linear product formation through the transition state **37**:

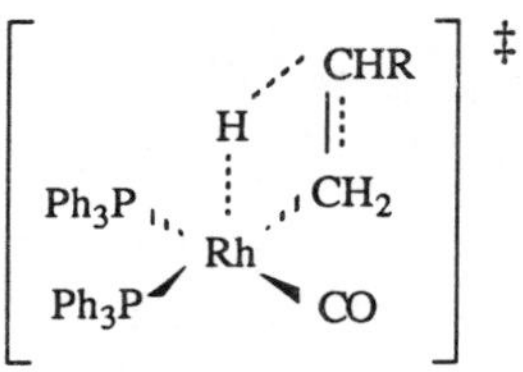

37

This is supported by the experimental observations that increasing the PPh_3-to-Rh ratio in the catalyst solution increases the selectivity of the system towards linear aldehyde formation (Table VII). However, this increase in selectivity is obtained at the expense of activity.

With phosphorus ligands smaller than PPh_3, the steric strain will be less, as will the tendency to be replaced by CO. Studies of the competition between CO and phosphanes for coordination to Ni(0), starting with $Ni(CO)_4$, have shown that the degree of substitution of CO by L decreases

TABLE VII
Influence of the PPh$_3$/**36** ratio on the
n/iso ratio of the products obtained from
the hydroformylation of propene
(T = 100°C, p = 35 bar, H$_2$/CO = 1/1)

PPh$_3$/**36**	*n/iso*
0/1	1/1
1/1	2/1
600/1[a]	15.3/1

[a] triphenylphosphine used as solvent,
125°C, 12.5 bar

linearly with increasing θ [2]For example, PPh$_3$ was unable to displace more than two carbonyl ligands, but smaller ligands such as P(OMe)$_3$ were able to displace all four ligands. On this basis, one can anticipate that rhodium systems containing very small ligands such as P(OMe)$_3$, especially if it is present in excess, should show little catalytic activity.

As already mentioned, phosphanes are decomposed under the reaction conditions used in hydroformylation. Therefore, the use of an excess of phosphanes prevents kinetic studies, especially in the case of rhodium catalysts. However, the use of phospholes such as **38** offered catalysts which are surprisingly not sensitive to **38**/Rh ratios higher than 2. Indeed, the regioselectivity in hydroformylation of styrene is not affected when this ratio varies from 1 to 20 and the yields of aldehydes do not change very much after this ratio reached 2 [43].

38   **39**

Moreover, the behaviour of this system allows the first kinetic study reported for a phosphane-rhodium(I) system which indicates that the rate determining step is the oxidative addition of hydrogen [44]:

$$v = k[styrene]_0^{-0,47} [styrene][Rh]pH_2pCO^{-1}.$$

This equation suggests also that it is a 14-electron acylrhodium species which is involved in the molecular hydrogen oxidative addition process. This apparently tricoordinated intermediate is presumably stabilized by further interaction with the double bonds of the phosphole ligands, *viz..*

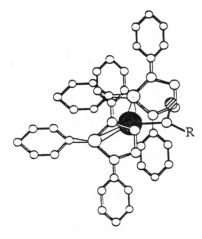

Additional examples of the application of «exotic» phosphanes are given with the use of phosphanorbornadienes which are also valuable ligands for rhodium catalysts for the hydroformylation of alkenes. Noteworthy is the demonstration that phosphanorbornadiene **39** is the first *monodentate* phosphane capable of converting ethyl acrylate into ethyl 2-formylpropionate under mild conditions [45].

Since the hydroformylation reaction is sensitive to structural changes in the added ligands, an interesting prospect will be asymmetric hydroformylation with catalysts substituted by chiral phosphines. If high (>90%), enantiomeric excesses could be obtained from prochiral olefins, such reactions could be useful in natural product synthesis and pharmaceutical manufacturing. Unfortunately, the highest ee value which has been obtained so far is 73%, Equation 8. A further difficulty exemplified by this equation is that for terminal alkenes, the chiral *iso*-aldehyde rather than the *n*-aldehyde is the desired product [46].

$$PhCH=CH_2 + CO + H_2 \xrightarrow[\text{DIPHOL}]{\text{PtCl}_2/\text{SnCl}_2} PhCH(CHO)CH_3 + PhCH_2CH_2CHO + PhCH_2CH_3 \quad (8)$$
$$\phantom{PhCH=CH_2 + CO + H_2 \xrightarrow[]{}} 56\% (73\%) \qquad (14)\% \qquad (20)\%$$

5.3. TWO-PHASE HYDROFORMYLATION

The economy of the rhodium LOP process rests on the fact that it is used mainly for light olefins leading to facile recovery of the catalyst by stripping of both reactants and products. Making homogeneous hydroformylation catalysts heterogeneous by binding them *via* their phosphine ligands to insoluble supports has not produced better catalysts.

A much more promising approach is to confine the active catalyst in a second liquid phase. By using the water-soluble phosphine TPPTSNa, **40**, with a rhodium oxo catalyst in the reactor containing two unmiscible phases, this «inverse phase-transfer» process facilitates the separation of the product (in the nonaqueous phase) from the catalyst (in the aqueous phase). Such a design provides virtually no catalyst leaching, good rates and excellent heat transfer and is now applied on the industrial scale by Ruhrchemie.

$$P \left(\underset{SO_3^-Na^+)_3}{\bigcirc} \right) \quad \textbf{40}$$

6. Conclusion

The great strength of homogeneous catalysis and especially when involving phosphine ligands is the opportunity it offers to tailor ligands so as to enhance the reactivity and selectivity of metal centred catalysts. Only slight changes in the ligand can produce considerable changes in selectivity. That the reaction occurs in liquid phase(s) also facilitates the study of the mechanistic steps involved in the catalytic cycle and, by changing the phosphine or metal centre, may allow the isolation of stable species which may represent intermediates of the catalytic cycle.

Although phosphanes may degrade under certain reaction conditions, they are quite unique in achieving selectivity control, especially at the stereoselectivity and enantioselective levels. Further research is needed for making more robust and more specific phosphorus ligands – at low cost if possible.

Homogeneous catalysis has the great drawback that the catalyst recovery is very often difficult. Therefore, new processes for the separation of the catalyst from the products are required and liquid-liquid biphasic systems may provide interesting applications.

Laboratoire de Chimie de Coordination du CNRS,
Unité n° 8241 liée par conventions à l'Université Paul Sabatier et à l'Institut National Polytechnique,
205, route de Narbonne, 31077 Toulouse Cedex (France)

References

1. McAuliffe, C.A.; Lewason, W. *Phosphine, Arsine and Stibine Complexes of Transition Elements*; Elsevier, New York; 1979.
2. Tolman, C.A. *Chem. Rev.* **1977**, *77*, 313.
3. Stelzer, O. *Topics in Phosphorus Chemistry*, 9; Wiley, New York, 1977; 1.
4. Kagan, H.B. *Asymmetric Synthesis*, vol. 5; Morrison, J.D., Academic Press, New York, 1985; p. 1.
5. Garrou, P.E. *Chem. Rev.* **1985**, *85*, 171.
6. McAuliffe, C.A. *Comprehensive Coordination Chemistry*, Vol. 2; Wilkinson, G.; Gillard, R.D.; McCleverty, J.A., Pergamon Press, Oxford, 1987; 989.
7. *Comprehensive Organometallic Chemistry*, Vol. 8; Wilkinson, G.; Stone, F.G.A.; Abels, E.W., Pergamon Press, Oxford, 1982.
8. *Homogeneous Catalysis with Metal Phosphine Complexes*, Pignolet, L.H., Plenum Press, New York, 1983.
9. *Principles and Applications of Organotransition Metal Chemistry*; Collman, J.P.; Hegedus, L.S.; Norton, J.R.; Finke, R.G., University Science Books, Mill Valley, CA., 1987.
10. Tolman, C.A.; Seidel, W.C.; Gosser, L.W. *J. Am. Chem. Soc.* **1974**, *96*, 53.
11. Mann, B.E.; Musco, A. *J. Chem. Soc., Dalton Trans.* **1975**, 1673.
12. Marynick, D.S. *J. Am. Chem. Soc.* **1984**, *106*, 4064.
13. Golovin, M.N.; Rahman, Md. M.; Belmonte, J.E.; Giering, W.P. *Organometallics* **1985**, *4*, 1981.
14. Bogdanovic, B. *Adv. Organometal. Chem.* **1979**, *17*, 105.
15. Pardy, R.B.A.; Tkatchenko, I.*J. Chem. Soc., Chem. Commun.* **1981**, *49*.
16. de Souza, R.F. *PhD Thesis*, Université Paul Sabatier, Toulouse, 1987 ; de Souza, R.F.; Neibecker, D.; Tkatchenko, I. *J. Organomet. Chem.* submitted.
17. Keim, W. *New J. Chem.* **1987**, *11*, 531.
18. Bogdanovic, B.; Henc, B.; Karmann, H.-G.; Nüssel, D.; Walter, D.; Wilke, G. *Ind. Eng. Chem.* **1970**, *62*, 34.
19. Wilke, G. *Angew. Chem., Int. Ed. Eng.* **1988**, *27*, 185.
20. *Coordination Polymerization*; Chien, J.C.W., Academic Press, New York, 1975.
21. Ascensó, J.R.; Carrondo, M.A.A.F. de C.T.; Dias, A.R.; Goes, P.T.; Piedade, M.F.M.; Romáo, C.C.; Tkatchenko, I. *Polyhedron* **1989**, *8*, 2449-2457.
22. Ascensó, J.; Dias, A.R.; Gomes, P.T.; Neibecker, D.; Revillon, A.; Ramao, C.C.; Tkatchenko, I. *Makromol. Chem.* **1989**, *190*, 2773-2787.
23. Ascensó, J.; Dias, A.R.; Gomes, P.T.; Ramao, C.C.; Pham, Q. T.; Neibecker, D.; Tkatchenko, I. *Macromolecules* **1989**, *22*, 998.
24. Grubbs, R.H.; Miyashita, A.; Liu, M.; Burk, P. *J. Am. Chem. Soc.* **1978**, *100*, 2418.
25. Guibert, I.; Neibecker, D.; Tkatchenko, I. *J. Chem. Soc., Chem. Commun.* **1989**, 1850.
26. Guibert, I.; Neibecker, D.; Tkatchenko, I. *New J. Chem.*, submitted.
27. Grenouillet, P.; Neibecker, D.; Tkatchenko, I. French Patent Appl. 86.046444 (Rhône-Poulenc).
28. Jolly, P.W.; Wilke, G. *The Organic Chemistry of Nickel*, Vol. II; Academic Press, New York, 1975.
29. Heimbach, P.; Schenklun, H. *Topics Current Chem.* **1980**, *92*, 46.
30. Barnett, B.; Büssemeir, B.; Heimbach, P.; Jolly, P.W.; Kruger, C.; Tkatchenko, I.;

Wilke, G. *Tetrahedron Lett.* **1972**, *15*, 1457.
31. Tkatchenko, I. *J. Organomet. Chem.* **1977**, *124*, C39; Ballivet-Tkatchenko, D.; Riveccie, M.; El Murr, N. *J. Am. Chem. Soc.* **1979**, *101*, 2763.
32. Medema, D.; Van Helden, R. *Rec. Trav. Chim. Pays-Bas* **1974**, *90*, 324.
33. Grenouillet, P.; Neibecker, D.; Poirier, J.; Tkatchenko, I. *Angew. Chem., Internat. Ed. Engl.* **1982**, *21*, 767.
34. Touma, M. *PhD Thesis* , Université Paul Sabatier, Toulouse, 1986.
35. Sen, A.; Lai, T.W. *J. Am. Chem. Soc.* **1981**, *103*, 4627.
36. Agbossou, S.; Bonnet, M. C.; Tkatchenko, I. *Nouv. J. Chim.* **1985**, *9*, 311.
37. Cornils, B. *in New Syntheses with Carbon Monoxide*; Falbe, J., Springer-Verlag, Berlin, 1980; p. 1.
38. Slaugh, L.H.; Mullineaux, R.D. *J. Organomet. Chem.* **1968**, *13*, 469.
39. Pruett, R.L. *Adv. Organomet. Chem.* **1979**, *17*, 1.
40. Kuntz, E. *Chemtech.* **1987**, *17*, 570; Jenck, J., French Patent 2,473,504, 1984.
41. Paulik, F.E. *Catal. Rev.* **1972**, *6*, 49.
42. Evans, D.; Osborn, J.A.; Wilkinson, G. *J. Chem. Soc. (A)* **1968**, *31*, 33.
43. Bergounhou, C.; Neibecker, D.; Réau, R. *J. Chem. Soc., Chem. Commun.* **1988**, 1370.
44. Réau, R. *PhD Thesis*, Université Paul Sabatier, Toulouse, 1988.
45. Neibecker, D.; Réau, R. *Angew. Chem., Internat. Ed. Engl.* **1989**, *28*, 500.
46. Ojima, I.; Hirai, K. *Asymmetric Synthesis*, Vol. 5; Morrison, J.D., Academic Press, New York, 1985; p. 104.

A. MORTREUX

LIGAND CONTROLLED CATALYSIS: CHEMO TO STEREOSELECTIVE SYNTHESES FROM OLEFINS AND DIENES OVER NICKEL CATALYSTS

During the last three decades, much efforts in the area of transition metal-catalyzed organic syntheses have been made on selective C–C bond formation. In this context, Wilke's group in Mulheim has extensively studied coupling reactions of butadiene over nickel catalysts [1], after the pioneering work of Foster [2] who discovered the cyclooctadiene synthesis from this diene as early as 1947.

Diene-olefin codimerization have also been the subject of thorough investigations at almost the same time, on nickel based catalytic systems, and most of these reactions have been studied in details, so that the mechanism are now generally well depicted, as several organometallic intermediates have been isolated and characterized. This paper will deal with this chemistry, and particular emphasis will be given to the role played by the ligand environment for the product selectivity. Special attention will be also given to aminophosphinephosphinites $PPh_2NMeCHR^1CHR^2OPPh_2$ (AMPP) and aminophosphinites $PPh_2OCHR^1CHR^2NH(CH_3)$ (AMP), which will show typical chemo, regio and even stereoselective behaviours.

1. Diene-olefin codimerization

1.1. CHEMO AND REGIOSELECTIVE BUTADIENE-ETHYLENE CODIMERIZATION

The simplest chemo and regioselective codimerization is performed from butadiene and ethylene. According to the catalytic system, the reaction can mainly be directed towards the production of linear or cyclic codimers, in which either 1 or 2 butadiene units are incorporated (Scheme 1).

47

A. F. Noels et al. (Eds.), Metal promoted selectivity in organic synthesis, 47–63.
© 1991 Kluwer Academic Publishers. Printed in the Netherlands.

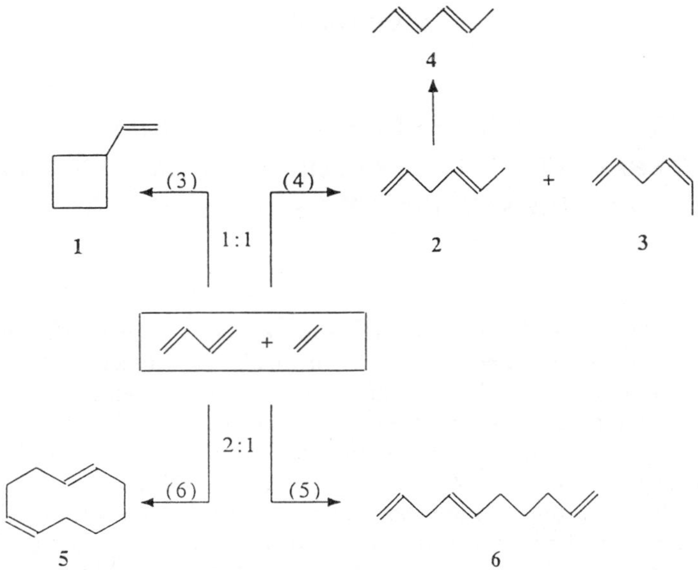

Scheme 1

Titanium, rhodium, palladium, cobalt, iron and nickel catalysts are used for these reactions [7, 8], some of them giving rise to the synthesis of commercial products (**2** and **3**).

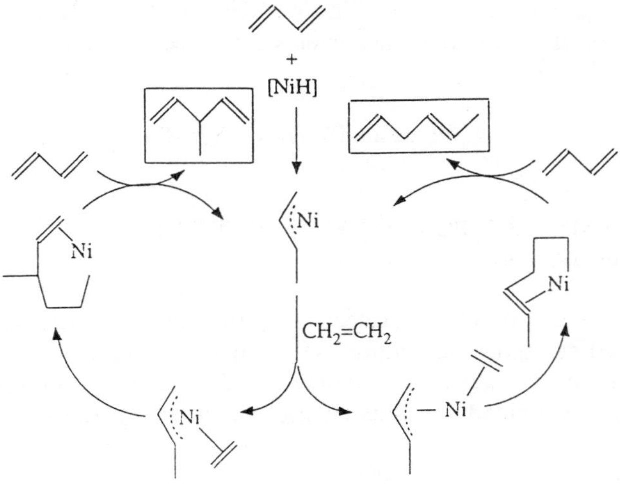

Scheme 2

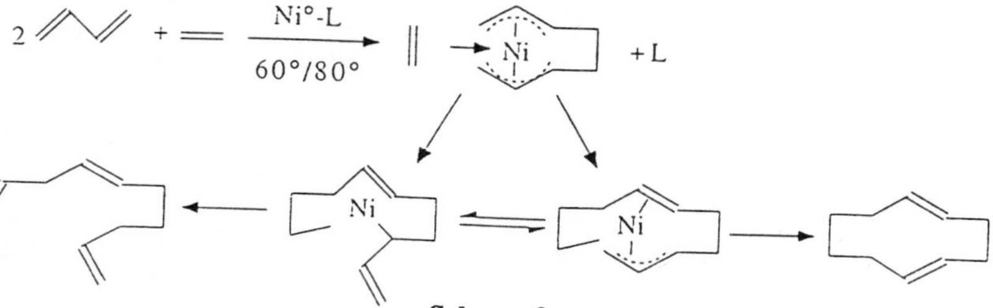

Scheme 3

In the case of nickel, NiH or NiHL have been proposed as the active species: the 1/1 adducts can be formed via π allyl intermediates (Scheme 2) and 2/1 adducts via production of bis π allyl complexes (Scheme 3).

1.2. BUTADIENE-FUNCTIONALIZED OLEFIN CODIMERIZATION

Diene-functionalized olefins codimerization can also be performed with production of either mono or bis dienic codimers. Table 1 summarizes some typical examples of codimerization with butadiene.

TABLE I

Coreactant	Catalyst	Products (Yield %)	Ref.
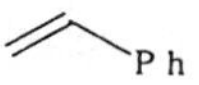 ⤳Ph	P(O-C6H4-Ph)3/Ni°	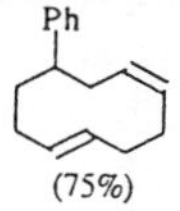 (75%)	[1]
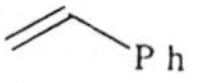 ⤳Ph	naked nickel	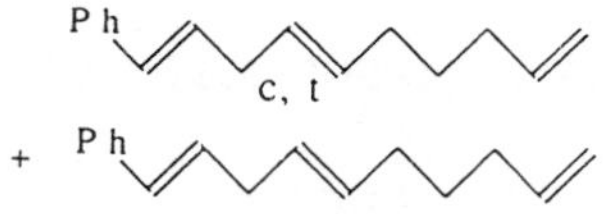 (97%)	[1,8]
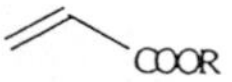 ⤳COOR	P(OPh)3/Ni°	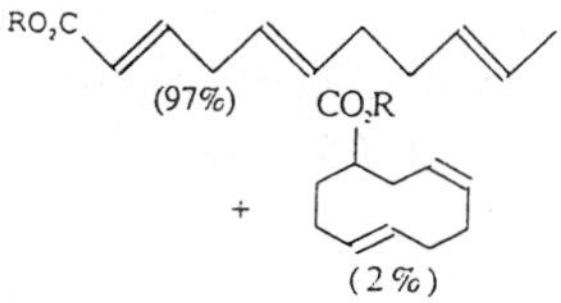 (2%)	[1]
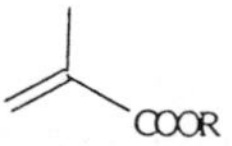 ⤳COOR	Ni(acac)2/AlEt3/AsPh3	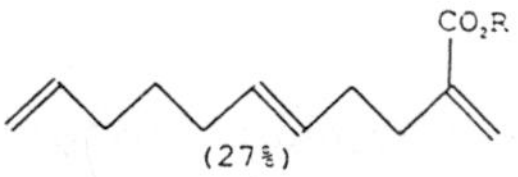 (27%)	[1,9]

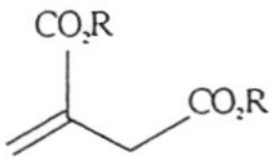	$Ni(acac)_2/AlEt_3/SbPh_3$	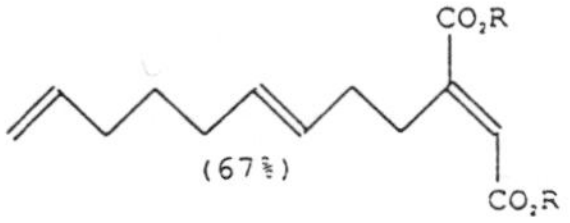	[10]
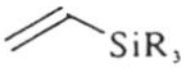	$Ni(COD)_2$/phosphine	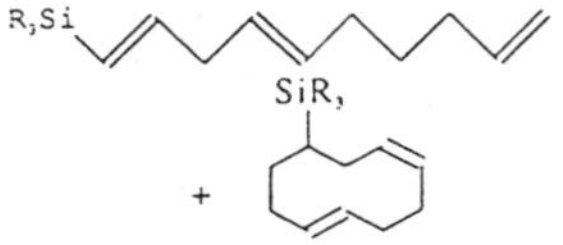	[11]
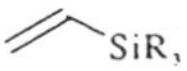	$Ni(COD)_2$/phosphine	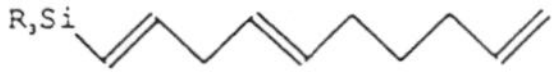	[11]

This table is generalized by the following scheme:

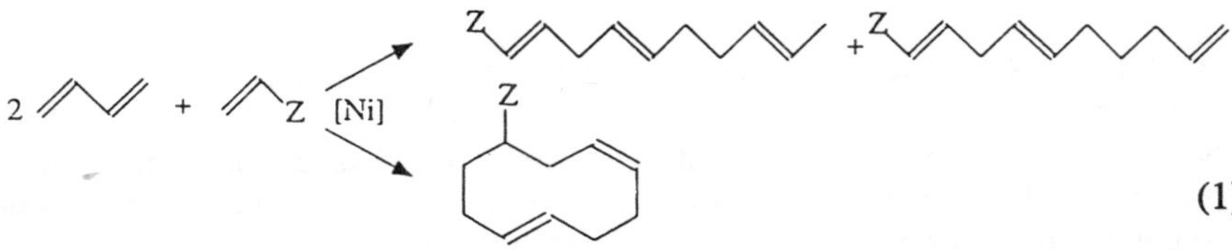

(1)

The cyclic/linear codimer ratio depends on several factors such as the temperature and the ligand (high temperature and a phosphite favour cyclic compounds) and also on the nature of the substituent: Ar and CO_2R groups favor hydrogen transfer reactions and therefore linear codimers. The mechanisms generally involve olefin insertion into a bis (π allyl) nickel to give a C_{10} chain complexed to the metal via a σ alkyl and a π allyl bond. Coupling between the alkyl group and the end chain carbon atom gives the cyclic compounds, and α H elimination followed by addition on the σ allyl group gives the linear products.

1.3. ASYMMETRIC DIENE-OLEFIN CODIMERIZATION

1.3.1. Scope of the reaction

Cyclic conjugated dienes codimerize with olefins on nickel catalysts to produce vinyl substituted cyclolefins:

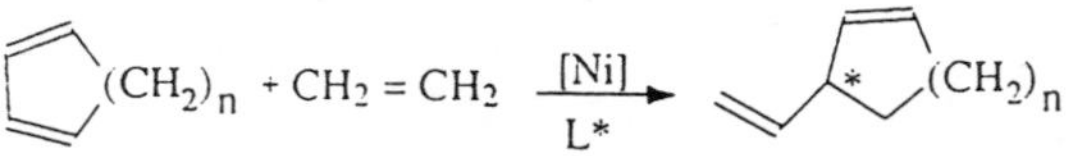

(2)

The observed regioselectivity during this reaction has been taken in

account in the first attempt of asymmetric codimerization, with cyclooctadiene and ethylene as substrates [12]: upon using (–) dimenthyl methyl
phosphine as ligand, S(–)-3-vinylcyclooctene is produced with 23,5% ee at 0
°C and 53% at –75°C, with a nickel system consisting of bis π allyl nickel
chloride and ethyl aluminium sesquichloride (Ni/Al/P = 1:1.5:1.2). The
highest optical purity at 0°C is obtained with a Ni : P ratio of 1:3:8.

$$
\text{(cyclooctadiene)} \quad + \quad CH_2{=}CH_2 \quad \xrightarrow[\text{CH}_2\text{Cl}_2]{\text{Ni} + \text{L*}} \quad \text{(vinylcyclooctene)} \tag{3}
$$

Extension of this diene-olefin reaction to norbornene has also given chiral
vinyl norbornane with ee's up to 77% [13]

$$
\text{(norbornene)} \quad + \quad CH_2{=}CH_2 \quad \xrightarrow{\text{Ni} + \text{L*}} \quad \text{(vinylnorbornane)} \tag{4}
$$

and more recently, codimerization between styrene and ethylene have been
claimed to give (–) 3-phenyl-1-butene with ee's > 95%:

$$
\text{(styrene)} \quad + \quad {=}\!= \quad \xrightarrow{\text{Ni} + \text{L*}}_{-70°C} \quad \text{(3-phenyl-1-butene)} \tag{5}
$$

In this last case, an aminophosphine in which a pinene group was incorporated was used as the chiral ligand [14].

1.3.2. Aminophosphinephosphinites chelating ligands for cyclohexadiene-ethylene codimerization

Apart from the above results, aminophosphine ligands have also been
previously used as chiral modifiers for the asymmetric synthesis of 3-
vinylcyclohexadiene and ethylene [15]:

$$
\text{(cyclohexadiene)} \quad + \quad \| \quad \xrightarrow[\text{Ph*CH(Me)N(Me)PPh}_2]{\text{Ni}° + \text{AlEt}_2\text{Cl}} \quad \text{(vinylcyclohexadiene)} \tag{6}
$$

and an optical rotation of 99° has been observed on the product at –70°C with L*/Ni = 10 (L* = PhMePPh$_2$). Much better results (Table II) were observed by using different chelating ligands, aminophosphinephosphinites, whose synthesis can be performed easily from the corresponding aminoalcohols:

$$\text{HOCHR}_1\text{CHR}_2\text{NHMe} + 2\text{PPh}_2\text{Cl} \xrightarrow[\text{-HCl}]{\text{NEt}_3} \text{Ph}_2\text{POCHR}_1\text{CHR}_2\text{NMePPh}_2 \tag{7}$$

In Table II are listed the results observed with different AMPP ligands arising from natural aminoacids [16].

TABLE II

Starting aminoacids	R	Ligand	$[\alpha]_D^{25}$, deg (cl.00, toluene)	config	T,°C	Optical yields, % ee
(2S, 3R)-Threonine	CH$_3$CH*(OPPh$_2$)	ThreoNOOP	+227,5	S	40	85
			+250		−30	93
(S)-Phenylalanine	PhCH$_2$-	PheNOP	−56	R	40	21
			−139		−25	52
(S)-Alanine	CH$_3$-	AlaNOP	−45	R	40	17
(S)-Valine	i-Pr-	ValNOP	−26,5	R	40	10
(S)-Aspartic acid	−CH$_2$CH$_2$OPPh$_2$	AspNOOP	−75	R	40	28
(S)-Glutamic acid	−(CH$_2$)$_2$OPPh$_2$	GlutaNOOP	−50	R	40	19

Such catalysts are very efficient as with a substrate/Ni ratio of 220, a total conversion is observed within 15 min. at 40°C. Ee's up to 93% are obtained with the tridentate ligand ThreoNOOP, in which two asymmetric centers within the hydrocarbon chelating chain are present. It has also to be noticed that the reaction must be stopped just after ethylene consumption, as an isomerization process into 3-ethylidene cyclohexene is observed after completion of the codimerization reaction.

Due to the good results obtained with this new series of ligands, the behaviour of these chelates during dimerization of dienes has also been examined.

2. Cyclodimerization of dienes

2.1. SCOPE OF THE REACTION

The following scheme gives the different possibilities of cyclodimers or cyclooligomer production:

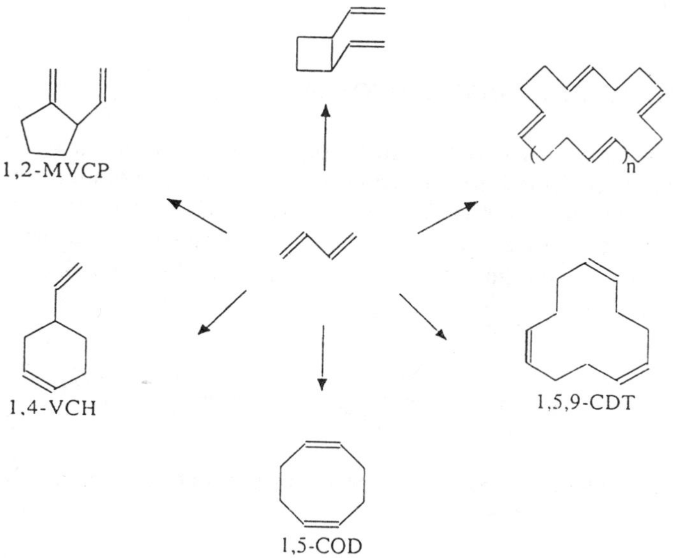

1,2-MVCP

1,4-VCH

1,5,9-CDT

1,5-COD

The cyclooligomers and 1,5,9-cyclododecatriene (1,5,9 CDT) have been observed by Reed [17]. These cyclooligomers contain 4 to 7 butadiene units and are prepared by catalysis on π allyl nickel complexes with low yield.

The cyclotrimer is produced in the absence of ligand on naked nickel. Different isomers are obtained (ttt, tcc and ttc) whose relative ratios are temperature and concentration dependant [18].

Upon using 1 equivalent of phosphine or phosphite per metal [1, 8, 9, 20], a mixture of 1,5-cyclooctadiene (1,5-COD) and 1,4-vinylcyclohexene (1,4-VCH) is obtained. The COD/VCH ratio depends upon the ligand: a phosphite (good π acceptor) favours bis allylic intermediates and consequently 1,5-COD as major product. A phosphine (good donor ligand) favors σ allyl moieties and therefore 4-vinylcyclohexene. Cis-1,2-divinylcyclobutane (cis-1,2-DVCB) is obtained at low conversion with a bulky phosphite ligand, but rearranges further during catalysis upon heating into 1,5-COD [1, 21]. If the reaction is conducted in the presence of an alcohol, 2-methylene-vinyl-cyclopentane has also been observed [8], the production of which has been studied by using deuterated alcohol [22].

Mechanistic studies on these oligomerizations have been performed by Wilke's group and a recent publication on nickel chemistry summarizes most of the details concerning the π-allylic intermediates involved in these reactions [23].

2.2. SUBSTITUTED CONJUGATED DIENES

In the presence of a «NiL» system, piperylene, isoprene and 2,3-dimethyl butadiene cyclodimerize into vinylcyclohexene or cyclooctadiene derivatives, but the reaction rate decreases, and the C_6 ring selectivity increases [1]. At low conversion, divinylcyclobutane derivatives are also observed, which allow to synthesize a precursor of grandisol via isoprene cyclo-dimerization [24].

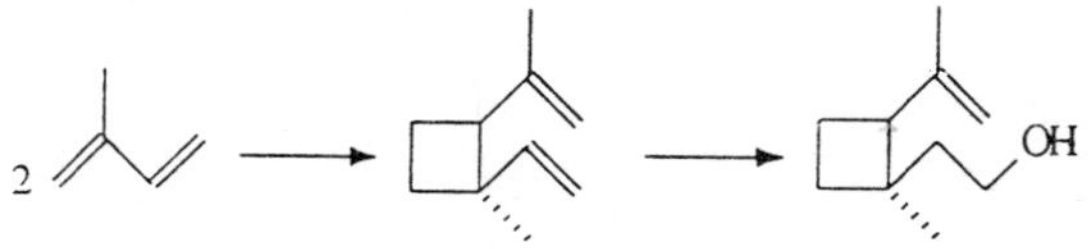

$$(9)$$

Limonene itself can be one of the major product of isoprene dimerization [25]:

$$(10)$$

The presence of CO might be the driving force for selective cyclodimerization in C_6 ring, via a Ni(PCy$_3$)CO 14 electron species as catalyst precursor, which avoid the catalytic cycle responsible for cyclooctadiene derivatives production, as the unprobable production of 20 electron species would be necessary for this reaction path.

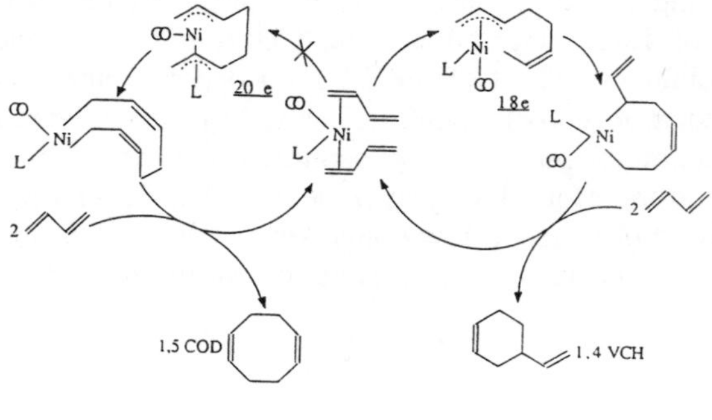

$$(11)$$

2.3. ENANTIOSELECTIVE CYCLODIMERIZATIONS

Butadiene cyclodimerization into optically active 1,4-vinylcyclohexene has only been studied by Richter with nickel catalysts modified by dioxaphospholane ligands [26].

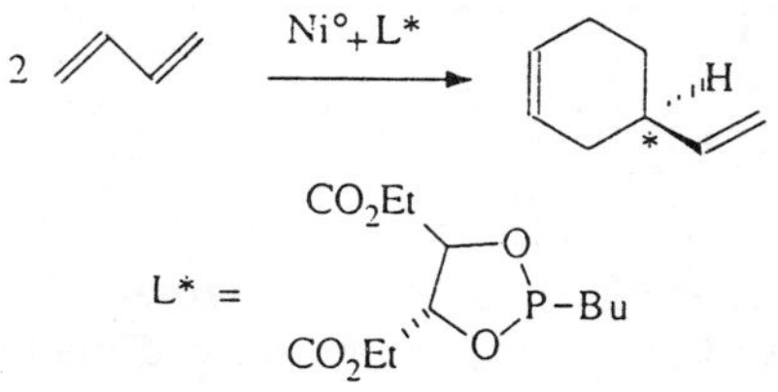

$$2 \diagup\diagdown \xrightarrow{Ni^\circ + L^*} \quad \text{(vinylcyclohexene)} \quad ...H$$

$$L^* = \quad \begin{array}{c} CO_2Et \\ CO_2Et \end{array} \diagup\diagdown \begin{array}{c} O \\ O \end{array} P-Bu$$

(12)

Chelating AMPP ligands have been used for the same reaction, and some of them found to be at least as efficient, although the maximum optical yield never exceed 30% [27] (Table III).

TABLE III

Ligand L_2	R^1	R^2	$\dfrac{\text{VCH}}{\text{COD}}$	ee% (Conf)
(S) AlaNOP	Me	H	1,7	6.4 (S)
(S) ValNOP	CHMe$_2$	H	2.5	8.9 (S)
(S) LeuNOP	CH$_2$CHMe$_2$	H	2.3	16.2 (S)
(S,S) IleNOP	CH$_2$CHMeEt	H	5.9	15.4 (S)
(S) PheNOP	CH$_2$Ph	H	2.7	10.4 (S)
(R) PheGlyNOP	Ph	H	1.6	7.7 (R)
(1S,2S) EPHOS	Me	Ph	0.7	20.7 (S)
(2R,3R) ThreoNOOP	MeC̊HOPPh$_2$	H	1.1	25.9 (R)

[a] Conditions : Butadiene/Ni(COD)$_2$ = 50; Solvent = toluene;
AMPP = PPh$_2$N(MeR^{1}CHCHR^{2}OPPh$_2$; L$_2$/Ni = 1; a 90% conversion is obtained within 24 h.

For this reaction, activity, optical yield and chemoselectivity are too low to be acceptable for large scale application so that more research in this field should be done to enhance the selective production of chiral vinyl-cyclohexene from readily available butadiene.

3. Linear dimerization of dienes

3.1. LINEAR DIMERIZATION OF BUTADIENE

Palladium catalysts are generally used to dimerise selectively butadiene into 1,3,7-octatrienes [28, 29] with selectivities up to 80%.

$$2 \quad /\!\!\!\diagup\!\!\!\diagup \quad \xrightarrow{\;[Pd]\;} \quad /\!\!\!\diagdown\!\!\!\diagup\!\!\!\diagdown\!\!\!\diagup\!\!\!\diagdown\!\!\!\diagup \tag{13}$$

The involved intermediates are the same as those depicted for cyclodimerization, but linear products are formed via hydrogen transfer by β elimination, which occurs readily with palladium:

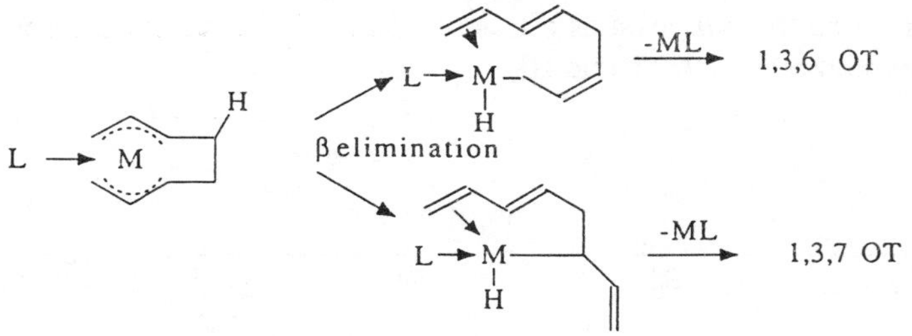

(14)

In the case of nickel, this hydrogen transfer is favoured by the presence of alcohols and gives mainly the 1,3,6-isomers when POEt$_3$ and PBu$_3$ are used as ligands [30]. A similar effect is observed with amines: with a Ni/POEt$_3$/Morpholine = 1/1/50 catalyst, Heimbach obtained a mixture of 1,3,6-octatrienes at 60°C [31]. The proposed mechanism involves a π-allyl nickel intermediate with the production of a Ni–N bond:

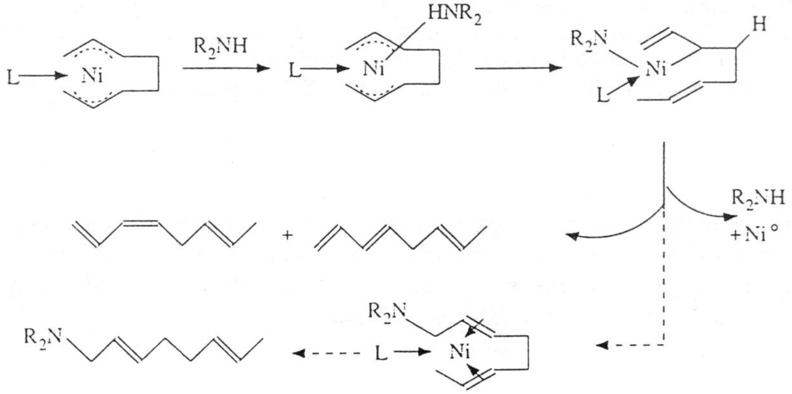

(15)

This mechanism is confirmed by production of octadienylamine at high conversion.

It has also to be noticed that in ethanol, 1,3,6-octatrienes could be produced on a $Ni(PPh_3)_2Br_2/NaBH_4$ system, at 100°C with a 95% selectivity [32].

3.2. LINEAR DIMERIZATION OF ALKYL SUBSTITUTED AND FUNCTIONALIZED DIENES

Dimerization of isoprene has been extensively studied for terpene synthesis. As this reagent is dissymmetric, four different coupling could occur (h = head, t = tail, one 1,3,7 isomer has only been represented).

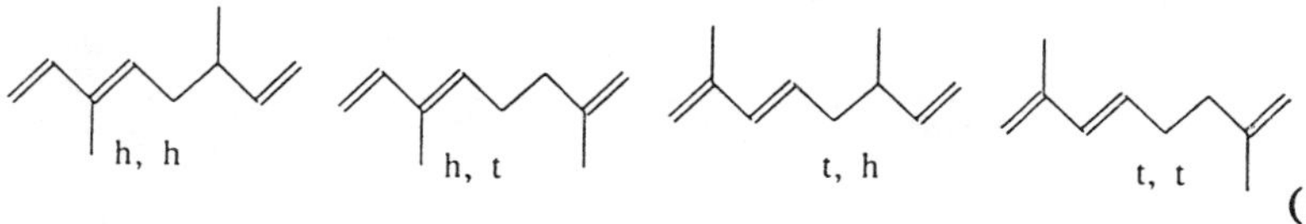

(16)

Palladium [33], titanium [34, 35] or zirconium [35, 36] based catalysts have been used successfully to produce several of these isomers, and nickel catalysts have been shown to give 2,6-dimethyl-1,3,6-octatrienes [37], but with activities far lower than with palladium. In every case, no head to tail linkage was observed, unfortunately, as this coupling could lead to natural terpenes.

A similar behaviour can be observed with functionalized dienes, *i.e.* the linear dimerization does not occur on nickel, whereas a palladium system catalyzes dienic esters dimerization [38].

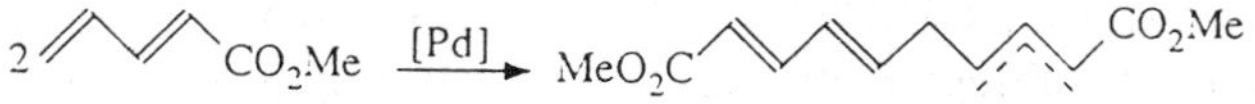

(17)

The reaction with methylsorbate $MeCH=CHCH=CHCO_2Me$ however failed in this case.

3.3. LINEAR DIMERIZATION ON NICKEL AMINOPHOSPHINITE COMPLEXES

3.3.1. Butadiene dimerization

From the observation that a combination of a zerovalent nickel precursor, a phosphine or a phosphite (Heimbach) ligand and a secondary amine as cocatalyst was required for selective dimerization of butadiene, the use of ancillary ligands bearing both P–O and N–H moities has been tried for this reaction. The synthesis of this new ligand was performed in one step by reaction of dimethylaminodiphenylphosphine PPh_2NMe_2 on an aminoalcohol [39].

$$HOCHR^1CHR^2NHR^3 + PPH_2NMe_2 \rightarrow$$
$$Ph_2POCHR^1CHR^2NHR^3 + NMe_2H \qquad (18)$$

With this ligand and $Ni(COD)_2$ as the source of nickel, selective production of 1,3,6-octatriene is observed, with a consecutive isomerization into 2,4,6-octatriene.

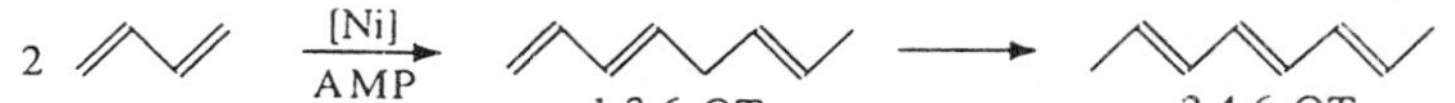

$$2 \xrightarrow[\text{AMP}]{[Ni]} \quad \text{1,3,6 OT} \longrightarrow \text{2,4,6 OT} \qquad (19)$$

This reaction occurs with high butadiene/nickel ratio (5000) and at room temperature or below. Particularly interesting also is the fact that only 1 mole of aminophosphinite ligand is required per nickel atom. These results were followed by a study of the mechanism of this reaction and its generalization to other dienes.

3.3.2. Labelled experiments [40]

The contribution of the NH group to the catalytic process, earlier suggested by Heimbach when morpholine was used in excess [31] has also been proved by deuterium experiments. Thus, upon using perdeuterated butadiene and the EPHOSNH $PPh_2OCHPhCHMeNHMe$ ligand with $Ni(COD)_2$, NMR analysis of the resulting octatrienes at the early stages of the reaction clearly shows that hydrogen is incorporated in the product in the methyl group of 1,3,6-octatriene.

A similar experiment conducted with C_4H_6 on $Ni(COD)_2$-EPHOSND gave 1,3,6-octatriene deuterated on the terminal methyl group:

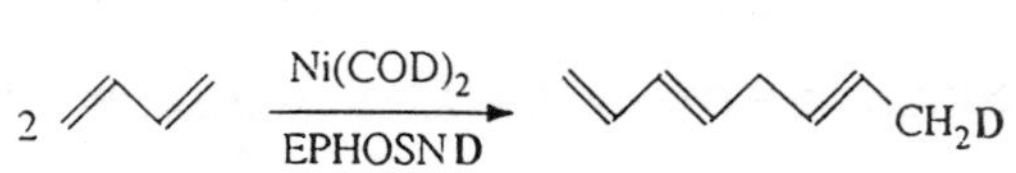

$$2 \xrightarrow[\text{EPHOSND}]{Ni(COD)_2} \quad CH_2D \qquad (20)$$

These experiments clearly proves that the NH group in the ligand is involved in the reaction mechanism and confirms the previous hypothesis of N-H oxidative addition according to the following scheme:

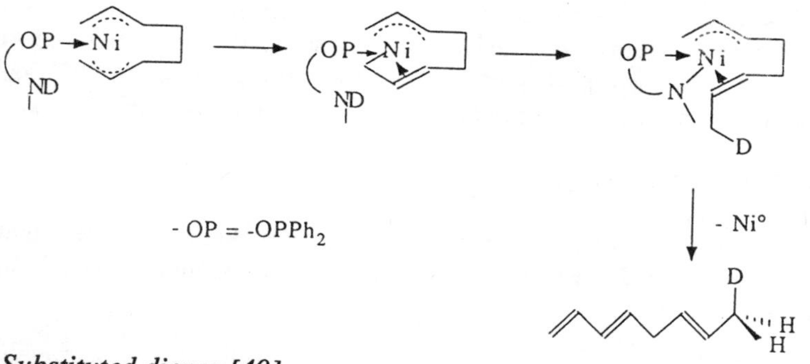

(21)

- OP = -OPPh$_2$

- Ni°

3.3.3. Substituted dienes [40]

Among several alkyl substituted dienes, only isoprene and piperylene are dimerized on this system. Moreover, isoprene gives a mixture of cyclic (40%) and linear (60%) dimers, the activity being reduced as compared with butadiene. Interestingly, the linear dimers are produced selectively in a tail to tail linkage (1,3,6 isomers):

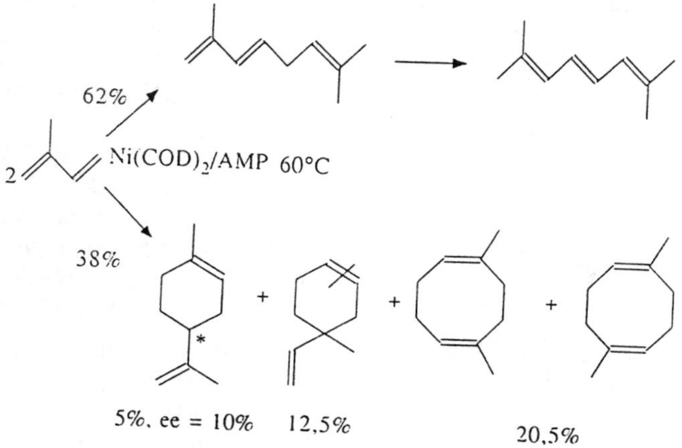

(22)

5%, ee = 10%   12,5%   20,5%

Piperylene is much more reactive than its isomer, as a 90% conversion is attained at 40°C (C$_5$H$_8$/Ni = 100), and only linear dimers are produced, with a remarkable regioselectivity in *head to head* linkage:

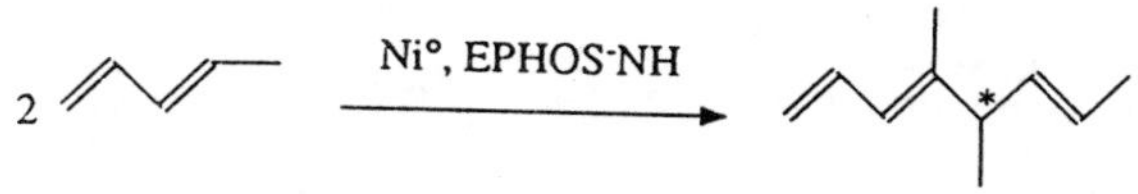

(23)

The 4,5 dimethyl-1,3,6-octatrienes are isomerized into their 2,4,6 conjugated isomers, but the reaction can be stopped selectively before this process to study the chirality of the 1,3,6 isomers: an ee up to 90% has been estimated for the (E, Z) 4,5-dimethyl-1,3,6-octatriene ($[\alpha]_D = -143°$)

3.3.4. Functionalized dienes [40]

Linear dimerization of functionalized dienes on nickel based systems was an unknown process. However, this Ni(COD)$_2$-EPHOSNH catalyst is able to catalyze such reactions, as depicted in Table IV.

The most interesting feature of these catalysis is that here again, linear products are obtained regioselectively: a *tail to tail* linkage is observed with methyl pentadienoate and a *head to head* linkage with methylsorbate, in which case again asymmetric C–C bond formation can occur.

4. Concluding remarks

AMPP and AMP ligands, easily available from natural aminoacids and aminoalcohols, have proved to be useful for diene reactions, especially during codimerization and linear dimerization with nickel catalysts. Due to the great variety and availability of starting reagents for ligand synthesis (in which no enantiomer separation is needed), it is found that at least one of these new ligands could give good enantioselectivities, as do ThreoNOOP for 3-vinylcyclohexene synthesis and (R,S)-EPHOS during piperylene dimerization. Although the research in asymmetric synthesis still remains somewhat empiric, the use of different AMPP or AMP also offers a good opportunity for studies involving stuctural and electronic effects in connection with the mechanism of asymmetric induction which are, in the context of C–C bond formation, far less developed than for asymmetric hydrogenation. Particulary interesting also would be the synthesis of natural products via regio and enantioselective dimerization of isoprene, to produce either cyclobutane derivatives in order to synthesize a precursor of grandisol as indicated in 2-2 or linear dimers from isoprene with a regioselective head to tail linkage. Catalysts for these reactions have yet to be found and this research area is certainly for the future one of the most interesting challenge for the synthesis of fine chemicals using homogeneous catalysis.

Laboratoire de Chimie Organique Appliquée,
ENSC Lille, URA CNRS 402,
USTLille Flandres Artois, BP108, 59652-Villeneuve d'Ascq (France)

TABLE IV

Diene	Diene/Ni	T°C Reaction time	Products	Yield
$CH_2=CHCH=CHCO_2Me$	100	40°C; 0.5h	$MeCO_2CH=CHCH=CHCH_2CH=CHCH_2CO_2Me$	98%
$MeCH=CH-CH=CHCO_2Me$	50	60°C; 25h	$MeCO_2CH=CHCH=CMeCHMeCH=CHCH_2CO_2Me$	22%
,,	50	80°C; 46h	,,	90%

References

1. Heimbach, P.; Jolly, P.W. and Wilke, G. *Adv. Orgamet. Chem.* **1970**, *8*, 29.
2. Foster, R.E. US Pat. Du Pont 2.504.016 (1947150).
3. Cannel, L.G. *J. Am.Chem. Soc.* **1972**, *94*, 6867.
4. (a) Ito, T. and Takami, Y. *Tetrahedron Lett.* **1972**, *47*, 4775; (b) Miller,R.G.; Kealy, T.G. and Barney, A.L. *J. Am. Chem. Soc.* **1967**, *89*, 3756; (c) Baker, R. *Chemical Reviews* **1973**, *73 (5)*, 487; Tolman, C.A. *J. Am. Chem. Soc.* **1970**, *92*, 4217.
5. Akhmedov, Y.M.; Mardanov, M.A. and Khanmetov, V. *J. Org. Chem. USSR* **1971**, *7*, 2610.
6. (a) Wilke, G.; Müller, F.W.; Kröner, M. and Bogdanovic, B. *Angew. Chem. Int. Ed. Engl.* **1963**, *2*, 105; (b) Heimbach, P. and Wilke, G. *Justus Liebigs Ann. Chem.* **1969**, *727*, 183.
7. (a) Keim, W. *Industrial Applications of Homogeneous Catalysis and Related Topics*, Gargnano, Oct 1984; (b) Taqui Khan, M. and Martell, A.E. *Homogeneous Catalysis by metal complexes*, Vol II; Acad. Press, N.Y. and London, **1974**; p 141.
8. Müller, H.; Wittenberg, D.; Seibt, H. and Scharf, E. *Angew. Chem. Int. Ed. Engl.* **1965**, *4*, 327.
9. Singer, H.; Umbach, W. and Dohn, M. *Synthesis* **1972**, 42.
10. Singer, H. *Synthesis* **1974**, 189.
11. Bartik, T.; Heimbach, P. and Himmler, T. *J. Organomet. Chem.* **1984**, *276 (3)*, 399.
12. Bogdanovic, B.; Henc, B.; Meister, B.; Pauling, H. and Wilke, G. *Angew. Chem. Int. Ed. Eng.* **1972**, *11*, 1023.
13. Bogdanovic, B. *Angew. Chem. Int. Eng.* **1973**, *12*, 954.
14. Wilke, G. *Angew. Chem. Int. Ed. Engl.* **1968**, *27*, 186.
15. Buono, G; Peiffer, G.; Mortreux, A. and Petit, F. *J. Chem. Soc. Chem. Comm.* **1980**, 938.
16. Buono, G.; Siv, C.; Peiffer, G.; Triantaphylides, C.; Denis, P.; Mortreux, A. and Petit, F. *J. Org. Chem.* **1985**, *50*, 1782.
17. Reed, H. B. W. *J.Chem. Soc. (London)* **1951**, 1931.
18. (a) Wilke, G. *Angew. Chem.* **1957**, *69*, 397; (b) Breil, H.; Heimbach, P.; Kröner, M.; Müller, H. and Wilke, G. *Makromol. Chem.* **1963**, *69*, 18.
19. Dubini, M.; Montino, F. and Chiusoli, G. P. *Chim. Ind. (Milan)* **1965**, *47*, 839.
20. Bremer, W.; Heimbach, P.; Hey, H.; Müller, E. W. and Wilke, G. *Ann. Chem.* **1969**, *727*, 161.
21. Heimbach, P. *Angew. Chem. Int. Ed. Eng.* **1973**, *12*, 975.
22. Kiji, J.; Masui, K. and Furukawa, J. *Tetrahedron. Lett.* **1972**, 1457.
23. Wilke, G. *Angew. Chem. Int. Ed. Engl.* **1988**, 1985.
24. Heimbach, P. and Hey, H.J. *Angew. Chem. Int. Ed. Engl.* **1970**, *9*, 528.
25. Barnett, B.; Büssemeier, B.; Heimbach, P.; Jolly, P.W.; Krüger, C.; Tkatchenko, I. and Wilke, G. *Tetrahedron Lett.* **1972**, 1457.
26. Richter, W.J. *J. Mol. Catal.* **1981**, *13*, 20; ibid **1983**, *18*, 145.
27. Gros, P.; Buono, G.; Peiffer, G.; Denis, P.; Mortreux, A. and Petit, F. *New J. Chem.* **1987**, *11*, 573.
28. Takahashi, S.; Shibano, T. and Hagihara, N. *Bull. Chem. Soc. Jap.* **1968**, *41*, 454; ibid, *Tetrahedron Lett.* **1967**, 2451.
29. Medema, D. and Van Heiden, R. *Rec. Trav. Chim. Pays Bas* **1971**, *90*, 324.
30. Müller, H.; Wittenberg, D.; Seibst, H. and Scharf, E. *Ann. Chem.* **1967**, *727*, 161;

ibid, *Angew. Chem.* **1964**, *77*, 320.

31. Heimbach, P. *Angew. Chem. Int. Ed. Engl.* **1968**, *7*, 882.
32. Pittman, C. U. Jr. and Smith, L. R. *J. Am. Chem. Soc.* **1975**, *97*, 341.
33. (a) Inoue, Y.; Sekiga, S.; Sasaki, Y. and Hashimoto, H., *Chem. Abstr.* **89** 42280 1978; (b)Takahashi, S.; Shibano, T. and Hagihara, N. *J. Chem. Soc. Comm.* **1969**, 161; (c) Musco, A. *J. Mol Catal.* **1976**, *1*, 443; (d) Takahashi, K.; Hata, G. and Miyake, A. *Bull. Chem. Soc. Jap.* **1973**, *46*, 600.
34. Itakura, I. and Tanaka, H. *Makromol. Chem.* **1969**, *123*, 274; Ger. Pat. 2.061.352 1971 to Mitsubishi Pet. Cie, cf *Chem. Abstr.* **75**, 63079.
35. (a) Yasuda, H. and Nakamura, A. *Reviews of Chemical Intermediates* **1986**, *6*, 365–389; (b) Yamamoto, H.; Yasuda, H.; Tatsumi, K.; Lee, K.; Nakamura, A.; Chen, J.; Kai, Y. and Kasaï, N. *Organometallics* **1989**, *8*, 105.
36. Misono, A.; Uchida, Y.; Furukata, K. and Yoshida, S. *Bull. Chem. Soc. Jap.* **1969**, *42*, 2303.
37. (a) Watanabe, S.; Suga, K. and Kikuchi, H. *Aust. J. Chem.* **1970**, *23*, 385; (b) Mochida, I.; Yusa, S. and Seiyama, T. *J. Catal.* **1976**, *41*, 101.
38. Tenaglia, A. Thesis Université Marseille III 1984.
39. Denis, P.; Mortreux, A.; Petit, F.; Buono, G. and Peiffer, G. *J. Org. Chem.* **1984**, *49*, 5274.
40. Denis, P.; Jean, A.; Croizy, P.; Mortreux, A. and Petit, F. *J. Am. Chem. Soc.* **1990**, *112*, 1292.

A. LECLOUX

CATALYTIC ACTIVATION OF HYDROGEN PEROXIDE IN SELECTIVE OXIDATION REACTIONS

1. Introduction

Over the past few years, the importance of hydrogen peroxide as an oxidising agent for organic compounds has grown considerably. These oxidations generally proceed under relatively mild conditions. Sensitive materials can therefore be handled with confidence and the reactions are often very specific [1].

Oxidations by hydrogen peroxide can be achieved in neutral as well as in acid or alkaline solution.

For some applications, the reactivity of H_2O_2 itself is inadequate so that it is used either in the form of a more active species, such as percaboxylic acid, or in the presence of a catalyst. In this paper, the various modes of H_2O_2 activation are briefly reviewed and some practical applications are given to illustrate the possibilities of catalytic activation.

2. General discussion on the activation of hydrogen peroxide

The activation of H_2O_2 corresponds to accentuating the cleavage of one of the bonds of the molecule. H_2O_2 exhibits a wide variety of action depending:
- on which of the 3 bonds is broken;
- on the fact whether the electron pair of the broken bond is shared or not.

If one of the OH bonds is broken, perhydroxyl groups are formed, while if the peroxidic linkage is broken, hydroxyl groups are produced.

In both cases, the reaction sequence can involve either an *ionic* or a *free radical* pathway, as shown on Scheme **1**.

65

A. F. Noels et al. (Eds.), Metal promoted selectivity in organic synthesis, 65–90.
© 1991 Kluwer Academic Publishers. Printed in the Netherlands.

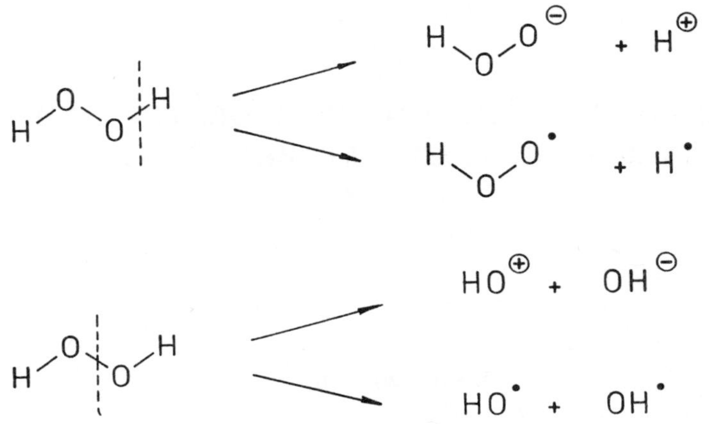

$$
H{-}O{-}O{-}H \quad \Bigg\langle
\begin{array}{l}
H{-}O{-}O^{\ominus} \quad + \quad H^{\oplus} \\[2em]
H{-}O{-}O^{\bullet} \quad + \quad H^{\bullet}
\end{array}
$$

(1)

$$
H{-}O{-}O{-}H \quad \Bigg\langle
\begin{array}{l}
HO^{\oplus} \quad + \quad OH^{\ominus} \\[2em]
HO^{\bullet} \quad + \quad OH^{\bullet}
\end{array}
$$

2.1. HOMOLYTIC DECOMPOSITION

The homolytic metal catalyzed decomposition of hydrogen peroxide (and also alkyl hydroperoxides) can be summarized as follows:

$$ ROOH + M^{(n-1)+} \longrightarrow RO^{\bullet} + M^{n+} + HO^{-} \tag{2} $$

$$ ROOH + M^{n+} \longrightarrow ROO^{\bullet} + M^{(n-1)+} + H^{+} \tag{3} $$

where R is H or alkyl group.

Very often, only one of these reactions is produced by the metal ion.

In this case, the metal ion acts as an initiator rather than a catalyst, because the non catalytic radical chain decomposition can occur via the following steps:

$$ 2\,ROO^{\bullet} \longrightarrow 2\,RO^{\bullet} + O_2 \tag{4} $$

$$ RO^{\bullet} + ROOH \longrightarrow ROO^{\bullet} + ROH \tag{5} $$

R is H or an alkyl group

- When the metal complexe is a strong oxidant, reaction **3** predominates (Pb^{IV}, Ce^{IV}, Ag^{II}, Tl^{III}).
- When the metal ion is a strong reducing agent, reaction **2** predominates

(Cr^{II}, Cu^{I}, Ti^{III}, V^{III}).

- When the metal has 2 oxidations states of comparable stability, reactions 2 and 3 occur concurrently and a true catalytic decomposition takes place. This is the case:
- with *iron* in the decomposition of H_2O_2 (Fenton reagent),
- with *cobalt* and *manganese* in the decomposition of alkyl hydroperoxides.

The electron transfer between metal ions and hydroperoxides occurs via an outer-sphere mechanism in polar solvents and via an inner-sphere covalently bonded complex in non polar solvents.

This kind of activation leads to electrophilic radical species. The reactivities of different C–H bonds towards these species is thus more dependent on the electron supply than on the C–H bonds strength. These electrophilic radical species attack organic compounds by hydrogen abstraction or addition to multiple bonds. The radicals thus formed undergo further reactions finally leading to oxidised products. But radical reactions are not selective, normally producing a mixture of products.

A classical example of this type of oxidation is the hydroxylation of phenols with Fenton's reagent (Fe^{2+}/H_2O_2). This reaction proceeds via the OH radical which attacks the aromatic nucleus forming a cyclohexadienyl radical. This, in turn, is oxidised by Fe^{3+} to form the hydroxylated product; the regenerated Fe^{2+} starts another redox chain.

The general scheme is summarized hereafter:

$$H\text{-}O\text{-}O\text{-}H + Fe^{2+} \longrightarrow Fe^{3+} + OH^- + HO^\bullet$$

$$R\text{-}H + {}^\bullet OH \longrightarrow R^\bullet + H_2O$$

$$R^\bullet + Fe^{3+} \longrightarrow R^+ + Fe^{2+}$$

$$R^+ + OH^- \longrightarrow R\text{-}OH$$

$$\overline{}$$

$$H_2O_2 + R\text{-}H \xrightarrow{\ Fe^{2+}\ } R\text{-}OH + H_2O \tag{6}$$

The Fenton's reagent is a very efficient source of OH radicals. It can initiate polymerization, hydroxylations and oxidative coupling for example.

2.2. HETEROLYTIC DECOMPOSITION

2.2.1. Natural activation

The ionic decomposition of H_2O_2 is primarily due to its weakly acidic character

$$H-O-O-H \quad \rightleftharpoons \quad H-O-O^{\ominus} + H^{\oplus} \tag{7}$$

This natural activation can generate, depending on the pH of the medium, different ionic species.

In basic medium, H_2O_2 reacts with hydroxyl ions to give perhydroxyl anions

$$H-O-O-H + HO^{\ominus} \quad \rightleftharpoons \quad H-O-O^{\ominus} + H_2O \tag{8}$$

This equilibrium is displaced to the right as the basicity of the medium increases.

The perhydroxyl anion may be considered as a super nucleophile, the reactivity of which is about 200 times that of the OH⁻ ion. The perhydroxyl anion reacts very rapidly with a substrate containing an electrophilic center, such as the double bond of an olefin activated by an electron withdrawing substituent, or the carbon atom of a carbonyl compound or of a nitrile group, or an organometallic cation. In the absence of any other reagent, HOO⁻ oxidises a second molecule of H_2O_2, via an unstable transition complex which decomposes with release of molecular oxygen.

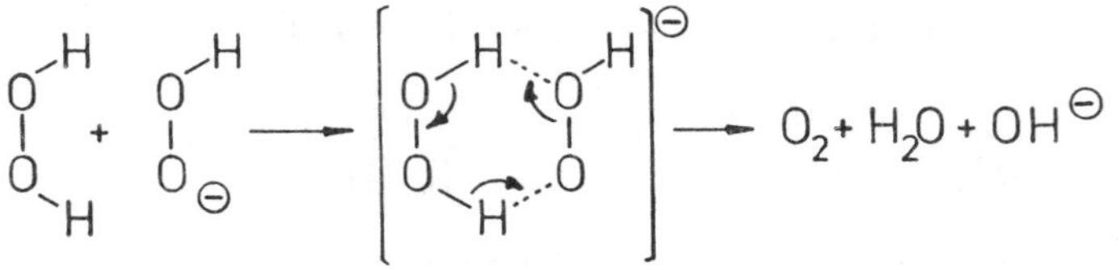

$$\tag{9}$$

This reaction explains the well known instability of H_2O_2 in alkaline media. This decomposition reaction of H_2O_2 can be largely or even entirely overridden in the presence of electrophilic substrate. One of the main applications of the activation of H_2O_2 in basic medium is the synthesis of amine oxides according to the scheme:

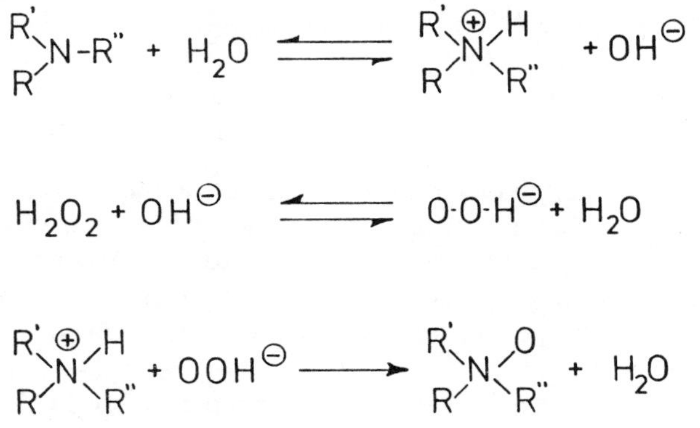

$$\underset{R}{\overset{R'}{>}} N\text{-}R'' + H_2O \;\rightleftharpoons\; \underset{R}{\overset{R'}{>}} \overset{\oplus}{\underset{R''}{N}}{\overset{H}{\diagup}} + OH^{\ominus}$$

$$H_2O_2 + OH^{\ominus} \;\rightleftharpoons\; O\text{-}O\text{-}H^{\ominus} + H_2O$$

$$\underset{R}{\overset{R'}{>}}\overset{\oplus}{\underset{R''}{N}}{\overset{H}{\diagup}} + OOH^{\ominus} \;\longrightarrow\; \underset{R}{\overset{R'}{>}}N{\overset{O}{\underset{R''}{\diagdown}}} + H_2O \tag{10}$$

Amino hydroperoxides or amino peroxides can be formed at room temperature starting from H_2O_2, ammonia and a carbonyl compound:

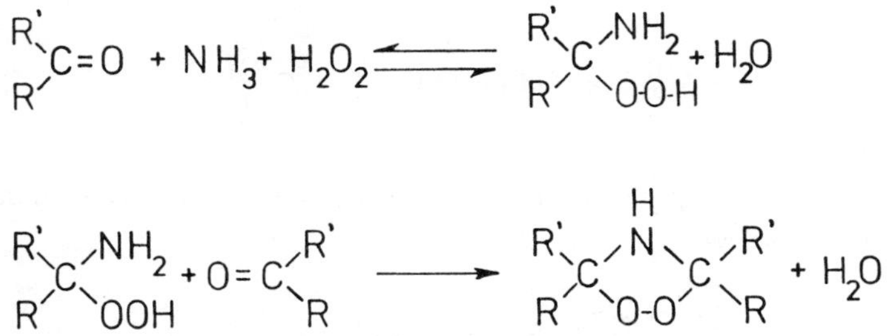

$$\underset{R}{\overset{R'}{>}}C=O + NH_3 + H_2O_2 \;\rightleftharpoons\; \underset{R}{\overset{R'}{>}}C{\overset{NH_2}{\underset{O\text{-}O\text{-}H}{<}}} + H_2O$$

$$\underset{R}{\overset{R'}{>}}C{\overset{NH_2}{\underset{OOH}{<}}} + O=C{\overset{R'}{\underset{R}{<}}} \;\longrightarrow\; \underset{R}{\overset{R'}{>}}C{\overset{\overset{H}{N}}{\underset{O\text{-}O}{<}}}C{\overset{R'}{\underset{R}{<}}} + H_2O \tag{11}$$

These peroxides are relatively stable and their pyrolysis results in a rearrangement to give N-substituted amides.

$$\underset{R}{\overset{R'}{>}}C{\overset{\overset{H}{N}}{\underset{O\text{-}O}{<}}}C{\overset{R'}{\underset{R}{<}}} \;\xrightarrow{\Delta T}\; R'\text{-}\underset{\overset{\|}{O}}{C}\text{-}N\text{-}R + \underset{R}{\overset{R'}{>}}C=O \tag{12}$$

In acidic medium, H_2O_2 is more stable because the equilibrium

$$H\text{-}O{\underset{O}{\diagdown}}\text{-}H \;\rightleftharpoons\; H\text{-}O\text{-}O^{\ominus} + H^{\oplus} \tag{13}$$

is displaced to the left. The formation of an oxonium structure is observed,

A. Lecloux

resulting from the solvatation of protons by hydrogen peroxide.

$$H-O_{O-H} \quad H^{\oplus} \quad \rightleftharpoons \quad H-O_{O<^H_H} \quad \oplus \tag{14}$$

This structure confer an electrophilic character on the H_2O_2 molecule. But when the reaction medium contains a substrate which is a more powerful nucleophile than H_2O_2 (acid, alcohol, ketone, oxometal group), this substrate is protonated, a new electrophilic intermediate is generated which reacts with H_2O_2 to give a new peroxy compound: peracid, hydroperoxide, metalloperoxidic complexe.

In this activation process, the oxidizing properties of H_2O_2 have been simply transfered to an other molecule acting as a support. In some cases this transfer may be a catalytic process. In this transfer, the oxidizing power might be slightly decreased or markedly increased depending on the peroxidic compound formed [2]. In general the reactivity increases in the order.

$$R'-O-O-H \; < \; H-O-O-H \; \ll \; R-C{<^O_{O-OH}} \tag{15}$$

Within the percarboxylic acid series, the following sequence of increasing reactivity is observed.

$$CH_3 \; < \; C_6H_5 \; < \; m\text{-}Cl\cdot C_6H_4 \; < \; H \; < \; CF_3 \tag{16}$$

The use of peracids, preformed or prepared in situ, is well known in epoxidation reaction for example. The epoxidation of olefins with peracids and its mechanism have been the subject of many reviews published over the past ten years. It implies an electrophilic attack by the peracid on the double bond of the olefin. The mechanism proposed by Bartlett (Scheme 17) is generally accepted.

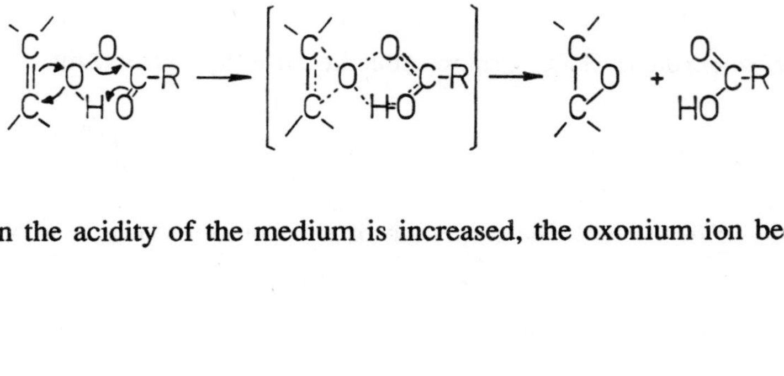

$$\tag{17}$$

When the acidity of the medium is increased, the oxonium ion becomes unstable and generates OH^+ cations. These OH^+ exhibit a very strong electrophilic character towards aromatics, alcohols or even alkanes.

Several selective methods of monohydroxylation of substituted aromatics by H_2O_2 in the presence of an acid have been developed [3]. The acid can be either protic or Lewis acid. Its role is a double one: on the one hand, it takes part in the formation of OH^+ cation and on the second hand it partially neutralizes the effect of the substituent present in the initial molecule (or of the OH group which is introduced), thus limiting the polyhydroxylation.

2.2.2. Catalytic activation

Some years ago, according to the work of Mimoun [4], the catalytic activation of H_2O_2 has been considered as leading to peroxo like species, similar to the active intermediates involved in oxygen activation. This approach was supported by the spectroscopic data showing that the O–O bond in the H_2O_2 molecule lies between superoxo and peroxo bond.

TABLE I
Oxygen activation

Name	Molecular	Superoxo		Peroxo	Oxo
Form	O_2	O_2^-		$O_2^=$	$2O^=$
Bond order	2	1,5		1	0
O–O distance, Å	1.21	1.33	1.4	1.49	–
O–O frequency, cm⁻¹	1580	1097	890	802	–
			↑		
			H_2O_2		

These peroxo complexes can act as shown hereafter (Figure 1).

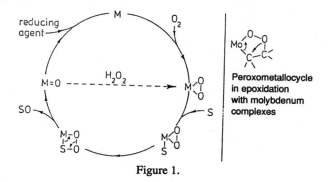

Peroxometallocycle in epoxidation with molybdenum complexes

Figure 1.

In this scheme, the substrate and the O–O bond are on the same metal atom M. There is a nucleophilic attack of the O–O bond on the substrate, the formation of a peroxometallocycle and the evolution of the oxidized substrate, leaving the catalyst in the form of an oxo complex. This complexe has to be regenerated: a reducing agent is needed to complete the cycle. In principle H_2O_2 must act as this reducing agent, but all efforts to use these *peroxo* compounds as catalyst failed. The main reason of this failure is probably the fact that in these complexes the metal is in its highest oxidation state and there is no more free electron to be back donated to the O–O bond. Consequently, the nature of the metal and the ligands has nearly no effect on the O–O distance and the activation of the peroxidic bond. These results indicate that such peroxo complexes are not directly related to the catalytic activation of H_2O_2 or hydroperoxides.

In fact, it is better to consider that the catalytic activation of H_2O_2 must proceed via the transfer of the perhydroxyl group to the catalyst metal ion.

Many acidic metal oxides such as OsO_4, WO_3, MoO_3, Cr_2O_3, V_2O_5, TiO_2 and SeO_2 activate H_2O_2 via the formation of inorganic peracid, where the M=O group plays a similar role to that of the C=O group in organic peracids.

The principal function of the catalyst is to withdraw electrons from the peroxidic oxygens, making them more susceptible to attack by nucleophiles such as olefins. In so doing the catalyst acts as a Lewis Acid. This type of activation of the peroxidic linkage is also observed with hydroperoxides.

A mechanism has been suggested by Shirmann [5] involving in a first step an hydroxy-hydroperoxy structure:

$$M = O + H_2O_2 \rightleftharpoons M \begin{array}{c} \diagup OH \\ \diagdown OOH \end{array} \tag{18}$$

where the activity highly depends on the nature of M and the ligands carried by M. The catalytic cycle could be described as follows, in the case of olefin epoxidation (Figure 2).

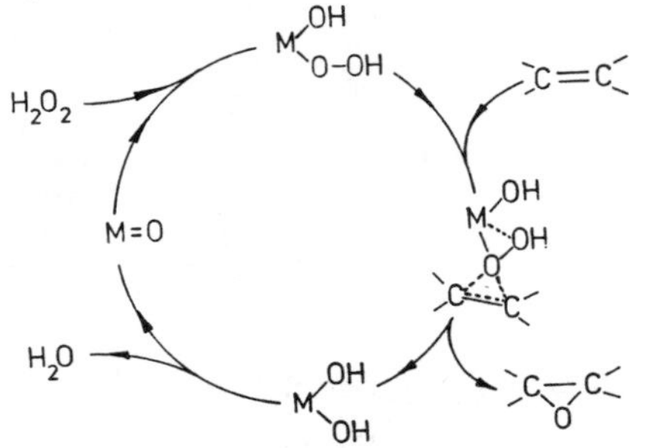

Figure 2.

In a recent review paper, Sheldon discusses the different proposed mechanisms related to the olefin epoxidation by hydroperoxides in the presence of a catalyst. We will return on this subject later. This general discussion on the various ways of activating hydrogen peroxide has shown the versatility of this compound as a source of active oxygen. In the second part of this paper these possibilities are illustrated by three different examples. These show that interesting results can be obtained by the combina-hydrogen peroxide and a catalyst.

3. Practical applications

In the laboratory and industrial practice, most of the oxidation reactions of organic substrates are carried out either with percarboxylic acids or with organic hydroperoxides and a catalyst. These processes generally require a two steps operation: the preparation of the peroxydic compound and the oxidation reaction itself. Moreover, it is interesting to point out that the use of H_2O_2 instead of other peroxidic compounds could reduce by a factor of 4 the cost of the «active oxygen».

It is the reason why, over the past few years, we have checked the possibilities of performing oxidation reactions, on organic substrates, in one step and using hydrogen peroxide as an oxidant.

3.1. OXIDATION OF KETONES TO ESTERS OR LACTONES

The first reaction we were interested in was the oxidation of ketones to esters or lactones [6]. We discovered that the direct oxidation of ketones by hydrogen peroxide is possible provided the carbonyl bond of the ketone is activated by a Friedel Crafts catalyst. The reaction procedure is a very simple one. The H_2O_2 is gently introduced into a reactor containing the carbonyl compound and the catalyst.

The temperature and the pressure in the reactor are monitored in such a way that the water is continuously removed by azeotropic distillation. The steady state concentrations of water in the reaction medium is maintained at a very low level during the reaction: about 1g/kg. The main characteristics of this process are the following:

-- Among the Friedel-Crafts catalysts, the Lewis Acids are much more effective than the protonic acids. Hydrogen fluoride is the only protonic acid which gives satisfactory results, but it is well known that it is not very dissociated. All the other protonic acids give a lot of peroxides as by products and favor the hydrolysis of the esters formed.

- With the best catalysts such as BF_3, SbF_3, SbF_5, $SnCl_4$, TaF_5, WF_6, $MoCl_5$, the rate of reaction is so high that hydrogen peroxide is consummed as soon as introduced.
  As a consequence, the concentration of H_2O_2 in the medium is never high and the process is very safe. In practice, the rate of introduction of H_2O_2 in the reactor is only limited by the possibility of removing from the medium, the heat and the water produced by the reaction.

- As already stated, the water concentration in the reaction medium must be kept at a low level essentially to limit the complexation or the hydrolysis of the catalyst by water in order to keep it free to complex the carbonyl

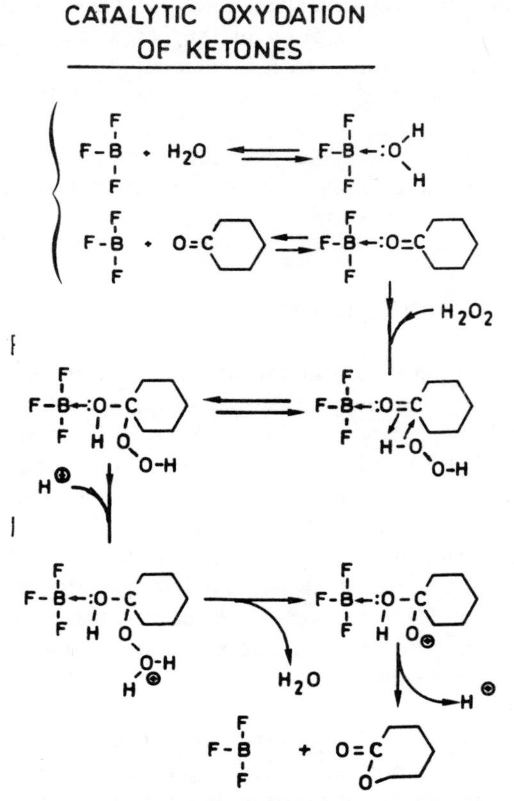

CATALYTIC OXYDATION
OF KETONES

Figure 3. Catalytic oxydation of ketones.

group of the reactant. The catalyst concentration is indeed much lower (about 10^{-4} on a molar basis) than the concentration generally used with protonic acid. The mechanism of reaction proposed on Figure 3 is a very classical one but it explains the inhibiting effect of water and the interest of limiting the protonic acidity to avoid by-product formation.

– An other interesting feature of the process is its high productivity: 200 to 500g of products per hour in an one liter reactor can be produced with a good selectivity. Figures as high as 90% have been obtained. In general, the selectivity towards H_2O_2 is slightly lower due to a secondary decomposition, especially in laboratory vessels where the surface/volume ratio is particularly high. Most of the experimental work we have done corresponds to the oxidation of cyclohexanone to ε-caprolactone. The Table II hereafter shows some of the results obtained in this case.

TABLE II
Oxidation of cyclohexanone

Catalyst		Productivity gε-CL/h.l.	Selectivity towards	
Name	Amount g/kg		H_2O_2 %	cyclohexanone %
SbF_5	0.08	231	75.9	84.7
$SnCl_4$	0.3	209	68.8	91.7
BF_3	0.7	233	76.7	90.8
SbF_3	1.4	234	76.9	92.1
$MoCl_5$	0.2	174	57.3	91.6
TaF_5	0.4	222	73.0	91.3
H_2So_4	0.2	50	17	–
HF	0.3	59	27	–
ZnO	0.3	9	13	–
HF + ZnO	0.3+0.3	189	62	–
Sb_2O_5	0.7	33	16	–
HF + Sb_2O_5	0.3+0.5	222	73	–

Reaction temperature: 80°C
Reaction pressure: 13.3 kPa (100 torr)
Reaction time: 30 min.

– Finally, this catalytic process of ketone oxidation presents two particularities which could be of interest in fine chemical synthesis.

1. On the one hand, the use of a catalyst introduces some interesting steric

effects which do not appear when a percarboxylic acid is used. The presence
of the catalyst in the reacting complex reduces in some way the accessibility
to the carbon atom on which the nucleophilic attack of H_2O_2 takes place.
The steric hindrance can modify the approach or even slow down the rate of
reaction, as shown on Figure 4.

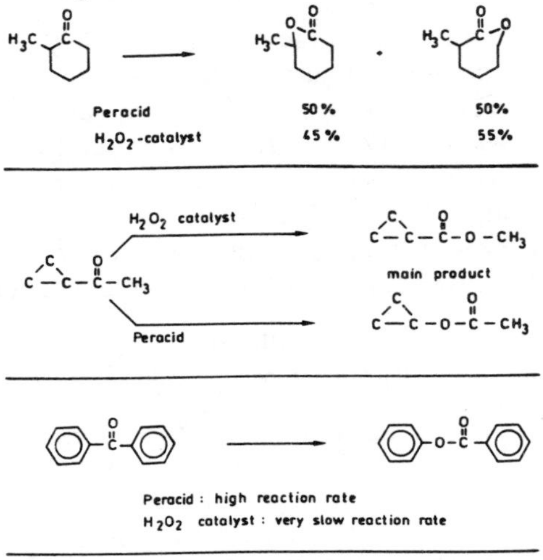

Figure 4.

2. On the other hand, by the system H_2O_2/Lewis Acid catalyst, it is possible
to prepare unsaturated lactones or esters by reacting an unsaturated ketone.
This is not possible by using percarboxylic acids as an oxidant because an
epoxidation reaction will occur at the same time as the esterification
proceeds (Figure 5).

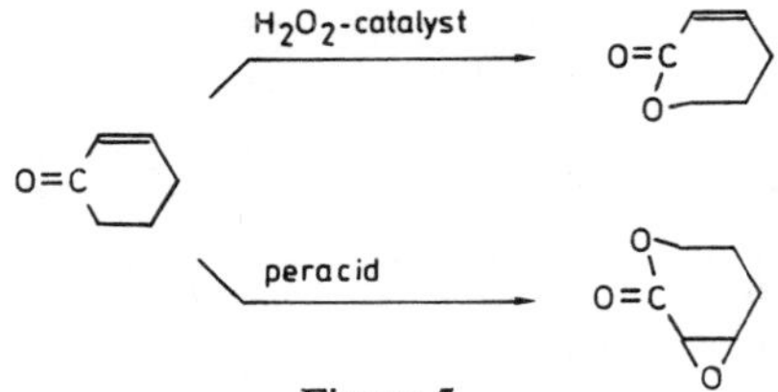

Figure 5.

Thus, by combining some steric hindrance and the specific activation of
the carboxylic group it could be possible to synthetise particular unsaturated
esters or lactones difficult to obtain by an other process.

3.2. OLEFIN EPOXIDATION

Over the past ten years, many studies have been published on the olefin epoxidation by peroxidic compounds, mainly oriented to the synthesis of propylene oxide. The Oxirane process is based on the catalytic activation of hydroperoxides, while the Interox and Degussa processes claim the use of perpropionic acid as oxidative agent. More recently, PCUK announced the development of a new process using hydrogen peroxide as an oxidant and boric acid as a catalyst. Union Carbide has proposed the use of an arsenic based catalyst to epoxidize olefins with H_2O_2.

In most cases, the catalytic oxidation of olefins by hydrogen peroxide is limited to low conversion and productivity. We discovered [7] that very good yields of epoxide can be obtained by oxidizing olefins with H_2O_2 in the presence of a selenium based catalyst. This is quite surprising because selenium is well known to give vicinal diols from olefins. The reaction procedure is a very simple one. The H_2O_2 is gently introduced in the reaction medium containing a solvent, the olefin, the dissolved catalyst and an organic base. The reaction temperature is generally set at the boiling point of the mixture, but it can be varied by reducing the pressure in the reaction vessel. The vapours are either directly refluxed in the reactor or condensed in a side vessel where a part of the distilled water is separated from the organic phase which is recycled to the reactor. Despite the simplicity of the process, various conditions have to be fullfilled in order to obtain good yields and hight selectivities in epoxide:

- the reaction medium must be *homogeneous*. When it is present, the solvent must be chosen so that only one phase is present in the reactor containing the olefin, the catalyst, an organic base, hydrogen peroxide and some water. For this reason, and for others that will be discussed later on, the solvent is most often an alcohol;
- the catalyst is most often selenium dioxide or selenious acid H_2SeO_3;
- an organic base must be present in order to improve the selectivity and the yield of epoxidation. The pK_a of the base is less than 8.

Before the discussion of the mechanistic aspects of the reaction, some figures are given in Table III. These show the practical possibilities of this catalytic system, which is particularly well suited to internal olefins and water soluble olefins.

It is worthwhile to point out the combination of high selectivities with high productivities and relatively high olefin conversion.

The ratios olefin/solvent and organic base/catalyst have to be adapted to each case. In general, when these two ratios are progressively increased:

TABLE III

Olefins epoxidation

Olefin	Solvent	Base	Conversion of		Selectivity towards		Productivity in epoxide
			H_2O_2 %	olefin %	H_2O_2 %	olefin %	g/h.l.
Cyclohexene	Benzyl alc.	Pyridine	96.8	31.3	86.7	92.6	467
x-Pinene	n-Butanol	Pyridine	98.5	34	88	89	210
Allyl alcohol	–	Quinoline	89	20	94	95	212
2,5-Dihydrofuran	benzene as stripping agent	Lutidine	85	19	90	>90	>200

1. The yield in epoxide and the rate of reaction pass through a maximum as shown on Figure 6.

2. The selectivity continuously increases (Figure 6).

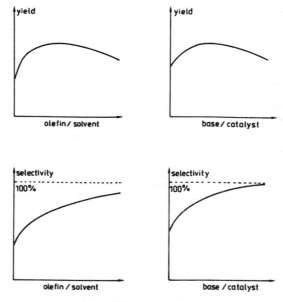

Figure 6.

Another interesting observation is illustrated on Figure 7. It shows the rate of H_2O_2 consumption during the epoxidation of a given olefin in two different solvents (isopropanol and n-butanol) in the same conditions.

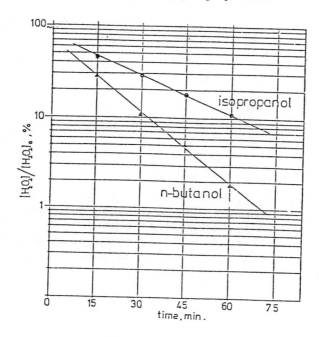

Figure 7.

This result clearly shows that, if an alcohol is used as solvent, it can modify the rate of reaction and is thus involved in some way in the catalytic complex. Actually, no solvent is needed when the olefin is itself an alcohol or a polar compound like an ether or an ester. In this case the role of the solvent at the catalyst level is played by the olefin. Finally, in our attempt to understand the role of the base in the catalytic system, we came to the conclusion that there are many different equilibria between the catalyst, the solvent, the base, the olefin and the water present in the medium. At first, we thought that the role of the base is to neutralize the acidity of the medium. It is the reason why we started to titrate the remaining acidity in the reacting medium. The titration was made by adding a strong base (n-butylamine) to the medium containing the solvent, the olefin, the catalyst, some water and the catalytic base. Bromophenol blue was used as indicator. Surprisingly, as shown on Figure 8 the acidity of the medium increases by increasing the amount of pyridine added. The measured acidity generally reaches a plateau, the level of which is lower than the value corresponding to the first acidity of selenious acid.

A. *Lecloux*

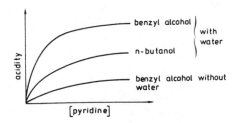

Figure 8.

Moreover the level of the plateau and the pyridine effect depends on the alcohol used as solvent. If no water is present in the medium, the apparent acidity decreases and the effect of pyridine is much smaller.

If no base is added, the apparent acidity of the medium increases with the amount of water added and the variation depends on the type of alcohol present, as shown on the Figure 9.

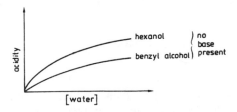

Figure 9.

It is important to note that the olefin modifies the apparent acidity of the medium in a way depending on its nature. The acidity seems to be taken up by the olefin as shown on Figure 10.

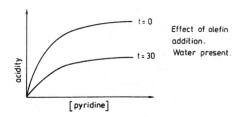

Figure 10.

The upper curve is obtained directly after the addition of the olefin in the medium and the lower curve, 30 minutes later. This effect is particularly marked in the case of substituted olefins and depends on the solvent used.

All these results indicate the formation and dissociation of various competing complexes between the catalyst, the base, the alcohol, the olefin and water.

We think the base plays three different roles:

— first, it assists the dissociation of selenium dioxide and selenious acid and, doing so, it increases its reactivity towards H_2O_2 and aids to the formation of a perselenious compound;
— second, the base protects the epoxide by catching free protons: limiting the hydrolysis of the oxirane bridge;
— third, the base reduces the secondary decomposition of hydrogen peroxide.

CATALYTIC EPOXIDATION : MECHANISM

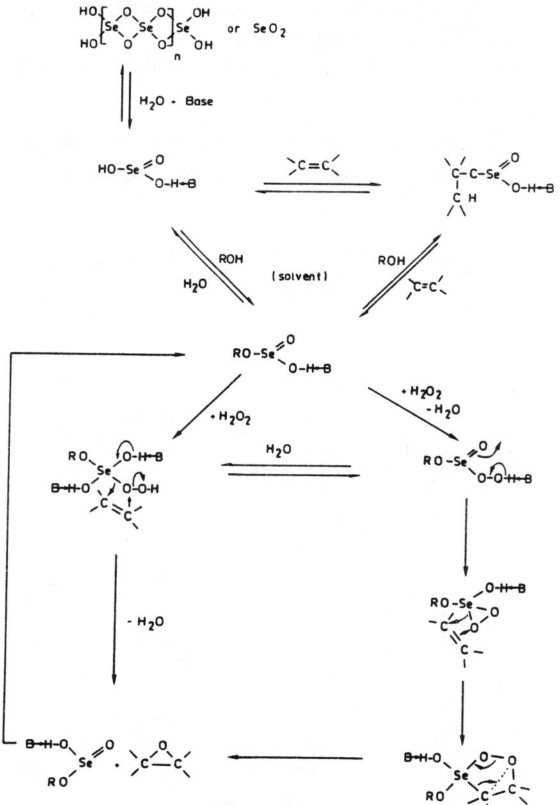

Figure 11.

The optimal yield in epoxide is obtained with an optimal quantity of base, this quantity depending on the reaction medium composition and on the pK_a of the base. With a lesser amount of base, the catalyst is poorly dissociated and the hydrolysis is favoured. With a too high quantity of base, the rate of reaction decreases (due to complexation) but the selectivity increases.

To explain all these observations, we propose the catalytic scheme shown on Figure 11. Some uncertainty remains because the perselenious compound cannot be considered neither as a true peracid nor as a transition metal complexe. The proposed scheme seems however to be able to explain most of the experimental observation.

To conclude, the main advantages of the catalytic process presented here are the facts that:
— a commodity chemical is used as oxidant;
— productivity and selectitivy figures are compatible with industrial requirements;
— highly substituted epoxides can be obtained in good yield despite the reactivity of this kind of oxirane bridge;
— aqueous soluble olefins (like unsaturated alcohols, ketones or ethers) can be easily epoxidized with a good selectivity.

These advantages open the way to new interesting and cheap synthesis of chemical compounds such as those presented on Figure 12.

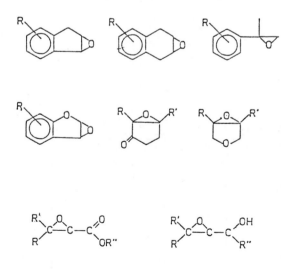

Figure 12.

3.3. AMINES OXIDATION

3.3.1. Aromatic amines

It is well known that the oxidation of aromatic amines to the corresponding oxides is a reaction requiring strong oxidative agent, such as peracids, and long reaction times. We discovered [8] that very good yields of aromatic amine-oxides can be obtained by oxidizing the corresponding amines with hydrogen peroxide in the presence of a selenium based catalyst.

The reaction procedure is a very simple one. The H_2O_2 is gently introduced into the reaction medium containing a solvent, the aromatic amine and the catalyst. The reaction temperature is generally set at the boiling point of the mixture, but it can be lowered by reducing the pressure in the reaction vessel. The vapours are, either directly refluxed in the reactor, or condensed in a side vessel where a part of the distilled water is separated from the organic phase which is recycled to the reactor.

In order to obtain good yields and high selectivities in amine oxide, various conditions have to be fullfilled:

- the steady state concentration of water in the reaction medium has to be maintained at a low level during the reaction: less than 10g/kg; it is the reason why the solvent is generally chosen in such a way that the water can be removed by azeotropic distillation; for the same reason highly concentrated hydrogen peroxide (70% wt or higher) is preferably used as reactant;
- the Se/amine molar ratio has to be set about 0.03 and the solvent/amine volume ratio at 3 or higher;
- the reaction temperature is preferably lower than 80 C and the reaction time between 1 and 2 hours;
- for safety reason, it is important to reach high hydrogen peroxide conversion; in order to do so the amine conversion has to be limited to max. 80%; the economics of the process implies of course the recycling of the unconverted amine.

In those conditions, pyridine and methylpyridines give the best results, typical values being:

- amine conversion: 65 to 80%
- selectivity vs the amine: 96 to 100%
- H_2O_2 conversion: 90 to 98%
- selectivity vs H_2O_2: 83 to 89%.

If these figures as well as the reaction rate observed at the lab scale are used to calculate the productivity in amine oxide, very high values are obtained, typically 700 to 800 tons of oxide per year and per m³ of reactor. Such values look very attractive at an industrial point of view.

In the case of the 2-chloropyridine, the nitrogen is desactived due to the electron withdrawing effect of the chlorine atom and, consequently, the yield and the selectivities are lower than in the case of pyridine, and the productivity of amine oxide is limited to 300 tons/year/m³.

As show on Figure 13, the H_2O_2 consumption rate appears to obey a first order law when n-butanol is used as solvent.

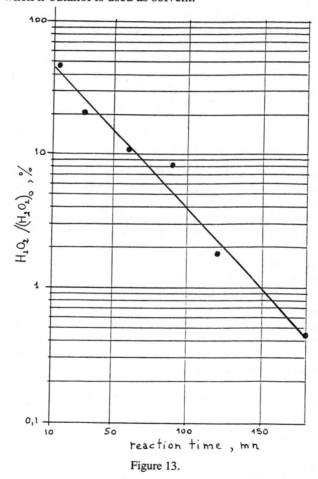

Figure 13.

To explain the various experimental results, perselenious acid is supposed to be formed as an intermediate. This per compound either reacts with the amine to produce the corresponding amine oxide (rate constant k) or decomposes, with a rate constant k_d, in selenious acid and oxygen according to the following scheme:

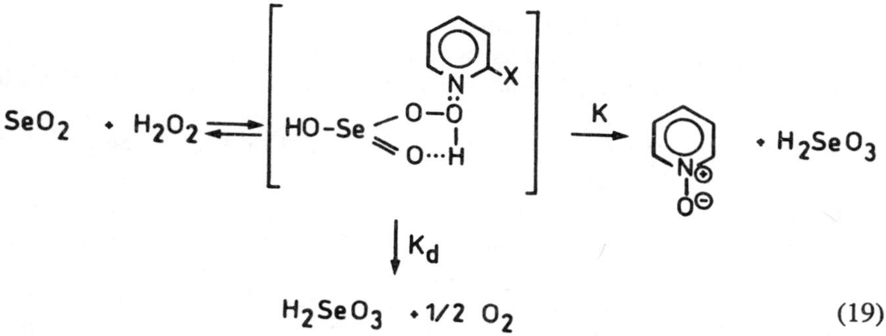

$$\text{SeO}_2 \cdot \text{H}_2\text{O}_2 \rightleftharpoons \left[\text{HO-Se} \underset{\text{O}\cdots\text{H}}{\overset{\text{O-O}}{<}} \right] \xrightarrow{\text{K}} \bigcirc + \text{H}_2\text{SeO}_3$$

$$\downarrow k_d$$

$$\text{H}_2\text{SeO}_3 \cdot 1/2 \ \text{O}_2 \tag{19}$$

Consequently, a high selectivity vs H_2O_2 is linked to a high value of the k/k_d ratio; a higher k value can be obtained by increasing the amine/peroxide ratio in the medium, while a low k_d value can only occur if the polarity of the medium is sufficiently weak to allow «internal» hydrogen bond in the active intermediate. This could explain the negative effect of water on the reaction selectivity.

It is also interesting to note that, in the case of picolines, no oxidation of the methyl group occurs even for a large H_2O_2 excess.

3.3.2. Aliphatic secondary amines

It is well known that the oxidation of aliphatic secondary amines by hydrogen peroxide leads to the corresponding hydroxylamines. The yields are however limited due to the «natural» decomposition of H_2O_2 in basic media.

A yield improvement is observed when complexing agents (such as EDTA or polyphosphonic acids) are added in the reaction vessel. These compounds reduce the catalytic hydrogen peroxide decomposition by chelating the metalic impurities present in the reaction medium. On the contrary, neither positive nor negative effect of the addition of radical scavengers in the medium is observed; this is an indication that radical species are not significantly involved in the reaction mechanism of amine oxidation.

By looking at possible catalytic effects in this type of reaction, we suprisingly discovered a significant promoting effect of zinc and cadmium salts increasing the selectivity vs H_2O_2 by about 15% even when they are

used at molar concentrations as low as 10^{-4}. The exact role of these compounds is still unclear:
- these two metals are not known as activating agent of hydrogen peroxide;
- the same effect is observed with different types of salts even if they are not soluble in the reaction medium.

It is believed that Zn or Cd acts more as promoting agent than as catalyst, probably by inhibiting in some way the H_2O_2 decomposition in basic medium.

Whatever the exact role played by Zn and Cd, high selectivities in oxidation of dialklamines are obtained under very different operating conditions from those found for pyridine. These are summarized hereafter:
- high polarity solvents are the most suitable: water, methanol and their mixtures are most often used; the ideal is to adjust the solvent composition at the highest polarity compatible with a monophasic reacting medium. In these conditions, diluted hydrogen peroxide can be used as reagent;
- as shown on Figures 14 and 15, the yield in hydroxylamine vs hydrogen peroxide increases with the amine concentration in the reaction medium and with the amine/H_2O_2 molar ratio. On the economical point of view there is thus a balance between a high yield vs H_2O_2 and a high amine conversion level.

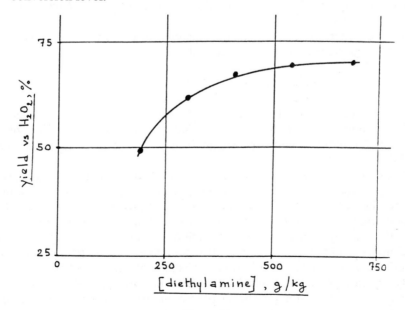

Figure 14.

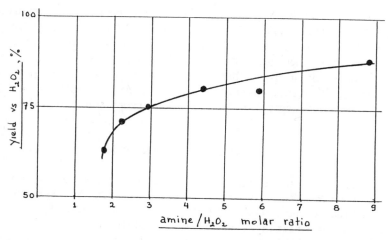

Figure 15.

To explain the positive effect of the polarity, it is supposed that the transition reaction state is a quaternary ammonium species similar to:

$$\begin{array}{l} R \\ {}^{R}\!\!\diagdown N\text{-}H \end{array} + H_2O_2 \rightarrow \left[\begin{array}{c} R \\ R\text{-}N^+\text{-}H\ldots\ldots\ ^-OH \\ OH \end{array} \right] \rightarrow \begin{array}{l} R \\ {}^{R}\!\!\diagdown N\text{-}OH \end{array} + H_2O \qquad (20)$$

However the role of the Zn or Cd compounds in such a mechanism is still unclear.

To illustrate the practical interest of this system some representative data are given in Table IV.

TABLE IV

Type of amine	diethylamine	di-n-propylamine
Solvent	water	methanol (65)/water (35)
$2ZnCo_3 \cdot 3Zn(OH)_2$	2,2 mmole	0.6 mmole
Reaction temperature	60°C	78°C
Reaction time	45 mn	90 mn
Amine/H_2O_2 (molar)	2,84	3
H_2O_2 conversion	100%	100%
Selectivity vs H_2O_2	75%	70%
Amine conversion	26%	23%
Selectivity vs amine	100%	97%
Productivity t/y/m^3	1400	600

These figures are industrially attractive, but the different steps involved in the product separation and their impact on the manufacturing cost have to be examined in detail before establishing the economical balance of such a process. We have checked, on the one hand, the feasibility of the isolation of diethylhydroxylamine by distillation and, on the other hand, the possibility of replacing the methanol/water solvent by a heavier polar solvent like butanediol or ethyleneglycol in the case of di-n-propylamine oxidation.

4. Conclusion and safety rules

The various oxidation reactions reviewed in this paper clearly show the interest of the catalytic activation of hydrogen peroxide. As these oxidations generaly proceed under relatively mild conditions and most often in a very selective way, sensitive materials can be handled with confidence.

Hydrogen peroxide is normally quite stable but may be rendered unstable by a wide variety of contaminants. Decomposition results in the liberation of a large quantity of gaz and energy. That is the reason why, to prevent incidents caused by the rapid decomposition of peroxygen chemicals in

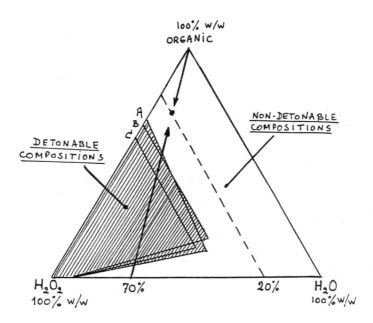

Figure 16.

reaction with organic compounds, a set of safety rules should be followed. These safety rules have been derived logically by reference to the typical triangular diagram shown on Figure 16; it illustrates the three component mixtures of hydrogen peroxide, organic compound and water which are highly representative of most organic reactions. (A, B, C indicate three different organic species).

that mixtures do not occur with compositions within the detonable area. It is recommended that reactions are carried out in such a way as to prevent the hydrogen peroxide content exceeding 20% w/w. If the reaction has two or more phases this should apply to each phase. Proper attention should be paid to ensure adequate mixing in all cases. As illustrated in the diagram, the order of addition of the component is of great importance.

To help people in handling hydrogen peroxide-organic mixtures in safe conditions, a check list of safety precautions is proposed hereafter:

1. Wear adequate personal protection;
2. Clean all glassware and working area;
3. Protect vessels from sources of contamination;
4. Store in a cool place, away from sunlight, in vented containers;
5. Vent all reactors and equipment;
6. Carry out new reactions on a small scale;
7. Use alcohol thermometers and stainless stell ball hydrometers;
8. Always plan reaction;
9. Always add peroxygen compound to the organic material;
10. Addition should be controlled;
11. Provide efficient agitation. Stop peroxygen feed if agitator fails;
12. Ensure peroxygen content does not exceed 20% w/w (as H_2O_2) during reaction;
13. Pre-select reaction temperature. Do not increase temperature after addition if no reaction results;
14. Supply reaction with adequate cooling facilities;
15. Analyse reactor product and destroy unreacted peroxide before distillation or evaporative crystallisation;
16. Never use acetone (or other lower aliphatic ketones) as a solvent or cleaning fluid.

In the vast majority of cases the precautions quoted will ensure that no incidents occur. There are however situations where the general triangular diagram present above does not hold. In such cases, specialized companies offer free consultation on all aspects of hasards and safety.

A. Lecloux

Acknowlegdements

This work was supported by INTEROX Group.

SOLVAY & Cie, Central Laboratory,
Rue de Ransbeek, 310, B-1120
Bruxelles (Belgium)

References

1. Sheldon, R.A.; Kochi, J.K. *Metal Catalyzed Oxidations of Organic Compounds*; Acad. Press: New-York, 1981.
2. Sheldon, R.A. *The Chemistry of Peroxides* in *The Chemistry of Functional Groups*; Patai, S., Wiley, New-York, 1983; pp 161–200.
3. di Furia, F.; Modena, G. *Pure and Appl. Chem.* **1982**, *54*, 1853.
4. Mimoun, H. et al. *J. Amer. Chem. Soc.* **1983**, *105*, 3101.
5. Schirman, J.P. et al., German Patent 2752626, 1978.
6. Lecloux, A.J. et al., European Patent 021525, 1980; in the name of Interox.
7. Lecloux, A.J. et al., European Patent 068564, 1982; in the name of Interox.
8. Lecloux, A.J. et al. European Patent 224662, 1986; in the name of Interox.

VALERIA CONTE, FULVIO DI FURIA,
GIULIA LICINI AND GIORGIO MODENA

ENANTIOSELECTIVE S-OXIDATION: SYNTHETIC APPLICATIONS

The homochiral sulfinyl group, owing to its ability to promote and to direct reactions taking place at the adjacent carbon atoms, has been widely used in synthetic procedures [1]. A few typical examples taken from the recent literature are shown in Scheme 1.

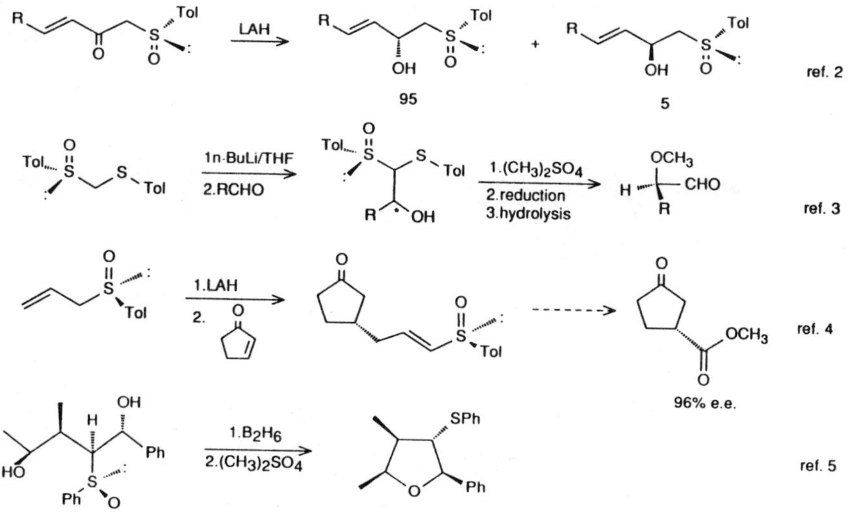

Scheme 1

The preparation of optically active sulfoxides is still mainly based on the method developed in the early sixties by Andersen involving the separation of diastereomeric menthyl sulfinates and their transformation to sulfoxides [6]. Some procedures for the resolution of racemic sulfoxides have been also reported [7].

At any rate, a direct oxidation of the parent thioethers would be clearly a

A. F. Noels et al. (Eds.), Metal promoted selectivity in organic synthesis, 91–105.
© 1991 Kluwer Academic Publishers. Printed in the Netherlands.

V. Conte et al.

more appealing route. In principle this could be achieved by using either an achiral oxidant and a homochiral thioether or a homochiral oxidant and a prochiral thioether. Methods based on both approaches have been published but their synthetic value is usually rather scarce because of the low optical yields obtained or of the very narrow field of application. A noticeable example is the reaction shown in Figure 1 where a high degree of enantioselection is obtained in the oxidation with *m*-chloroperbenzoic acid (MCPBA) of a thioether containing a chiral, not racemic group [8].

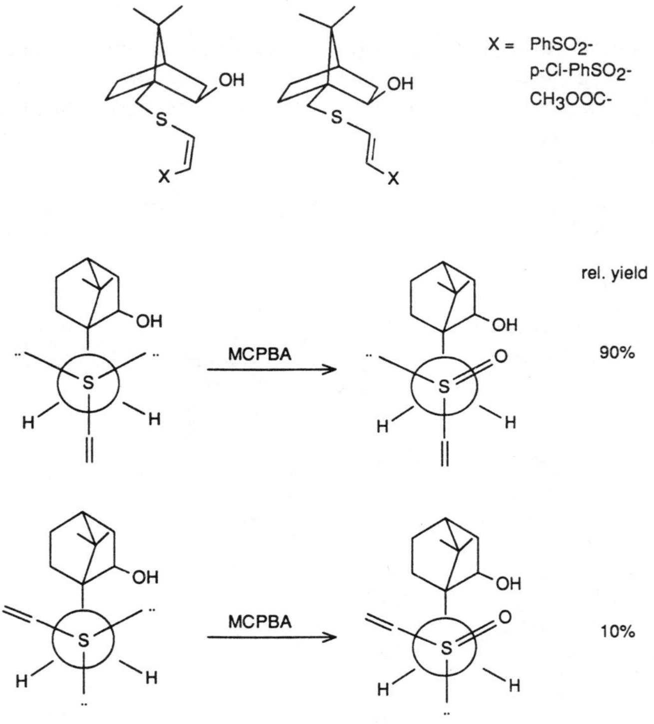

X = PhSO2-
     p-Cl-PhSO2-
     CH3OOC-

rel. yield

90%

10%

Figure 1.

As far as the enantioselective oxidation with chiral reagents is concerned, the attention was initially focused on percarboxylic acids [9]. These systems, however, provide values of optical yields not exceeding 10% with a variety of sulfides. Better results (e.e. 40–60%) had been obtained by using optically active oxaziridines [10a,b]. Very recently a reagent of this class, *i.e.* (-)-α,α-dichlorocamphorsulfonyloxaziridine, has been found to afford e.e. values up to 85–95%, for rather wide variations of the sulfide structure [10c]. Several other systems have been tested including enzymatic oxidations [11]. These provide in some cases satisfactory optical yields, which however are counterbalanced by the very well known drawbacks of such reagents.

The development of transition metal catalyzed systems employing alkylhydroperoxides or hydrogen peroxide opened new routes to the enantioselective oxidations of sulfides since the high coordination number of the metal ions usually employed, *i.e.* Ti(IV), V(V), Mo(VI) and W(VI), allows an easy access to chiral peroxo complexes through the contemporary coordination of the oxidant and of the chiral ligand(s) to the metal center. Indeed, since the early attempts, promising enantioselectivities were obtained. In Scheme 2 some typical examples are presented.

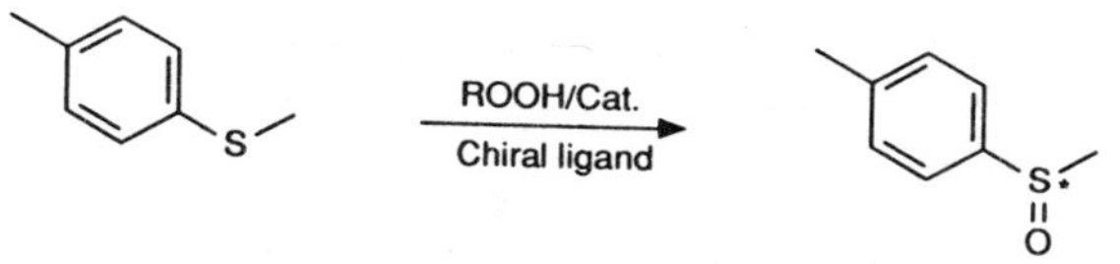

CATALYST	Chiral Ligand	ROOH	e.e. %	Ref.
VO(acac)$_2$	(-)-menthol	t-BuOOH	10	12
VO(3MeO-Sal-(R,R)-C$_6$H$_{10}$)	/	CHP	35	13
MoO$_5$HMPT	(+)-DET	t-BuOOH	28	14

Scheme 2

In this field, the real turning point may be considered the discovery by Sharpless that the titanium-(+)-DET-alkylperoxo complexes are extremely efficient asymmetric oxidants of allylic alcohols [15].

Although the Sharpless reagent as such does not show a significant enantioselectivity in sulfide oxidations [16], two modifications, independently developed in our [14] and in the Kagan's laboratory [16], proved to be of synthetic value, as indicated by the data shown in Table 1.

In the two oxidizing systems the active species formed in solution are likely different. In fact, the Kagan's procedure requires the presence of just one equivalent of water which, on the contrary, is excluded in our system. Nevertheless, they show a very similar enantioselectivity and also very similar sensitivity to the structure of the sulfide.

This makes it possible to utilize all the data available obtained either with the Kagan's or with our method in an attempt to establish some structure-reactivity correlations which would provide an useful guide in selecting the most promising substrates.

TABLE 1

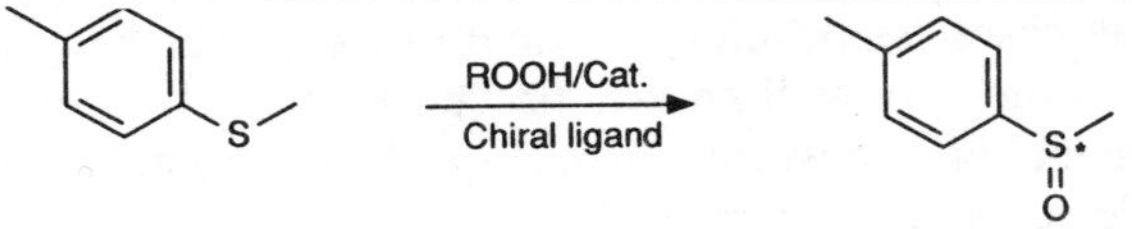

Ti(IV), eq	DET, eq	t-BuOOH, eq	H₂O, eq	Sulfide, eq	T,°C	Solvent	e.e,%	Ref.
1	1	2	-	1	-20	CH₂Cl₂	-	16
1	4	2	-	1	-20	DCE	88	14
1	2	1	1	1	-20	CH₂Cl₂	91	16

TABLE 2
Asymmetric oxidation of selected sulfides

Sulfide	e.e., %	confg. (ref. 14)	e.e., %	confg. (ref. 16)
p-Tolyl-S-Me	88	R	91	R
PhCH₂-S-Me	46	R	58	R
p-Tolyl-S-t-Bu	34			
p-Tolyl-S-n-Bu			20	R
p-Tolyl-S-i-Pr			63	
p-Cl-Ph-S-(CH₂)₂OH	14			

As far as open chain sulfides are concerned, the only apparent trend which may be observed is a tight connection between the optical yields and the difference in size of the two substituents at sulfur. When one of the two residues is as small as possible, *i.e.* a methyl, and the other one large enough, either a large alkyl group or a Π system, better an aromatic one, the enantioselectivity is satisfactorily high. On the contrary, when these conditions are not fulfilled, the enantioselectivity drops even if the sulfur center is congested by the presence of two bulky substituents.

A typical case history is that of the asymmetric oxidation of β-hydroxysulfides. We became interested in such substrates, potentially very useful in asymmetric synthesis, because they may resemble the allylic alcohols in the possibility of forming alkoxo-derivatives with the titanium catalyst. It should be recalled that the high e.e. values obtained with allylic alcohols are

TABLE 3

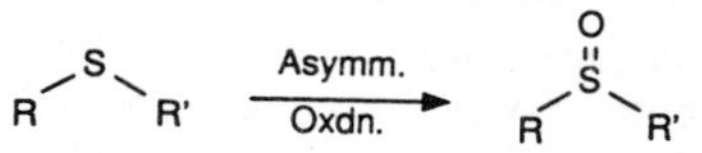

R	R'	e.e. %
Methyl	X-Phenyl (X= H, p-CH₃, p-Cl, p-Br, p-OCH₃, p-NO₂)	80-90
Cyclopropyl	Phenyl	95
n-Butyl	p-Tolyl	20
Benzyl	p-Tolyl	7
Methyl	t-Butyl	53
Methyl	n-Octyl	71
Methyl	Benzyl	58

ref. 16

rationalized just on the basis of the formation of such alkoxo-complexes. The data of Table 4 indicate however that the OH group rather than improving the enantioselectivity of the oxidation compared with that attainable with unfuctionalized thioethers, appears to play a negative role [17].

TABLE 4

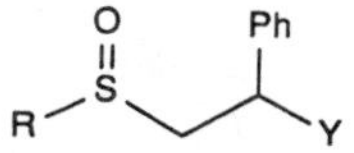

		diast.ratio (a:b)	e.e.% (a)
R = Ph	Y = OH	67 : 33	18
R = Ph	Y = OCOCH₃	55 : 45	21
R = t-Bu	Y = OH	74 : 26	47
R = t-Bu	Y = OSi(CH₃)₃	58 : 42	58

The logic consequence of such an observation has been to protect the OH group and at the same time to make as small as possible the other substituent at sulfur. The results of an investigation of the oxidative behavior of S-methyl-β-hydroxysulfides variously protected at the OH group, reported in Table 5, confirm the effectiveness of such an approach [18].

TABLE 5

X	Y	Z	yield,%	diast. ratio[b] a:b	e.e.,%[c] a	e.e.,%[c] b
H	Ph	OH	20	68:32	3	5
H	Ph	OSiPh$_3$	78	56:44	70	64
H	Ph	OSiPh$_3$	85	50:50	80	75
H	Ph	OCOCH$_3$	81[d]	50:50	76	76
Ph	H	OSitBuPh$_2$	90	87:13	74[e]	76
Ph	Ph	OH	84	97: 3	20	n.d.
Ph	Ph	OCOCH$_3$	79	87:13	18	67
Ph	CH$_3$	OH	22	100:0	18	--
Ph	CH$_3$	OSiPh$_3$	90	88:12	70	n.d.
Ph	CH$_3$	OSiPh$_3$	91[d]	91: 9	78	70
Ph	CH$_3$	OCOCH$_3$	80[d]	86:13	73	n.d.
Et	Et	OSiPh$_3$	95[d]	80:20	65	n.d.

a) isolated yields based on the oxidant; [substrate]:[oxidant] = 2:1.
b) obtained by ^{1}H-NMR (CDCl$_3$).
c) obtained by ^{1}H-NMR in the presence of (+)-(S)-2,2,2-trifluoro-1-(9-antryl)-ethanol.
d) obtained by using cumyl hydroperoxide as oxidant.
e) obtained by ^{1}H-NMR in the presence of (-)-(R)-N-(3,5-dinitrobenzoyl)-α-ethyl-phenyl amine.

It may be noticed that, as previously observed, the hydroxy group has indeed a large negative effect on the enantioselectivity. However, after its protection, sulfoxides characterized by significant enantiomeric excesses are obtained so that synthetically useful products become available. In fact, at least for some of the compounds that we have so far investigated, such as for example the one presented in Scheme 3, not only the e.e. values are fairly good but an upgrading of the optical purity up to e.e. >98% may be obtained by simple crystallization [19]

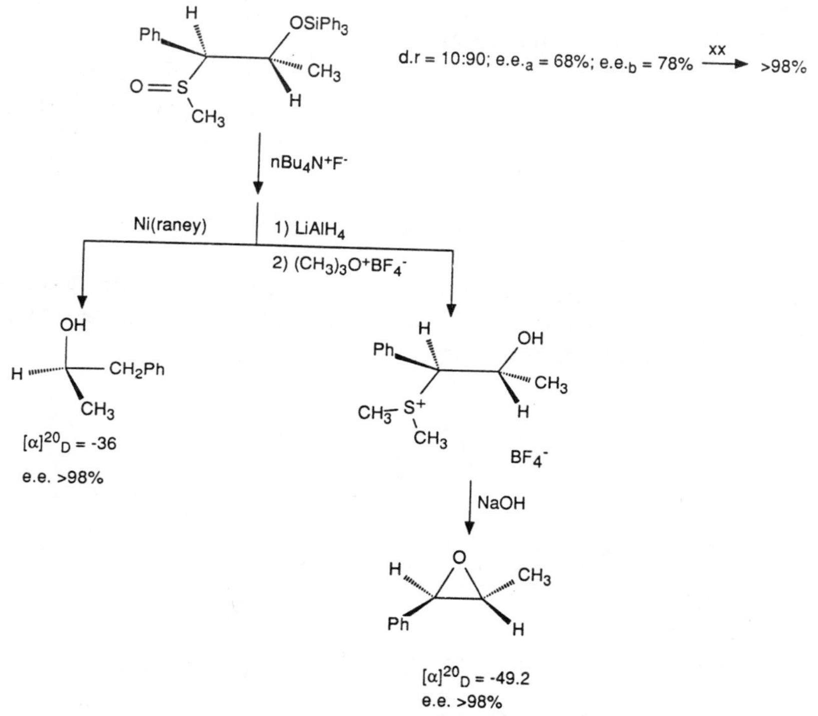

Scheme 3

The Scheme 3 shows the series of reactions which allow to obtain optically pure alcohols and epoxides starting from β-hydroxysulfoxides, as exemplified by (1R, 2S, SR)-1-methyl-2-(methylsulfinyl)-benzeneethoxy-triphenylsilane.

Our oxidation procedure was also applied to 1,1′-binaphtalene derivatives [20] aiming at obtaining optically pure sulfoxides which may be utilized as chiral synthons in several reactions.

The series of reactions leading to [1,1′-binaphtalene]-2,2′-dithioethers starting from the corresponding dithiol are reported in Scheme 4:

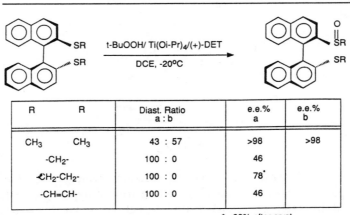

Scheme 4

In Table 6 are reported the results of the asymmetric monooxidation of the thioethers to sulfoxides. It is still observed that the methyl derivatives give the best results, with e.e. values larger than 98%.

TABLE 6

R	R	Diast. Ratio a : b	e.e.% a	e.e.% b
CH$_3$	CH$_3$	43 : 57	>98	>98
-CH$_2$-		100 : 0	46	
-CH$_2$-CH$_2$-		100 : 0	78*	
-CH=CH-		100 : 0	46	

* >98% after cryst.

An immediate application of these results is the resolution of [1,1'-binaphthalene]-2,2'-dithiol which can be accomplished, starting from the monosulfoxides both diasteromerically and enantiomerically pure, *via* the reactions outlined in Schemes 5 and 6 [20].

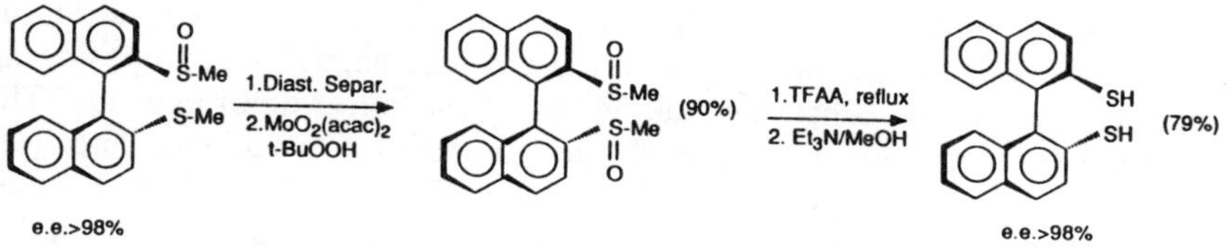

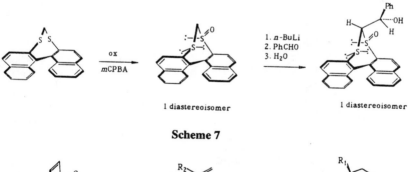

Scheme 5

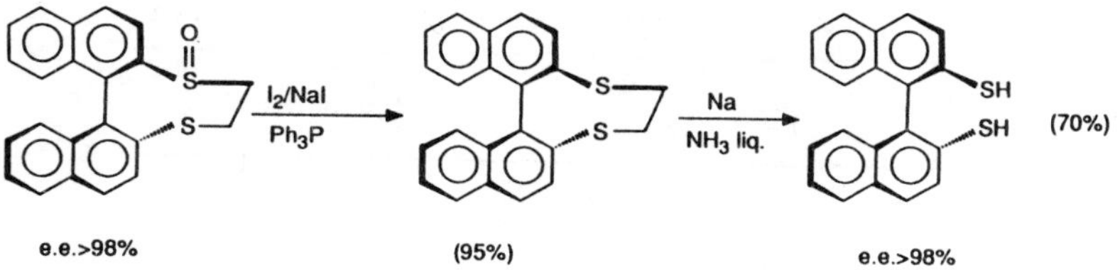

Scheme 6

As far as the application of sulfoxides as chiral synthons are concerned, two typical examples are presented in Schemes 7 [21] and 8 [22] concerning a C-C bond forming reaction and a cycloaddition respectively, both characterized by a very high selectivity.

Scheme 7

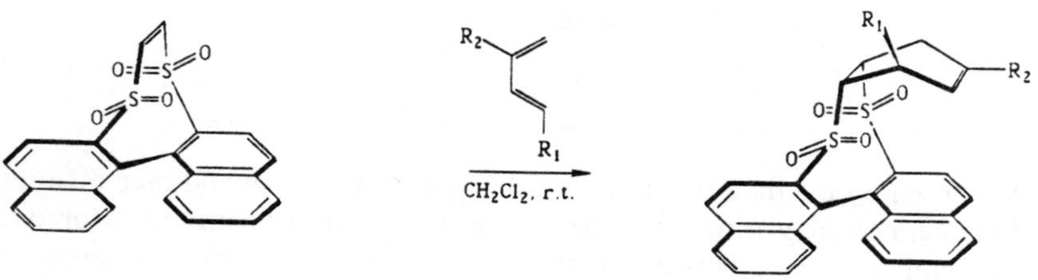

Scheme 8

Aiming at improving the efficiency of the asymmetric oxidation of sulfides we investigated the effect on the enantioselectivity of imposing an extra-constrain to the substrates, i.e. by fusing the sulfur atom in a ring. The first set of compounds investigated was that of 1,3-dithiolanes and 1,3-dithianes [23, 24]. The pertinent results, Table 7, show that 1,3-dithiolanes are particularly suitable substrates so that we concentrated our efforts on this system.

TABLE 7
Asymmetric oxidation of 1,3-dithiolanes and 1,3-dithianes

cis:trans		e.e., % (major diast.)
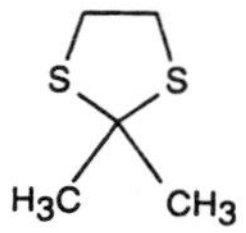	-	40
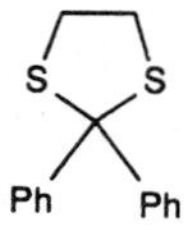	-	64
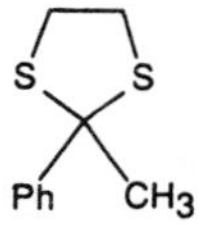	97:3	83
	-	20 (Kagan Reagent)
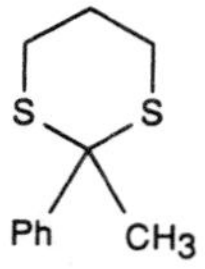	85:15	39

As shown by the data reported in Table 7 both ring size and face differentiation seem to play a role in maximizing the enantioselectivity. Moreover the difference in size between the two substituents at sulfur, as discussed above for the open chain sulfides, has a remarkable effect. We further investigated the effect of increasing the rigidity of the system by fusing carbons 4 and 5 of the dithiolane ring in either a saturated or an unsaturated ring, Table 8 [24]. These compounds are also of interest because, contrary to the simpler dithiolanes reported above [25], they resist strongly alkaline conditions and hence would be useful for synthetic purposes, such as the C-C bond forming reactions that involve carbanion intermediacy [26].

TABLE 8
Asymmetric oxidation of byciclic dithioacetals

	Diast. Ratio a : b : c : d	e.e., %
	100 : 0	24
	55 : 45 : 0 : 0	39(a) 25(b)

Although the diastereoselectivity (facial selectivity) appears to be significantly improved, the enantioselectivity shows a dramatic drop. Very likely the increase in size of the «small group» has, also in this system, a negative effect on the chiral recognition of the oxidant.

Returning now to the asymmetrically substituted dithiolanes that on the basis of the preliminary results looked promising, we have confirmed, Table 9, that, within a large variability of 2,2 substitution the diastereoselectivity and the enantioselectivity are both high enough to render the method a synthetically useful one [23, 24, 27].

On the other hand the fragility of the dithiolane ring system particularly under basic conditions allows for only a few chemical transformations.. As mentioned earlier, the simpler 1,3-dithiolane-S-oxides cannot be used in synthetic procedures involving carbanions. However, they have been utilized for the resolution of racemic aldehydes and ketones [24], see Scheme 9.

At any rate, the intrinsic fragility of 1,3-dithiolanes may be reduced by increasing the EWP of one of the substituents in the 2-position [28], thus making the carbanion mediated synthesis a viable procedure, as shown in a very general way in Scheme 10 [19].

V. Conte et al.

TABLE 9
Asymmetric oxidation of various 1,3-dithiolanes

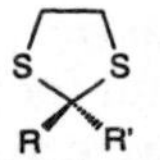

Substrate R	R'	yield,%	diast. ratio.	e.e.,%	e.e.,%after cryst.
CH₃	Ph	66	97: 3	83	98
H	Ph	76	94: 6	76	98
CH₃	t-Bu	61	>99: 1	68	70
H	t-Bu	80	>99: 1	70	liquid
H	COOEt	62	85:15	85	liquid
H	COOBz	81	92: 8	n.d.	98

TABLE 10

ASYMMETRIC OXIDATION OF

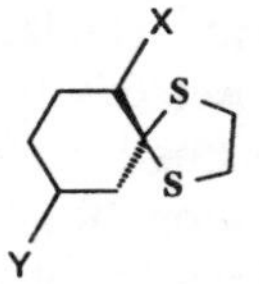

X	Y	Diast. Ratio	e.e., %	e.e., %	e.e.,%	e.e., %
CH₃	H	49:29:14:8	51	51	40	n.d.
Ph	H	63:37:0:0	78	80	--	--
t-Bu	H	75:25:0:0	72	74	--	--
i-Pr	CH₃	87:13:0:0	93	83	--	--
		40:10:10:40	44	48	32	46

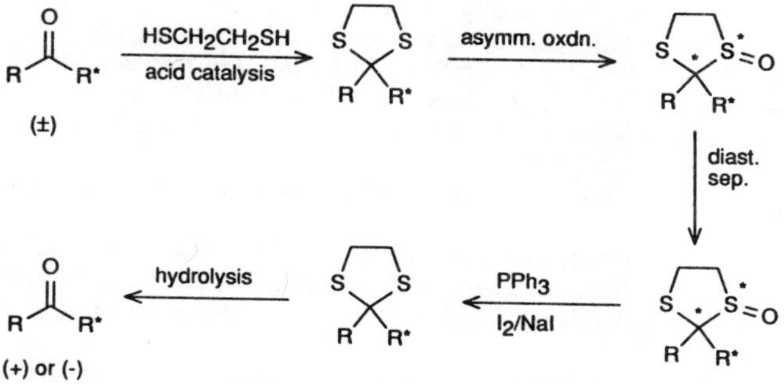

Scheme 9

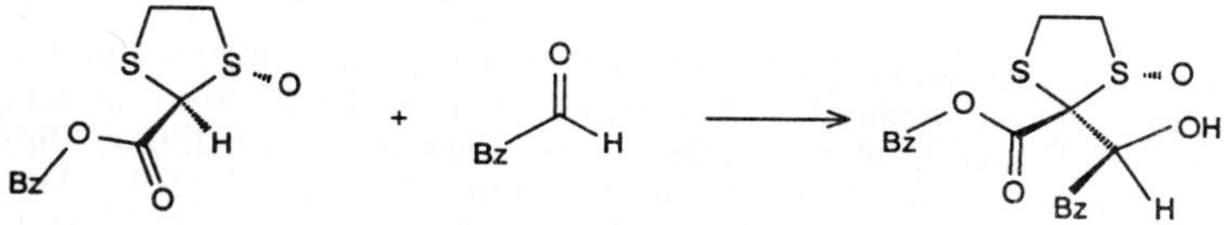

Scheme 10

In summary we have been able to show that enantioselective oxidation of sulfides reaches e.e. values synthetically useful whenever the two substituents at the sulfur atom are different enough to allow chiral recognition. In particular this has been verified to be the case for CH_3–S–R open chain sulfides and for the 1,3-dithiolane system. In the latter case, the unsubstituted –CH_2–CH_2– group appears to be small enough in respect to the substituted carbon 2 to differentiate the two sulfur atoms.

The drawbacks intrinsic in the dithiolane system, *i.e.* its fragility, may be overcome by the cooperation of an extra electron withdrawing group. The preliminary results so far obtained are very encouraging and we hope to be able to extend this novel approach to the enantioselective C–C bond formation.

CNR, Centro Meccanismi Reazioni Organiche,
Department of Organic Chemistry,
University of Padova,
Via Marzolo 1, I-35131 Padova (Italy)

References

1. Barbachyn, C. R.; Johnson, C. R.: «Optical Activation and Utilization of Compounds Containing Chiral Sulfur Center», *Asymmetric Synthesis*, vol. 4, chap. 2 , Morrison, J.D. Ed., Academic Press, Inc., New York, USA, (1983).

2. Solladié, G.; Greck, C.; Demailly, G.; Solladié-Cavallo, A. *Tetrahedron Lett.* **1982**, *23*, 5047; Solladié, G.; Damailly, G.; Greck, C. *ibid*, **1985**, *26*, 435.

3. Colombo, L.; Gennari, C.; Scolastico, C.; Guanti, G.; Narisano, E. *J. Chem. Soc. Chem. Commun.* **1979**, 591; *J. Chem. Soc. Perkin I* **1981**, 1278.

4. Hua, D.H.; Sinai-Zingde, G.; Venkataraman, S. *J. Am. Chem. Soc.* **1985**, *107*, 4088; *J. Org. Chem.* **1987**, *52*, 719.

5. Williams, D. R.; Phillips, J. G,; White, F. H.; Huffman, J. C. *Tetrahedron* **1986**, *42*, 3003.

6. Andersen, K.K. *Tetrahedron Lett.* **1962**, 93; Andersen, K.K. Andersen *J. Org. Chem.* **1964**, *29*, 1953; Andersen, K.K.; Gaffield, W.; Papanikolaou, N.E.; Foley, J.W.; Perkins, R.I. *J. Am. Chem. Soc.* **1964**, *86*, 5637; Andersen, K.K. *Int. J. Sulfur Chem.* **1971**, *6*, 69.

7. Mikolajczyk, M.; Drabowicz, J., Chiral Organosulfur Compounds, *Topics in Stereochemistry*, vol. 13, 333, Allinger, N.L.; Eliel, E.L.; Wilen, S.H. Eds., John Wiley & Sons, Inc., New York, USA, (1983).

8. De Lucchi, O.; Lucchini, V.; Marchioro, C.; Valle, G.; Modena, G. *J. Org. Chem.* **1986**, *51*, 1457 .

9. Mayr, A.; Montanari, F.; Tramontini, M. *Gazz. Chim. Ital.* **1960**, *90*, 739; Balenovic, K.; Bregant, N.; Francetic, D. *Tetrahedron Lett.* **1960**, 4637; Maccioni, A.; Montanari, F.; Secci, M; Tramontini, M. *Tetrahedron Lett.* **1961**, 607; Mislow, K.; Green, M.; Raban, M. *J. Am. Chem. Soc.* **1965**, *87*, 2761; Folli, U.; Iarossi, D.; Montanari, F.; Torre, G. *J. Chem. Soc. C* **1968**, 1317; Bucciaelli,F.; Forni,F.; Marcaccioli,S., Moretti, I., Torre, G. *Tetrahedron* **1983**, *39*, 187.

10. a) Davis, F.A.; Mc Cauley, J.P.; Chattopadhyay, S.; Hazakal, M.E.; Towson, J.C.; Watson, W.H.; Tavanaiepour, I. *J. Am. Chem. Soc.* **1978**, *109*, 3370; b) Davis, F.A.; Jenkins, Jr., R.H.; Awad, S.B.; Stinger, O.D.; Watson, W.H.; Galloy, J. *J. Am. Chem. Soc.* **1983**, *104*, 5412; c) Davies, F. A.; ThimmaReddy, R.; Weismiller, M. C. *J. Am. Chem. Soc.* **1989**, *111*, 5964.

11. Auret, B.J.; Boyd, D.R.; Henbest, H.B.; and Ross., S.J. *J. Chem. Soc. C*, **1968**, 2371; Abushanabab, E.; Reed, D.; Suzuki, F.; Sih, C.J. *Tetrahedron Lett.*, **1978**, 3415; May, S.W.; Phillips, R.S. *J. Am. Chem. Soc.* **1980**, *102*, 5981; Poje, M.; Nota, O.; Balenovic, K. *Tetrahedron* **1980**, *36*, 1895; Auret, B.J.; Boyd, D.R.; Breen, F.; Greene, M.E.; Robinson, P.M. *J. Chem. Soc. Perkin Trans. I*, **1981**, 930; Auret, B.J.; Boyd, D.R.; Cassidy, S.D.; Turley, F.; Drake, A.F.; Mason, S.F. *J. Chem. Soc., Chem. Commun.* **1983**, 283; Auret, B.J.; Boyd, D.R.; Cassidy, E.S.; Hamilton, R.; Turley, F. *J. Chem. Soc., Perkin Trans I* **1985**, 1547; Fujimori, K.; Haekim, Y. *Bull. Chem. Soc. Jpn.* **1983**, *56*, 2310; Kagan, H.B., Asymmetric Oxidation Mediated by Organometallic Species, *Stereochemistry of Organic and Bioorganic Transformations, Workshop Conferences Hoechst*, vol. 17, 31, Bartmann, W. and Sharpless, K.B. Eds., VCH, Weinheim, FRG, (1987).

12. Di Furia, F.; Modena, G.; Curci, R. *Tetrahedron Lett.* **1976**, 4637; Curci, R.; Di Furia, F.; Edwards, J.O.; Modena, G. *Chim. Ind.* **1978**, *60*, 595;

13. Nakajima, K.; Kojima, M.; Fujita, J. *Chem. Lett.* **1986**, 1483.

14. Di Furia, F.; Modena, G.; Seraglia, R. *Synthesis* **1984**, 325.

15. Katsuki, T.; Sharpless, K.B. *J. Am. Chem. Soc.* **1980,** 102, 5974.

16. Pitchen, P.; Dunach, E.; Deshmukh, M.N.; Kagan, H. B. *J. Am. Chem. Soc.* **1984**, *106*, 8188.

17. Bortolini, O.; Di Furia, F.; Licini, G.; Modena, G. *Phosphorus and Sulfur* **1988**, *37*, 171.

18. Conte, V.; Di Furia, F; Licini, G.; Modena, G. *Tetrahedron Lett.* **1989**, *30*, 4859.
19. Licini, G.,PhD Thesis, University of Padova, Italy (1988).
20. De Lucchi, O; Di Furia, F; Licini, G; Modena, G. *Tetrahedron Lett.* **1989**, *30*, 2575.
21. Delogu, G.; De Lucchi, O.; Licini, G. *J. Chem. Soc. Chem Commun.* **1989**, 411.
22. Cossu, S,; Delogu, G,; De Lucchi, O.; Fabbri, D.; Licini, G.; *Angew. Chem. Int. Ed. Eng.* **1989**, *28*, 766.
23. Bortolini, O.; Di Furia, F.; Licini, G.; Modena, G.; Rossi, M. *Tetrahedron Lett.* **1986**, 6257.
24. Bortolini, O.; Di Furia, F.; Licini, G.; Modena, G., Structure-Behavior Relationship in the Enantioselective Oxidation of Sulfides, *Reviews on Heteroatoms Chemistry,* vol. 1, 66, Oae, S. Ed. , MYU, Tokyo, Japan, (1988).
25. Oida, T.; Tanimoto, S.; Terao, H.; Okano, M. *J. Chem Soc. Perkin Trans I* **1986**, 1715; Tanimoto, S.; Terao, H.; Oida, T.; Ikehira, H. *Bull. Inst. Chem. Res., Kioto Univ.* **1984**, *62*, 54; Schaumann, E. *Bull. Soc. Chim. Belg.* **1986**, *95*, 312.
26. Carey, F.A.; Dailey, O.D. *Phosphorus and Sulfur* **1981**, *10*, 169.
27. Bortolini, O.; Di Furia, F.; Licini, G.; Modena, G., Enantioselective Oxidation of Sulfides, *The Role of Oxygen in Chemistry and Biochemistry, Studies on Organic Chemistry,* vol. 33, 193, Ando, W., Moro-oka, Y. Eds., Elsevier Sci. Publ., Amsterdam, The Netherlands, (1988).
28. Hermann, J.L.; Richman, J.E., Schlessinger, R.H. *Tetrahedron Lett.* **1973**, 2598.

J. C. FIAUD

MECHANISMS IN STEREO-DIFFERENTIATING METAL-CATALYZED REACTIONS. ENANTIO-SELECTIVE PALLADIUM-CATALYZED ALLYLATION

The palladium-catalyzed allylation of carbo- and heteronucleophiles is one of the most widely used transition-metal catalyzed transformation of organic material [1].

1. The palladium-catalyzed substitution of allylic substrates by nucleophiles. The reaction. The catalytic steps

Following the pioneering work of Tsuji on addition of carbonucleophiles to η^3-allylpalladium complexes [2], of Atkins *et al.* [3] and Hata *et al.* [4] on Pd-catalyzed allyl transfer from allylic electrophiles (acetates, phenates ...) to nucleophiles (active hydrogen organic compounds, amines), Trost developed this reaction showing the catalytic properties of Pd(0) complexes in the reaction of allylic acetates with stabilized carbanions (malonates, sulfonylacetates ...) [5]

1.1. THE η^3-ALLYLPALLADIUM FORMING STEPS (OXIDATIVE ADDITION OF THE SUBSTRATE TO A PALLADIUM(0) COMPLEX) THE REACTIVE ALLYLIC COMPOUNDS

The scope of the active substrates has since ever increased, and the Table I collects some allylic compounds which have been shown to be active in Pd-catalyzed allylic substitution. Palladium(0) complexes activate these substrates by ionization of the leaving group to give a cationic η^3-allylpalladium(II) complex, through the breaking of a C–O, C–N or C–S bond.

The electrophilic complex may then react with a nucleophile (neutral or anionic) to give the allyl nucleophile and liberate back a Pd(0) complex, according to Scheme 1.

A. F. Noels et al. (Eds.), Metal promoted selectivity in organic synthesis, 107–131.
© 1991 *Kluwer Academic Publishers. Printed in the Netherlands.*

TABLE I
Some anionic moieties obtained through the Palladium-induced ionization of allylic compounds

(allyl)G	+	Nu⁻	$\xrightarrow{\text{Pd catalyst}}$	(allyl)Nu	+	G⁻
allylic substrate						ionized group

ionized group G	allylic substrate	leading or representative reference
From derivatives of allylic alcohols:		
R–C(=O)–O⁻	carboxylate	6
PhO⁻	phenoxide	7
$(EtO)_2P(O)O^{\ominus}$	phosphate	8
RO–C(=O)–O⁻ ⟶	CO_2 + RO⁻	9
	carbonate ⟶ alkoxide	
R_2N–C(=O)–O⁻ ⟶	CO_2 + $R_2N^{\ominus}$	10
	urethane ⟶ amidure	
Ph-S(=O)–O⁻	sulfinate	11
From other allylic derivatives:		
Ph-S(=O)(=O)⁻	phenyl sulfone	12
PhS⁻	phenyl sulfide	13
$NO_2^{\ominus}$	nitrite	14
From vinyl epoxydes:		

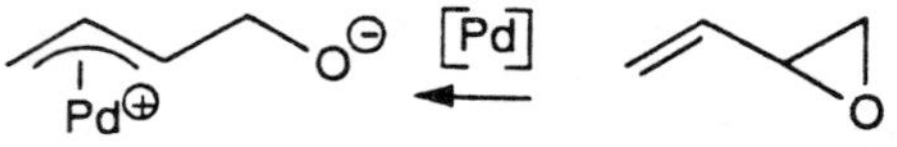

(allyl)G   +   NuM $\xrightarrow{\boxed{Pd}\ catalyst}$ (allyl)Nu   +   GM

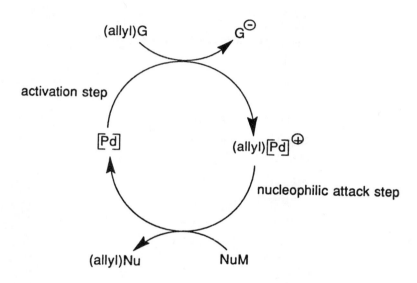

Scheme 1.
Catalytic cycle for the palladium-catalyzed substitution of allylic compounds.

1.2. THE NUCLEOPHILIC ATTACK STEP. THE ACTIVE NUCLEOPHILES

An even wider variety of active nucleophiles has been utilized up to now. A number are collected in Tables II, according to the nature of the bond to be formed (C–C, C–S, C–N, C–O, C–P ...). These nucleophiles are either neutral (amines or amine derivatives) or anionic.

J.C. Fiaud

TABLE IIa
The active nucleophiles in the palladium-catalyzed substitution of allylic sustrates

---

$$\text{(allyl)G} \quad + \quad \text{NuM} \quad \xrightarrow{\text{Pd catalyst}} \quad \text{(allyl)Nu} \quad + \quad \text{GM}$$
$$\text{nucleophile}$$

- CARBONUCLEOPHILES
  - Stabilized carbanions ("soft carbanions").

Nu is the fragment to be attached to the allylic frame by the reaction

NuM	Nu	Reference[a]
$NaCH(CO_2CH_3)_2$	$-CH(CO_2CH_3)_2$	6
$NaCH(COCH_3)_2$	$-CH(COCH_3)_2$	16
$NaCH(SO_2Ph)_2$	$-CH(SO_2Ph)_2$	17
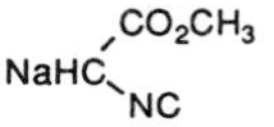 $NaHC\begin{smallmatrix}CO_2CH_3\\NC\end{smallmatrix}$	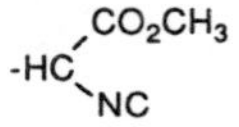 $-HC\begin{smallmatrix}CO_2CH_3\\NC\end{smallmatrix}$	18
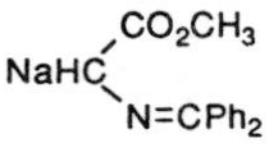 $NaHC\begin{smallmatrix}CO_2CH_3\\N=CPh_2\end{smallmatrix}$	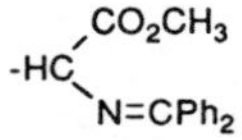 $-HC\begin{smallmatrix}CO_2CH_3\\N=CPh_2\end{smallmatrix}$	19
$H_3C-\underset{OLi}{C}=CH_2$	$H_3C-\underset{O}{\overset{\|}{C}}-CH_2^-$	20
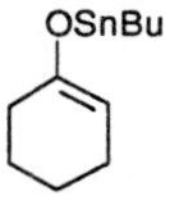	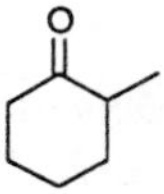	21
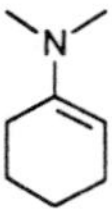	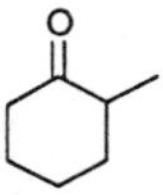	22
 Na^+	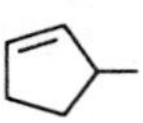	23

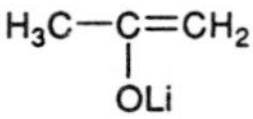

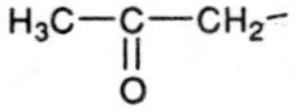

a) stands for a representative paper describing the use of the nucleophile

TABLE IIb
The active nucleophiles in the palladium-catalyzed substitution of allylic sustrates

$$(allyl)G \quad + \quad \underset{\text{nucleophile}}{NuM} \quad \xrightarrow{\text{Pd catalyst}} \quad (allyl)Nu \quad + \quad GM$$

▪ CARBONUCLEOPHILES

- Organometallics (or "hard" carbanions)

NuM	Nu-	Reference[a]
 (indenyl anion) Li[+]	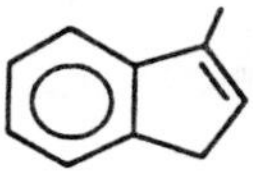 (methylindene)	24
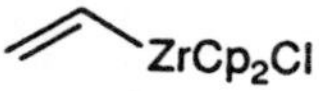 ZrCp$_2$Cl		25
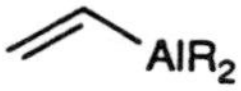 AlR$_2$	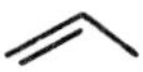	26
PhZnCl	Ph-	27
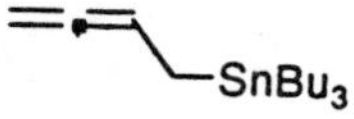 SnBu$_3$	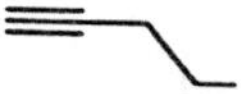	28
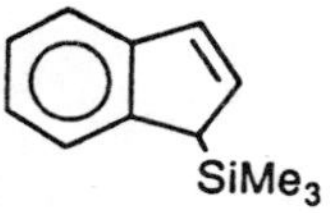 SiMe$_3$	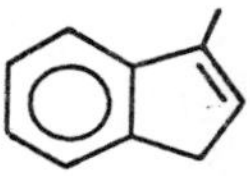	29
 [allyl]$_4$Sn		29

a) stands for a representative paper describing the use of the nucleophile

TABLE IIc
The active nucleophiles in the palladium-catalyzed substitution of allylic sustrates

$$(allyl)G \; + \; \underset{nucleophile}{NuM} \; \xrightarrow{Pd \; catalyst} \; (allyl)Nu \; + \; GM$$

- **HETERONUCLEOPHILES**

NuM	Nu-	Reference[a]
- N-nucleophiles		
R_2NH	R_2N-	3
NaN_3	$-N_3$	30
RNHOH	R(OH)N-	31
- O-nucleophiles		
ROLi	RO-	32
$HCOONH_4$	H-	33
Bu_3SnOPh	PhO-	34
- S-nucleophiles		
$PhSO_2Na$	$PhSO_2$-	35
$RS-SiMe_3$	RS-	36
- P-nucleophiles		
$Ph_2P(S)Li$	$Ph_2(S)P$-	37
- Si-nucleophiles		
$Al(SiMe_3)_3$	Me_3Si-	38
$Me_3Si-SiMe_3$	Me_3Si-	39
- reducing nucleophiles		
$NaBH_4$, $NaBH_3CN$	H-	40
$HCOONH_4$	H-	41

a) stands for a representative paper describing the use of the nucleophile

The anionic nucleophiles may be prepared extemporaneously from the corresponding acids which are deprotonated by an externally added base, or by the anionic leaving group (or a decomposition fragment) produced through oxidative addition of the allylic substrate to the Pd(0) catalyst. In such a way, particularly interesting are:

- the carbonate which affords through decarboxylation the basic methylate anion;
- the phenate ion, which is basic enough to produce sufficient concentration of carbanions from malonates, β-ketoesters and β-diketones. Ionization of vinyl epoxides also produces basic alcoholates.

2. The stereochemical course of the reaction; analysis of the stereochemistry of each individual catalytic step

It has been recognized that the substitution is stereoselective and proceeds either with overall retention of configuration of the allylic part (as for malonates) [16] or with overall inversion (as for phenylzinc chloride) [27] (Scheme 2). A reaction scheme has thus been proposed in agreement with these observations which correlates the stereochemical course of the reaction with the stereochemical behaviour of the palladium complexes towards the substrate and the nucleophiles (Scheme 3). The first step (ionization step, η^3-allylpalladium(II) complex forming step or oxidative addition step) would proceed through an attack of the palladium *anti* to the leaving group (*e.g.* the carboxylate). This is followed by attack of the cationic η^3-allylpalladium(II) complex by the nucleophile:

- either at the η^3-allyl ligand, *anti* to the palladium (Nu_L nucleophile);

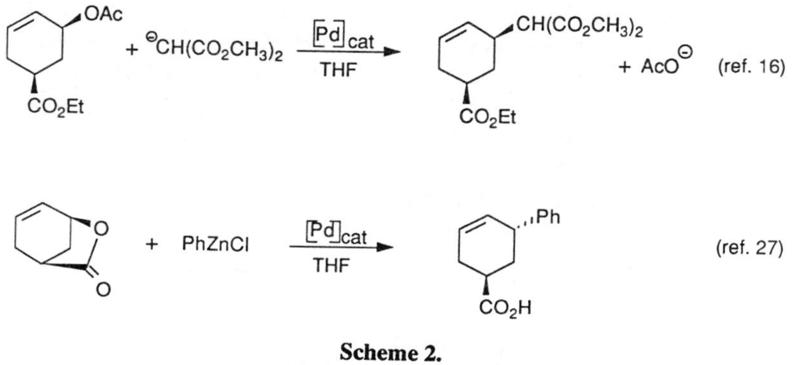

Scheme 2.
Stereochemical course of the Palladium-catalyzed allylation of nucleophiles.

- or at the palladium atom (Nu_M nucleophile), to give a neutral η^3-allylpalladium(II) complex, from which the product is liberated by *cis* ligand coupling.

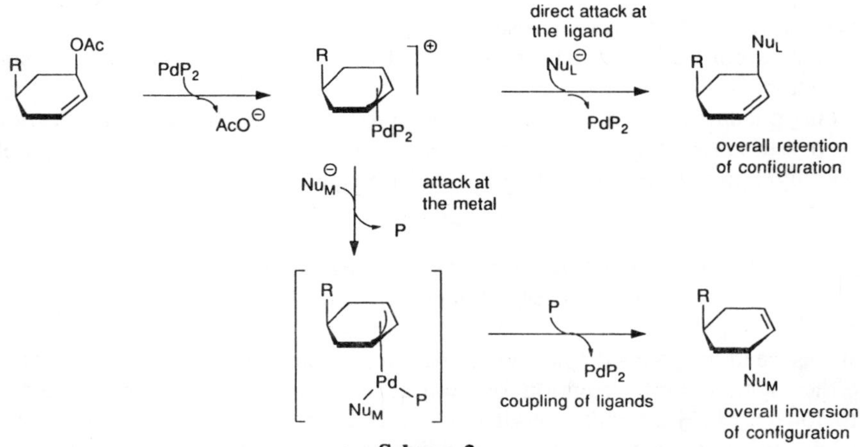

Scheme 3.

Stereochemical course of the palladium-catalyzed allylation of nucleophiles according to the nature Nu_L or Nu_M of the nucleophile.

2.1. THE η^3-ALLYL FORMING STEP IS UNDER STEREOELECTRONIC CONTROL; STEREOCHEMICAL REQUIREMENTS FOR THE ALLYLIC SUBSTRATE TO BE REACTIVE

Allylic substrates of proper stereochemical structure have been devised to get insight into the stereochemical course of the palladium-catalyzed reaction. Sodium dimethyl malonate and phenylzinc chloride (two representatives of Nu_L and Nu_M nucleophiles, respectively) were shown to be unreactive with substrate **1** [41]. This is due to the inability of the palladium(0) complex to approach the *endo* face of the molecule, i.e. the face of the olefin opposite to the C-O bond to be broken, indicative of an approach of the palladium *trans* to the acetate bond.

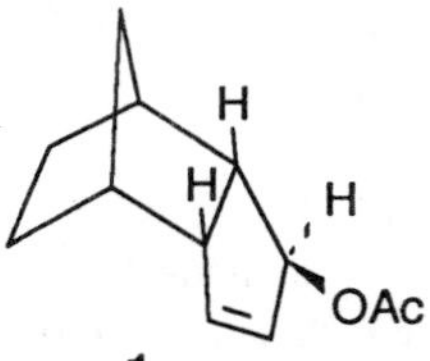

1

Competitive reactions of sodium dimethyl malonate or tri-n-butylvinyl aluminium with diastereomers **2** and **3** showed that **2** is > 200 times more reactive than dia **3**. This indicates a strong stereoelectronic control in the intermediate complex where the Pd-C bond to be formed and the Pd-O bond to be broken should be antiperiplanar (Scheme 4) [42]. This control which has been evidenced with rigid structures, also operates with acyclic, flexible substrates.

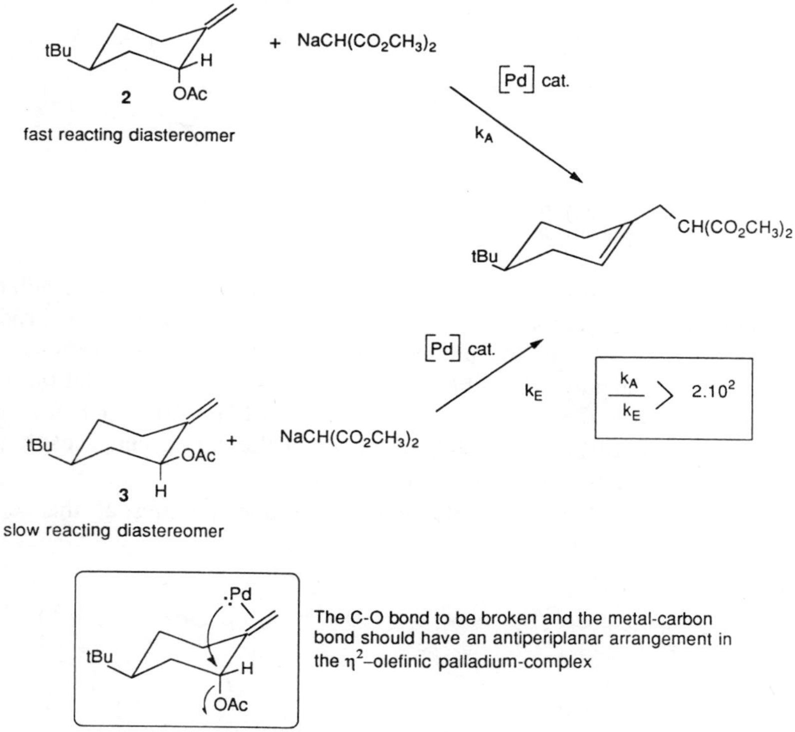

Scheme 4.
Stereochemical requirements for the allylic substrate to be reactive
in the palladium-catalyzed allylic substitution [42].

2.2. THE USE OF PROPERLY DEVISED MODELS TO APPRECIATE THE WAY OF ATTACK OF THE NUCLEOPHILE

The behaviour of the nucleophiles and hence the stereochemistry of the

nucleophilic attack may be determined according to different ways:
- reacting the nucleophile with an optically active η^3-allylpalladium(II) complex of known configuration and determining the absolute configuration of the product (Scheme 5) [43];

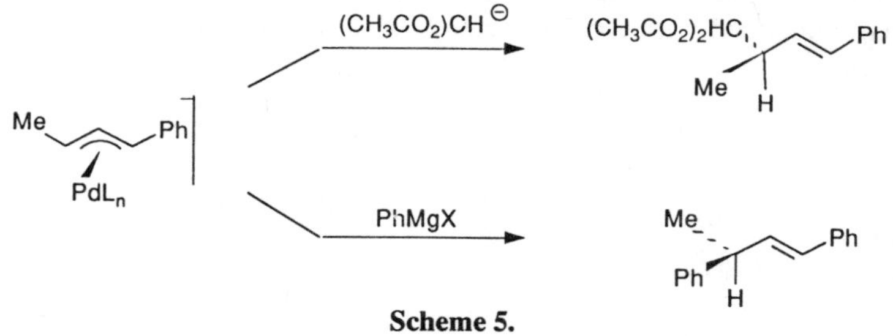

Scheme 5.
Determination of the stereochemistry of nucleophiles with the use of optically active cationic palladium complexes [43].

- reacting the nucleophile with a cyclic allyl carboxylate of known relative stereochemistry, and looking at the relative stereochemistry of the product [16, 29, 37, 44, 45, 46]. A number of allylic substrates have been used as models for that purpose (Scheme 6). Difficulties for interpretation arise will result when the relative stereochemistry of the product from these experiments is difficult to establish or when the experiments result in a scrambling of stereochemistry in the product;
- reacting the nucleophile with substrate **4** and looking at the regiochemistry of the substitution.

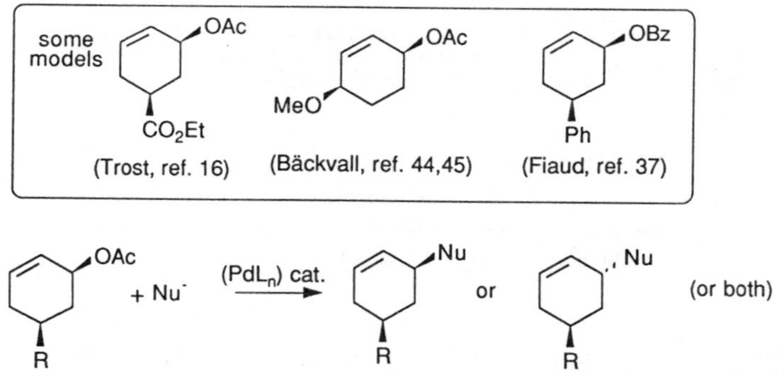

Scheme 6.
Stereochemical model substrates for the determination of the stereochemistry of the reaction of nucleophiles.

Keinan established an empirical correlation between the structure of the product and the behaviour of the nucleophiles (Scheme 7) [34];
– reacting the nucleophile with substrate **5** and looking at its reactivity.

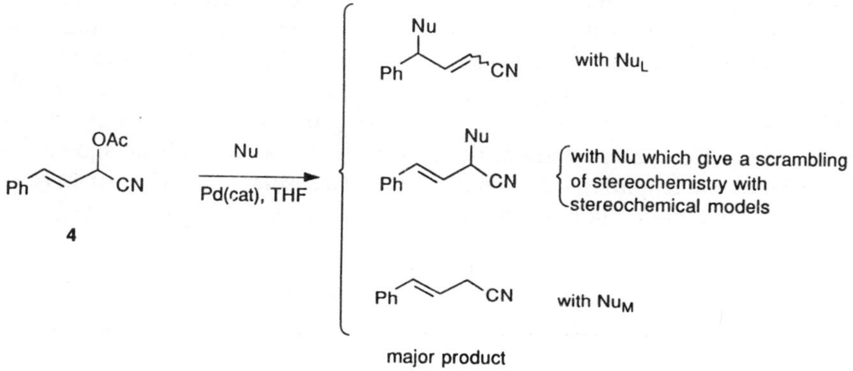

Scheme 7.
Determination of the stereochemistry of the reaction of nucleophiles through examination of the regioselectivity of reaction on a model substrate [34].

Substrate **5** should be readily ionized through palladium substitution to give the η^3-allylpalladium(II) complex **6** (Scheme 8). This complex would be reactive or unreactive according to the nature of the opposed nucleophile. Nu_M nucleophiles would give the *exo* product whereas Nu_L nucleophiles would be unreactive [41].

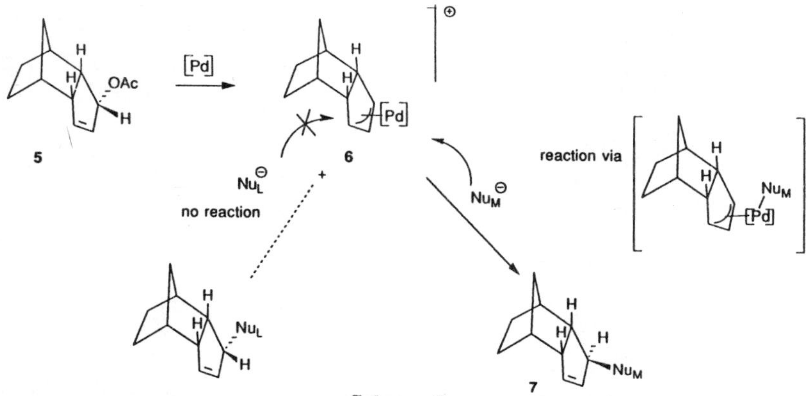

Scheme 8.
Determination of the stereochemistry of the reaction of nucleophiles through examination of the reactivity of reaction on the model substrate **5** [41].

Results in Table III are in agreement with these predictions. Sodium dimethyl malonate and lithium diphenyl phosphide which have been previously shown [16, 37] to react as Nu_L nucleophiles do not react with **5** whereas phenylzinc chloride (yet known as being a Nu_M nucleophile) [27] gives the exo-phenyl substituted product **7** (Nu = Ph) [41]. Cyclopentadienyl lithium reacts as a Nu_L nucleophile (to compare with a stabilized carbanion) and indenyl sodium as a Nu_M nucleophile (to compare with an organometal). These results may be compared to the pK_A of the corresponding acids, respectively 15 (cyclopentadiene) and 21 (indene). As it has been shown that these latter nucleophiles react with a high stereoselectivity [24], it may be concluded that these nucleophiles are strictly Nu_L and Nu_M, respectively.

TABLE III
Reactivity of various nucleophiles with **5** [41]

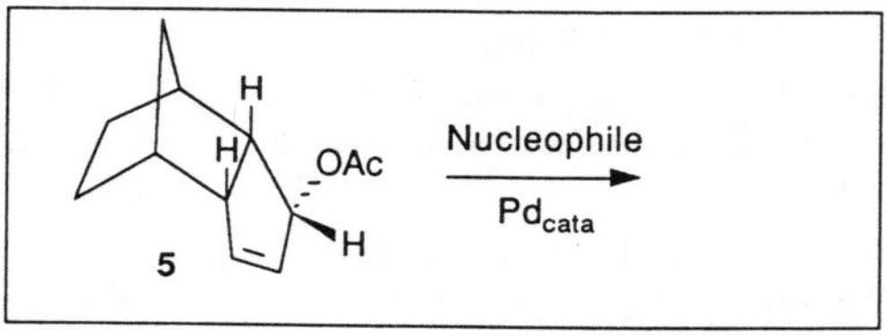

nucleophile	reaction	type[a]
$NaCH(CO_2CH_3)_2$	-	Nu_L
PhZnCl	+	Nu_M
$LiP(S)Ph_2$	-	Nu_L
 (cyclopentadienyl) Na^+	-	Nu_L
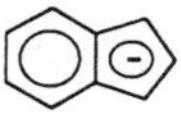 (indenyl) Li^+	+	Nu_M
$PhSO_2Na$	-	Nu_L
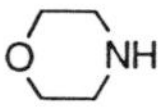 (morpholine) O⌢NH	-	Nu_L
$CH_3CO_2^-$	-	Nu_L
HCO_2NH_4	+	Nu_M

a) Nu_L is a nucleophile which attacks the ligand of the η^3-allylpalladium intermediate complex; Nu_M is a nucleophile which attacks the metal of the η^3-allylpalladium intermediate complex

With the help of stereochemical models, Trost showed that both sodium p-toluenesulfinate [47] and amines [48, 49] led to a scrambling of the stereochemistry in the product. Use of the reactivity model **5** indicates that these reagents are strictly Nu_L nucleophiles. Trost invoked an epimerization of the starting allylic acetate as a possible explanation for the scrambling of the stereochemistry that would occur through a readdition of the acetate anion to the metal. Use of model **5** indicates however that the acetate anion has no dual behaviour and strictly acts as a Nu_L nucleophile (Scheme 9). Dia **1** – that would be produced through re-attack of the acetate anion onto the η^3-allylpalladium(II) complex – should be stable and accumulate, as it cannot be ionized back through an *anti* attack of the palladium(0) complex (see above) [41].

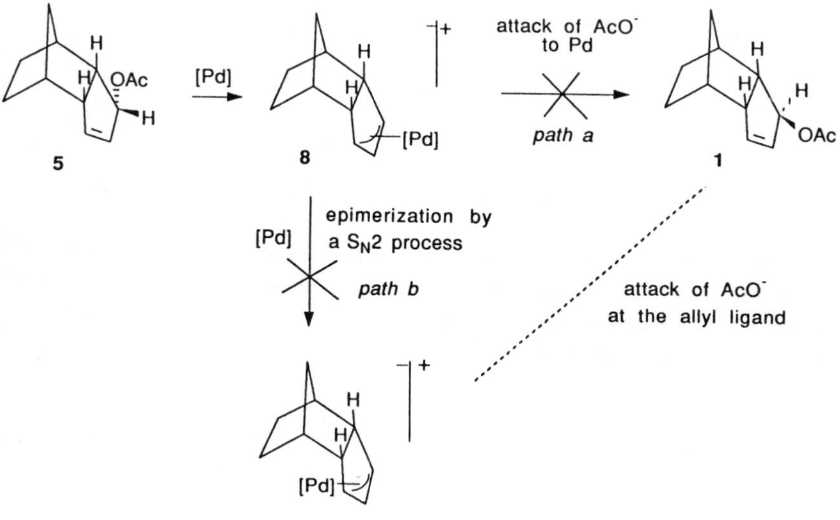

Scheme 9.
Substrate **5** (endo) does not epimerize to **1** (exo) [41].

The scrambling of the stereochemistry in the products may be likely due to an epimerization of the η^3-allylpalladium(II) complex through a S_N2 mechanism (displacement of the palladium from the η^3-allylpalladium(II) complex by a palladium(0) complex). That process cannot however operate on the *exo* palladium cationic complex **8** (Scheme 9, path b) where the *endo* face is hindered towards approach of a bulky Pd(0) complex.

The S_N2 mechanism appears to be the sole process able to enantiomerize a η^3-allylpalladium(II) complex in which the η^3-allyl ligand is dissymmetrically substituted (Scheme 10, c); a $\eta^3-\eta^1-\eta^3$ process will only epimerize the complex (Scheme 10, a) (except for a η^3-allyl ligand with two identical substituents at one allylic end) (Scheme 10, b).

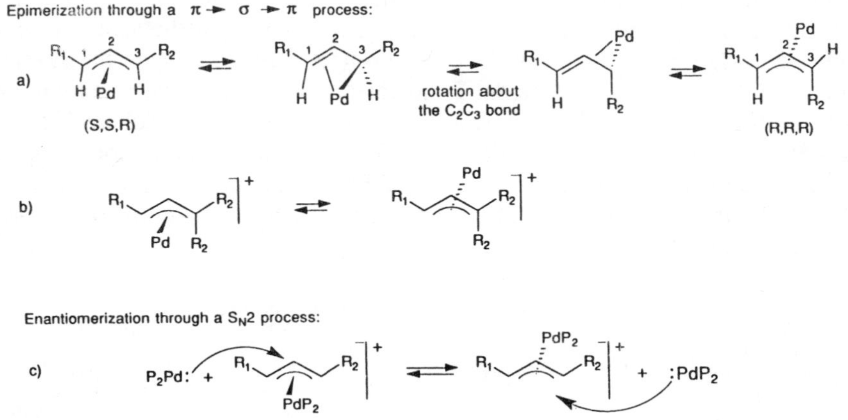

Epimerization through a $\pi \rightarrow \sigma \rightarrow \pi$ process:

Enantiomerization through a S_N2 process:

Scheme 10.

Processes for epimerization of η^3-allylpalladium complexes.

As the formate anion is reactive to give the reduced product, it is therefore concluded that this nucleophile attacks at the metal, then is decarboxylated to produce a palladium-hydride complex which then undergoes a hydride *cis* migration to the allyl ligand (Scheme 11).

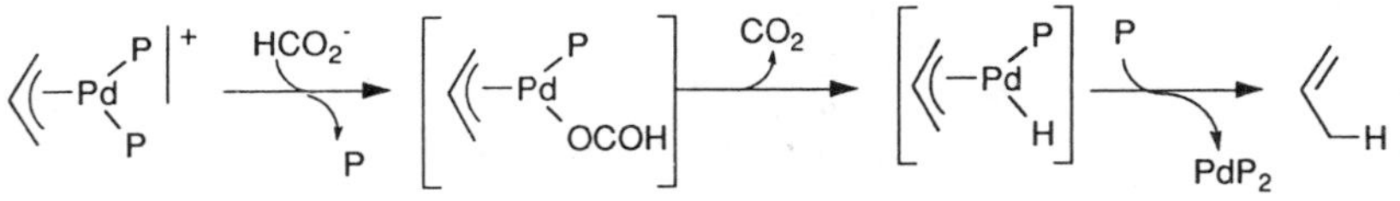

Scheme 11.

Mode of attack of the formate anion.

Up to now, all the nucleophiles tested with model substrate **5** appear to behave as strictly Nu_L or Nu_M nucleophiles.

3. Enantioselective synthesis. Asymmetric induction

Among the various ways to produce optically active organic material through palladium catalysis, several cases may be distinguished according to the location of the chiral inducer moiety in the substrate (as the leaving group or in the allylic frame), in the reagent or in the ligand of the catalyst.

3.1. OPTICALLY ACTIVE SUBSTRATES THE CHIRAL DIRECTOR IS LOCATED IN THE SUBSTRATE

3.1.1. As the leaving group

Few examples have been reported, for which the asymmetric induction results in a transfer of asymmetry from the leaving group to the product;

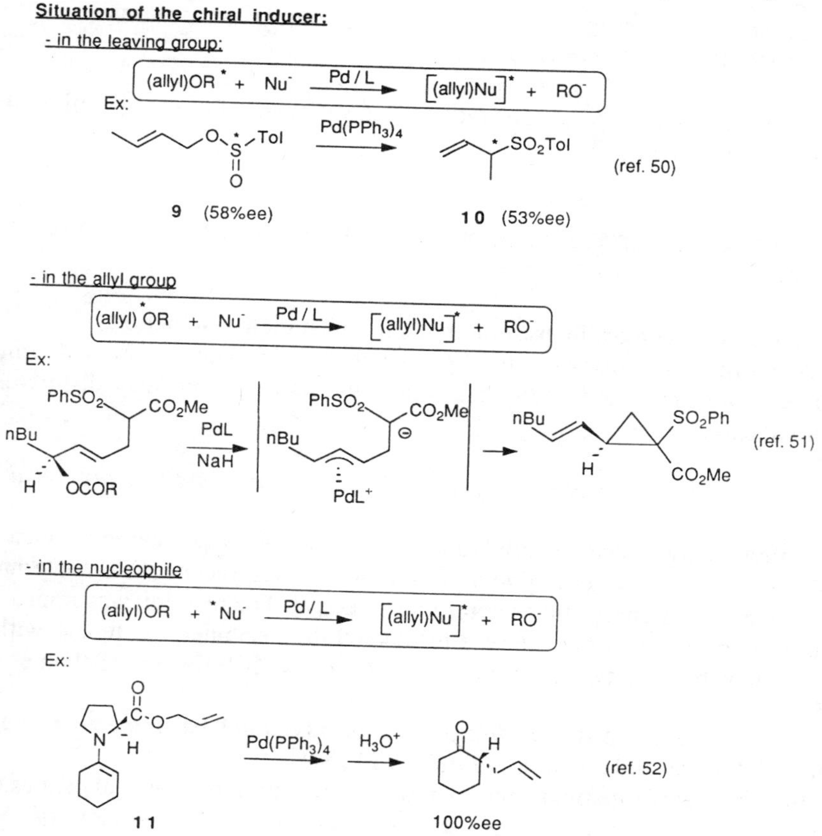

Situation of the chiral inducer:

- in the leaving group:

9 (58%ee)   **1 0** (53%ee)

- in the allyl group

- in the nucleophile

1 1   100%ee

Scheme 12.
Asymmetric induction processes according to the location of the chiral inducer.

among these, the palladium-catalyzed isomerization of chiral allylsulfinates **9** produce optically active sulfones **10** [50].

3.1.2. In the allylic frame

This more frequently encountered process accounts for asymmetry transfers; the success for such a transfer of asymmetry requires both stereoselective ionization and nucleophilic attack steps, together with a good optical stability of the intermediate η^3-allylpalladium complex. An example of such a successful transfer is depicted in Scheme **12** [51].

3.2. OPTICALLY ACTIVE NUCLEOPHILES. THE CHIRAL INDUCER IS LOCATED IN THE NUCLEOPHILE

Examples are scarce. A representative has been reported as the allylation of a proline enamine **11** [52].

3.3. OPTICALLY ACTIVE CATALYSTS. THE CHIRAL INDUCER IS LOCATED IN THE LIGAND OF THE CATALYST

This is by far the most interesting mode for asymmetric induction.

According to the process through which the chiral information is transmitted from the ligand of the catalyst to the product, one may distinguish several cases:

3.3.1. The substrate is achiral, non-prochiral; the nucleophile is prochiral

The first example was the allylation of β-ketoesters or β-diketones (such as **12**) by allyl phenoxide to give a product with a non racemizable quaternary carbon as new stereogenic center, in a low ee [7]. The reaction was improved (up to 81% ee) with the use of a ferrocenyl diphosphine-type ligand with a side chain bearing two primary hydroxy groups (BPPFA **13**) (Scheme 13) [53].

Optically active allyl-substituted aminoacids could be prepared through chiral Pd- catalyzed allylation of imines **14** [54].

In these transformations, the asymmetric induction may be analyzed as the selection, by an optically active η^3-allylpalladium(II) complex of two enantiotopic faces of an achiral nucleophile.

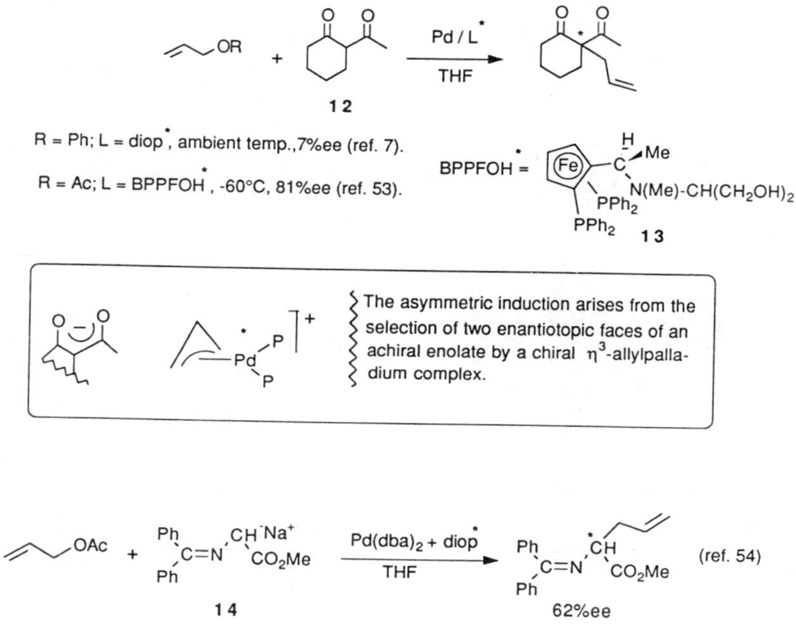

R = Ph; L = diop*, ambient temp., 7%ee (ref. 7).

R = Ac; L = BPPFOH*, -60°C, 81%ee (ref. 53).

BPPFOH* =

13

The asymmetric induction arises from the selection of two enantiotopic faces of an achiral enolate by a chiral η^3-allylpalladium complex.

(ref. 54)

14

62%ee

Scheme 13.
Selection by an optically active η^3allylpalladium complex
of two enantiotopic faces of an achiral nucleophile.

3.3.2. The substrate is prochiral, chiral or achiral; the nucleophile is not prochiral

– the allyl ligand of the η^3-allylpalladium(II) complex is symmetrically substituted.

In this process, the asymmetric induction arises from the selection, by an achiral nucleophile of two diastereosites of a chiral η^3-allylpalladium complex. Examples from acyclic allylic acetates have been reported (Scheme 14). The ee could be improved by a proper choice of the chiral inducing ligand.

– the allyl ligand of the η^3-allylpalladium(II) complex is dissymmetrically substituted.

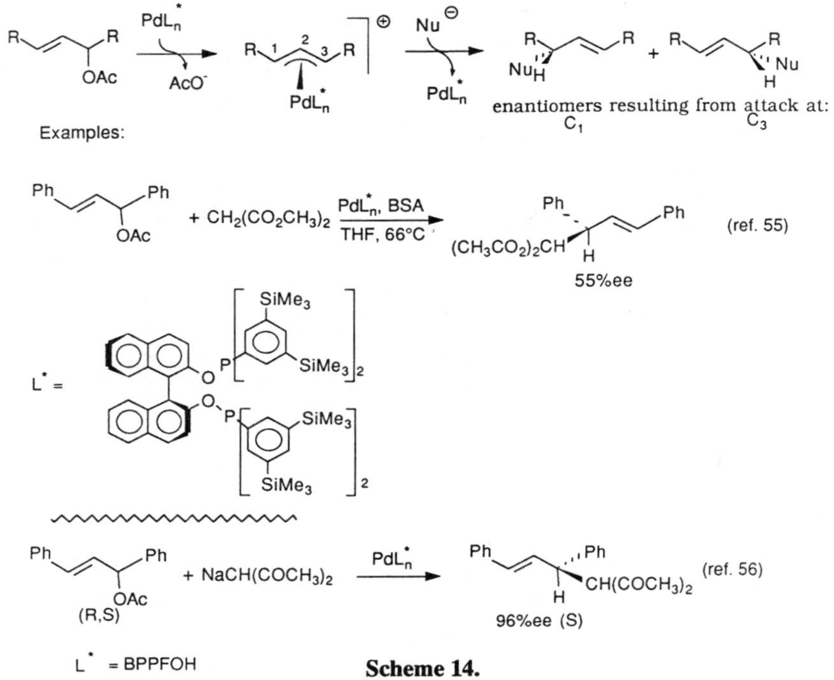

Examples:

(ref. 55)

55%ee

$L^* =$

(ref. 56)

96%ee (S)

$L^* = $ BPPFOH

Scheme 14.

Selection by an achiral nucleophile of two disastereotopic sites of a
chiral η^3allylpalladium complex.

The cationic η^3-allylpalladium intermediate complexes are diastereomeric. In the absence of any efficient interconversion process for the
diastereomers **15** and **16** (Scheme 15), different amounts of enantiomeric
compounds can be produced through:

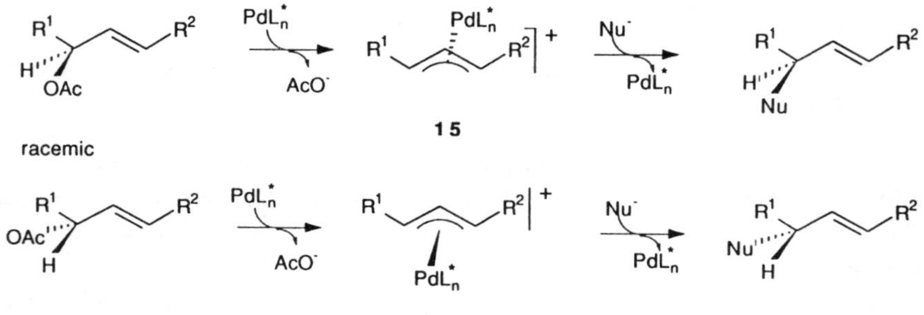

racemic

15

16

Scheme 15.

Dissymmetrically substituted η^3-allyl ligand of intermediate allylpalladium complexes.

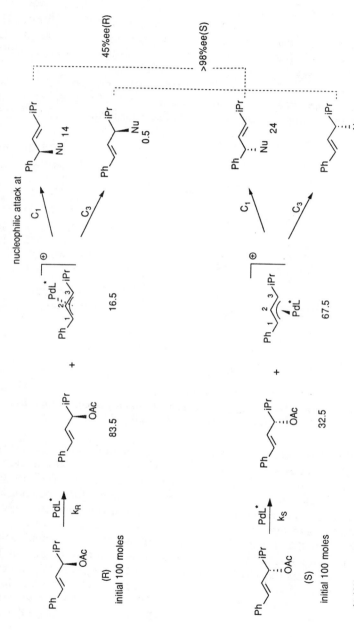

Scheme 16.

Selection by a chiral palladium complex of two enantiomers of the substrate (kinetic resolution) [58].

- kinetic resolution [57] of the substrate, *i.e.* selection by the optically active palladium(0) complex of the two enantiomers of the substrate. The process requires to carry out an incomplete reaction. One example has been reported by Hayashi [58] where both the substrate and the products were recovered under optically active form (Scheme 16).
- selection by the optically active palladium(0) complex of the two enantiotopic faces of the achiral substrate or of two enantiomeric conformations of the reacting substrate in fast equilibrium (Scheme 17).

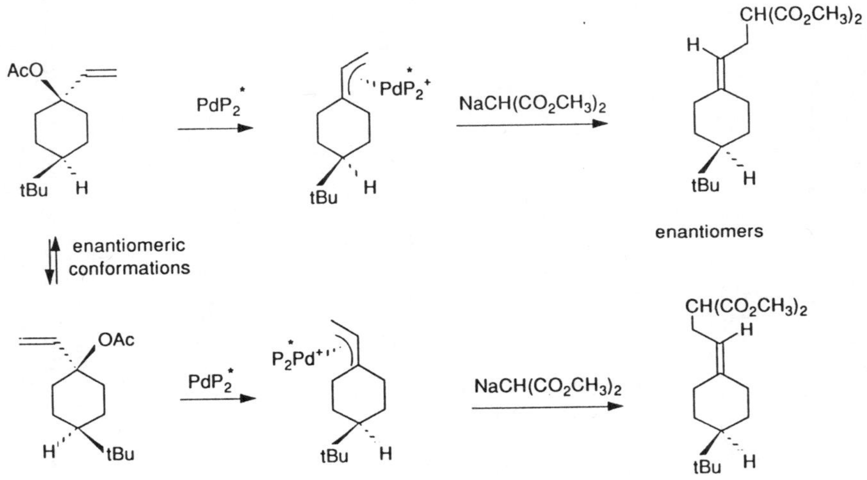

Scheme 17.

Selection by an optically active palladium(0) complex of the two enantiotopic faces of an achiral substrate [59, 60].

An example has been described where the achiral 4-t-butyl-1-vinyl-cyclohexen-1-yl acetate **17** through reaction with sodium dimethylmalonate gave the axially chiral cyclohexylidene compound **18** [59, 60]. In this reaction, the enantioselectivity has been shown to depend upon the nature of the inducing chiral phosphine ligand in the catalyst, the geometry (*cis* or *trans*) (Table IV) and the nature of the leaving group of the substrate, and upon the nature of the solvent. The latter dependences of the enantioselectivity upon the structure of the substrate, and more precisely that of the leaving group suggests that the Pd-induced ionization step is involved in the asymmetric induction process. Indeed the substitution reaction from a chiral, optically active (60% ee) linear allylic acetate **19** with an achiral Pd-catalyst in THF led to the optically active product **18** with a small loss of enantiomeric purity. This means that the η^3-allyl ligand in the palladium intermediate complex is enantiomerically stable and its rate of enantiomerization is slow compared with the rate of nucleophilic attack.

TABLE IV
Dependence of the ee of the product upon the phosphine and the geometry of the substrate [60]

phosphine*

$\left\{ R = Ac, \; solvent = THF \right\}$

substrate	phosphine*	ee (%)	rdt (%)
cis	(+)-DIOP	20 ± 5	60
cis	(S,S)-CHIRAPHOS	<10	73
cis	(R)-BINAP	25 ± 5	82
trans	(+)-DIOP	10 ± 5	82
trans	(S,S)-CHIRAPHOS	33 ± 5	68
trans	(R)-BINAP	47 ± 5	74

The enantioselective step is then more likely the η^3-allylpalladium forming step, and it is thus understood that the enantioselectivity is under the influence of the structure of the substrate. This is indeed the case *cis*-and *trans*-**17** afforded products with different ee and, more interestingly, the enantioselectivity is highly dependent upon the nature of the leaving group (Table V).

By a proper choice of the substrate geometry (*trans*), of the nature of the leaving group (p-anisoyl), of the chiral inducing ligand (BINAP) and of the solvent (dioxane), product **18** may be obtained with as high as 90%ee [61].
– selection by the achiral nucleophile of two diastereomeric η^3-allylpalladium(II) complexes in rapid interconversion.
Starting from a racemic substrate and in the absence of a kinetic selection by the palladium complex of the enantiomeric substrates, optically active products may be obtained provided that an efficient enantiomerization process for the η^3-allylpalladium intermediate complex takes place:

- a S_{N2} process in the case where the η^3-allylic ligand is dissymmetrically substituted.
- a S_{N2} or a $\eta^3-\eta^1-\eta^3$ process (or both) in the case where two substituents at one allylic terminus are identical.

One representative example of each situation is shown in Schemes 18 [62] and 19 [63] Bosnich could show with ^{31}P NMR analysis of the two cationic η^3-allylpalladium complexes that these were held in ratio 4:1 all along the reaction, to give the product with 83% ee. This means that the asymmetric induction process is mainly thermodynamic in origin, arising from the rapid interconversion of the two different-in-energy diastereomeric complexes.

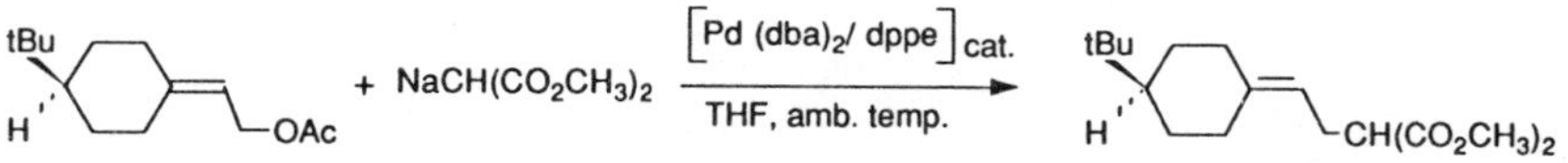

19 (S)   **18 (S)**

60 ± 6%ee   52 ± 4%ee

TABLE V

Dependence of the ee of the product upon the nature of the leaving group
and the solvent [63]

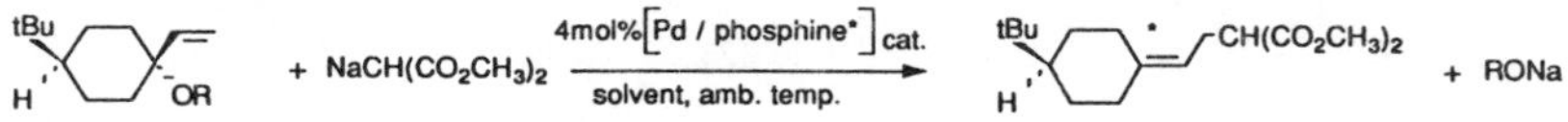

Leaving group {phosphine* = BINAP , solvent = THF}			Solvent {phosphine* = BINAP , solvent = dioxane}		
R	ee (%)	rdt (%)	R	ee (%)	rdt (%)
$-\overset{O}{\underset{\|}{C}}-OCH_3$	27 ± 3	74			
$-\overset{O}{\underset{\|}{C}}-CH_3$	47 ± 5	89	$-\overset{O}{\underset{\|}{C}}-CH_3$	49 ± 5	74
$-\overset{O}{\underset{\|}{C}}-\bigcirc$	63 ± 5	79	$-\overset{O}{\underset{\|}{C}}-\bigcirc$	76 ± 5	77
$-\overset{O}{\underset{\|}{C}}-\bigcirc_{OMe}$	72 ± 5	68	$-\overset{O}{\underset{\|}{C}}-\bigcirc_{OMe}$	80 ± 5	77
$-\overset{O}{\underset{\|}{C}}-\bigcirc-OMe$	78 ± 5	79	$-\overset{O}{\underset{\|}{C}}-\bigcirc-OMe$	90 ± 5	63

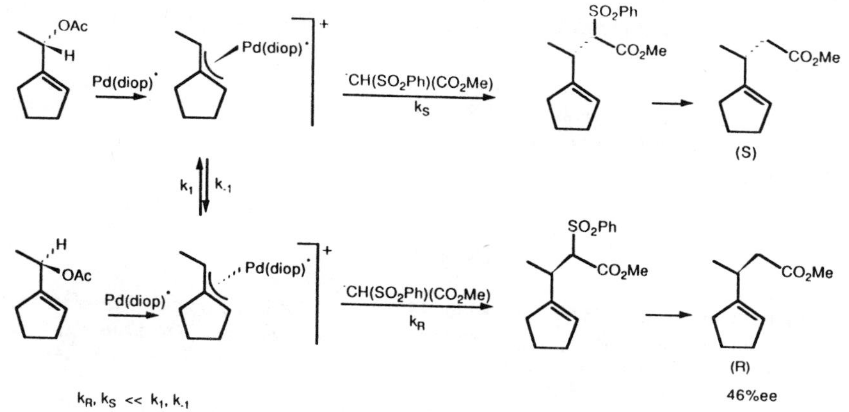

Scheme 18.

Selection by an achiral nucleophile of two diastereometric η^3-allylpalladium complexes in rapid equilibrium [62].

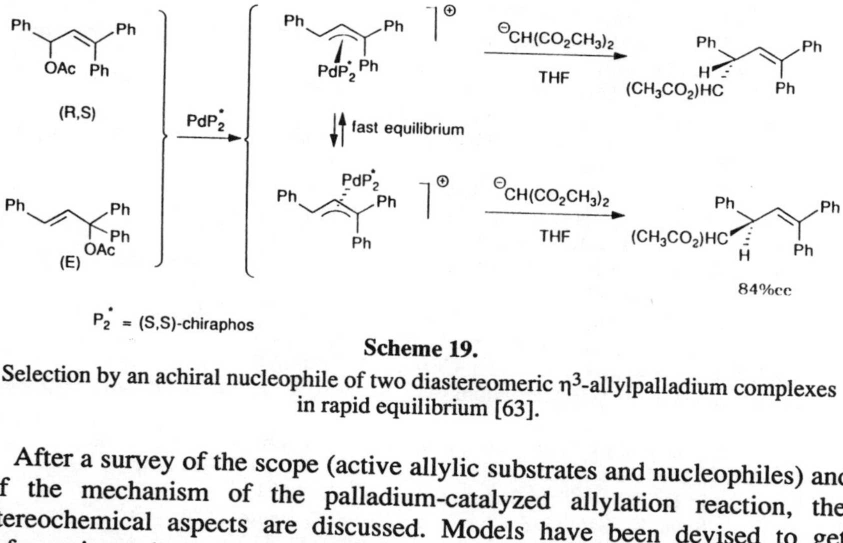

$P_2^* = (S,S)\text{-chiraphos}$

Scheme 19.

Selection by an achiral nucleophile of two diastereomeric η^3-allylpalladium complexes in rapid equilibrium [63].

After a survey of the scope (active allylic substrates and nucleophiles) and of the mechanism of the palladium-catalyzed allylation reaction, the stereochemical aspects are discussed. Models have been devised to get informations about the stereochemical course of each step of the catalytic process, and to inquire into the mode of attack of the nucleophiles.

Different processes for asymmetric induction have been described. The collection of informations about the reaction mechanism, the search for the enantioselective step(s) help in understanding the way in which the chiral

130   J.C. Fiaud

information contained in the ligand is transmitted to the product. and hence
how to set up the different parameters or partners in order to improve the
enantioselectivity.

Laboratoire de Synthèse Asymétrique,
Bât. 420, Université Paris-Sud,
91405 Orsay (France)

References

1. (a) Trost, B.M. *Acc. Chem. Res.* **1980**, *13*, 385; (b) Tsuji, J. *Organic Synthesis with Palladium Compounds*; Springer, Berlin 1980; (c) *Tetrahedron Lett.* **1986**, *42*, 4361; (d) Tsuji, J.; Minami, I. *Acc. Chem. Res.* **1987**, *20*, 140; (e) Trost, B.M. *Angew. Chem. Int. Ed. Eng.* **1989**, *28*, 1173.
2. Tsuji, J.; Takahashi, H.; Morikawa, M. *Tetrahedron Lett.* **1965**, *6*, 4387.
3. Atkins, K.E.; Walker, W.E.; Manyik, R.M. *Tetrahedron Lett.* **1970**, *11*, 3821.
4. (a) Hata, G.; Takahashi, K.; Miyake, A. *J. Chem. Soc, Chem. Commun.* **1970**, 1392; (b) Takahashi, K.; Miyabe, A.; Hata, G. *Bull. Chem. Soc. Jpn* **1972**, *45*, 230.
5. Trost, B.M.; Verhoeven, T.R. *J. Org. Chem.* **1976**, *41*, 3215.
6. Trost, B.M. *Acc. Chem. Res.* **1980**, *13*, 385
7. Fiaud, J.C.; Hibon de Gournay, A.; Larchevêque, M.; Kagan, H.B. *J. Organomet. Chem.* **1978**, *154*, 175.
8. Negishi, E.; Chatterjee, S.; Matsushita, H. *Tetrahedron Lett.* **1981**, *22*, 3737.
9. Tsuji, J. *J. Organomet. Chem.* **1986**, *300*, 281.
10. (a) Trost, B.M. *Acc. Chem. Res.* **1980**, *13*, 385; (b) Hayashi, T.; Yamamoto, A.; Ito, Y. *Tetrahedron Lett.* **1988**, *29*, 99.
11. Hiroi, K.; Makino, K. *Chem. Lett.* **1986**, 617.
12. Trost, B.M.; Schmuff, N.R. *J. Am. Chem. Soc.* **1985**, *107*, 396.
13. Hutchins, R.O.; Learn, K. *J. Org. Chem.* **1982**, *47*, 4380.
14. Tamura, R.; Kai, Y.; Kakihana, M.; Hayashi, K.; Tsuji, M.; Nakamura, T.; Oda, D. *J. Org. Chem.* **1986**, *51*, 4375.
15. Trost, B.M.; Molander, G.A. *J. Am. Chem. Soc.* **1981**, *103*, 5969.
16. Trost, B.M.; Verhoeren, T.R. *J. Org. Chem.* **1976**, *41, 3215.*
17. *Trost, B.M.; Verhoeren, T.R. J. Am. Chem. Soc.* **1980**, *102*, 4730.
18. Ito, Y.; Sawamura, M.; Matsumoto, M.; Hayashi, T. *Tetrahedron Lett.* **1987**, *28*, 4849.
19. Ferroud, D.; Genêt, J.P.; Kiolle, R. *Tetrahedron Lett.* **1986**, *27*, 23.
20. Fiaud, J.C.; Malleron, J.L. *J. Chem. Soc., Chem. Commun.* **1981**, 1159.
21. Trost, B.M.; Keinan, E. *Tetrahedron Lett.* **1980**, *21*, 2591.
22. Hiroi, K.; Suya, K.; Sato, S. *J. Chem. Soc, Chem. Commun.* **1986**, 469.
23. Fiaud, J.C.; Malleron, J.L. *Tetrahedron Lett.* **1980**, *21*, 4437.
24. Fiaud, J.C.; Denner, B.; Malleron, J.L. *J. Organomet. Chem.* **1985**, *291*, 393.
25. Hayasi, Y.; Riediker, M.; Temple, J.S.; Schwartz, J. *Tetrahedron Lett.* **1981**, *22*, 2629.
26. Matsushita, H.; Negishi, E. *J. Am. Chem. Soc.* **1981**, *103*, 2882.
27. Matsushita, H.; Negishi, E. *J. Chem. Soc., Chem. Commun.* **1982**, 160.

28. Keinan, E.; Peretz, M. *J. Org. Chem.* **1983**, *48*, 5302.
29. Trost, B.M.; Keinan, E. *Tetrahedron Lett.* **1980**, *21*, 2595.
30. Murahashi, S.I. ; Tanigawa, Y.; Imada, Y.; Taniguchi, Y. *Tetrahedron Lett.* **1986**, *27*, 227.
31. Murahashi, S.I.; Imada, Y.; Taniguchi, Y.; Kodera, Y. *Tetrahedron Lett.* **1988**, *29*, 2973.
32. Stanton, S.A.; Felman, S.W.; Parkhurst, C.S.; Godleski, S.A. *J. Am. Chem. Soc.* **1983**, *105*, 1964.
33. Tsuji, J.; Yamakawa, T. *Tetrahedron Lett.* **1979**, *20*, 613.
34. Keinan, E.; Roth, Z. *J. Org. Chem.* **1983**, *48*, 1769.
35. Inomata, K.; Yamamoto, T.; Kotabe, H. *Chem. Lett.* **1981**, 1357.
36. Trost, B.M.; Scanlan, T.S. *Tetrahedron Lett.* **1986**, *27*, 4141.
37. Fiaud, J.C. *J. Chem. Soc., Chem. Commun.* **1983**, 1055.
38. Trost, B.M.; Yoshida, J.; Lautens, M. *J. Am. Chem. Soc.* **1983**, *105*, 4494.
39. Hutchins, R.O.; Learn, K.; Fulton, R.P. *Tetrahedron Lett.* **1980**, *21*, 27.
40. Tsuji, J.; Yamakawa, T. *Tetrahedron Lett.* **1979**, *20*, 613.
41. Fiaud, J.C.; Legros, J.Y. *J. Org. Chem.* **1987**, *52*, 1907.
42. Fiaud, J.C.; Aribi-Zouioueche, L. *J. Chem. Soc, Chem. Commun.* **1986**, 390.
43. Hayashi, T.; Konishi, M.; Kumada, M. *J. Chem. Soc, Chem. Commun.* **1984**, 107.
44. Bäckvall, J.E.; Nordberg, R.E.; Björkman, E.E.; Moberg, C. *J. Chem. Soc., Chem. Commun.* **1980**, 943.
45. Bäckvall, J.E.; Nordberg, R.E. *J. Am. Chem. Soc.* **1981**, *103*, 4959.
46. Keinan, E.; Greenspoon, N. *Tetrahedron Lett.* **1982**, *23*, 241.
47. Trost, B.M.; Schmuff, N.R. *J. Am. Chem. Soc.* **1985**, *107*, 396.
48. Trost, B.M.; Keinan, E. *J. Org. Chem.* **1979**, *44*, 3451.
49. Trost, B.M.; Verhoeven, T. .; Fortunak, J.M. *Tetrahedron Lett.* **1979**, *20*, 2301.
50. Hiroi, K.; Kitayama, R.; Sato, S. *Chem. Lett.* **1984**, 929.
51. Colobert, F.; Genêt, J.P. *Tetrahedron Lett.* **1985**, *26*, 2779.
52. Hiroi, K.; Suya, K.; Sato, S. *J. Chem. Soc, Chem. Commun.* **1986**, 469.
53. Hayashi, T. ; Kanehira, K.; Hagihara, T.; Kumada, M. *J. Org. Chem.* **1988**, *53*, 113.
54. Genêt, J.P.; Jugé, S.; Ruiz-Montès, J.; Gaudin, J.M. *J. Chem. Soc, Chem. Commun.* **1988**, 718.
55. Trost, B.M.; Murphy, D.J. *Organometallics* **1985**, *4*, 1143.
56. Hayashi, T.; Yamamoto, A.; Hagihara, T.; Ito, Y. *Tetrahedron Lett.* **1986**, *27*, 191.
57. Kagan, H.; Fiaud, J.C. *Kinetic Resolution, Topics in Stereochemistry*, 18; Eliel and Wilen, Interscience, 1988; 249.
58. Hayashi, T.; Yamamoto, A.; Ito, Y. *J. Chem. Soc., Chem. Commun.* **1986**, 1090.
59. Fiaud, J.C.; Legros, J.Y. *Tetrahedron Lett.* **1988**, *29*, 2959.
60. Fiaud, J.C.; Legros, J.Y. *J. Organomet. Chem.* **1989**, *370*, 383.
61. Fiaud, J.C.; Legros, J.Y. *J. Org. Chem.* **1990**, *55*, 4840.
62. Trost, B.M.; Strege, P.E. *J. Am. Chem. Soc.* **1977**, *99*, 1649.
63. Auburn, P.R.; Mackenzie, P.B.; Bosnich, B. *J. Am. Chem. Soc.* **1985**, *107*, 2033.

A. HEUMANN

REGIO-, STEREO-, AND ENANTIOSELECTIVITY
IN PALLADIUM AND PLATINUM
CATALYZED ORGANIC REACTIONS

1. Introduction

There is a growing interest in metal catalyzed organic reactions as a tool for selective organic synthesis [1]. Among the metals which can be used, the group VIII elements, especially palladium and platinum, have found very widespread application [2]. This feature is best illustrated by the fact that even short syntheses of complex molecules may be accomplished using, more than one palladium catalyzed organic transformation. Let us cite as an illustrative example the «Total Synthesis of the N-acetyl methyl ester of (±)-Clavicipitic Acid», described by L.S. Hegedus and coworkers [3]. Starting from commercially available 2-bromo-6-nitro-toluene **1**, the desired tricyclic ester **10** can be obtained in overall 18% yield. It is remarkable that 4 of the 12 reaction steps are conducted with high selectivities in the presence of palladium catalysts:

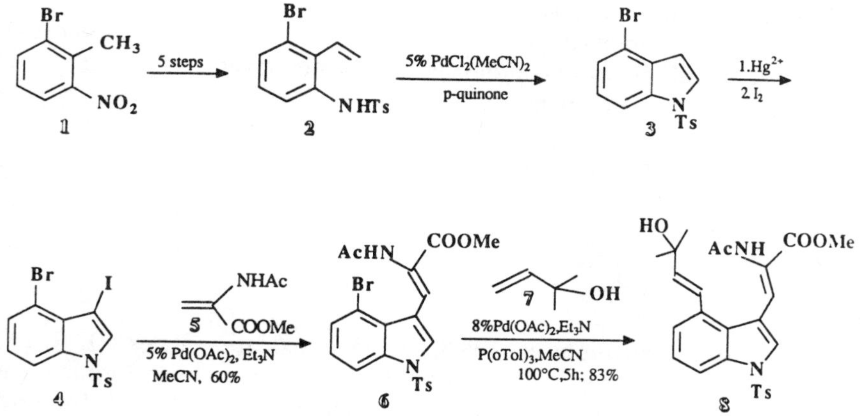

A. F. Noels et al. (Eds.), Metal promoted selectivity in organic synthesis, 133–159.

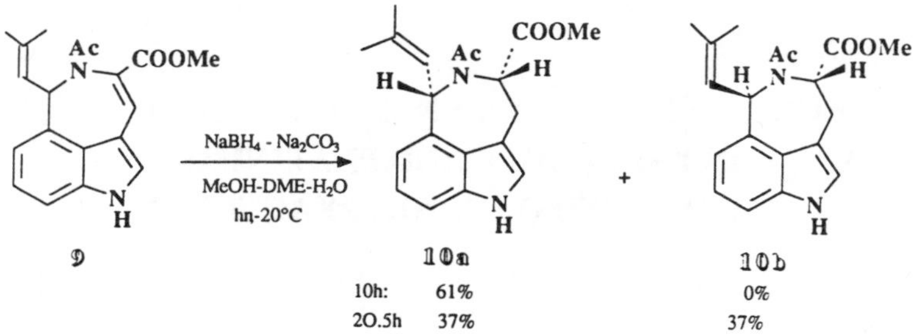

9

10a

10h: 61%
20.5h   37%

10b

0%

37%

Two different Pd(II) mediated cyclization reactions give rise to the formation of a five- **3** and seven membered ring **9**, respectively. Furthermore Pd(O) species are used to introduce, stereoselectively, the two C-side chain precursors at 2 different aromatic carbons via olefin-arylation (Heck) reactions [4]. This approach makes available compounds such as **10** with a «40 fold increase in yield over existing methods».

After this impressive demonstration of the potential efficacy of transition metals, and more precisely of palladium, in organic synthesis, the question arises: what makes palladium so predestinate to be the *reagent of choice* for so many selective reactions in organic synthesis ?

It has been pointed out [5], that *seven fundamental reactions* might be considered in order to rationalize the unique role of palladium compared to other transition metals. These reactions might be divided into three categories:

– formation of C–Pd bonds:
  1. oxidative addition (Pd°)
  2. nucleophilic attack to coordinated Pd(II) complexes
– modification of C–Pd bonds:
  3. insertion of an unsaturated ligand
  4. metathetical replacement
– cleavage of C–Pd bonds:
  5. reductive elimination
  6. β-hydride elimination
  7. reductive solvolysis.

In other words, palladium has readily variable oxidation states (O, II, IV), thus releasing or accepting electrons during different bond-forming and bond-breaking steps. It is also able to bind various kinds of ligands from soft π-acceptors (olefins, CO) to hard σ-donors (Cl, OH, OAc) and its catalytic activity is enhanced by the ease of coordinating several (different) ligands, thus enabling reactions between these ligands. Although the most common coordination number of palladium is 4 (square planar) the lower coordination number 3 and the higher ones, 5 and 6, are easily and reversably attained. This is important for the activation of stable complexes in solution $(Pd(PPh_3)_4$, for example) or for the renewal of «consumed» ligands (reaction product). On the other hand, the oxidative cleavage of Pd-C bonds,which is the product forming step in many Pd-catalyzed oxidation reactions (Wacker reaction and organic analogues), is supposed to proceed via an addition-elimination process.

We have already shown most of these fundamental reactions in the introductory reaction sequence by Hegedus [3], and as will be demonstrated in the examples throughout this lecture, most of the problems relative to selectivity will be inextricably interweaved with the stereochemistry of these reactions.

The following type of reactions will be discussed:
2. Rearrangements as a consequence of palladium(II) and palladium(O) catalysis
3. 1.4-Difunctionalization of 1,3-dienes
4. Cyclization of non-conjugated dienes
5. Asymmetric alkylation via π-allyl Pd complexes
6. Asymmetric hydroformylation

2. Rearrangements

Reactions which lead to a skeletal rearrangement of the organic substrate have continuously attracted the attention of the organic chemists [6]. This is not surprising since the migration of atoms or functional groups generally changes the topology of the molecule profoundly, thus offering the possibility of direct and rapid access to quite different structures.

Palladium easily forms σ-bonds with carbon, these may react, depending on the substrate *and* the reaction conditions via a heterolytic cleavage, and give rise, at the same time, to carbonium ion-like transformations.

The Wagner Meerwein Rearrangement observed by Baird [7] when he treated norbornene **11** with a mixed $(PdCl_2-CuCl_2)$ oxidation system in buffered acetic acid, is an attractive example to be cited:

$$ \textbf{11} \quad + \quad CH_3COOH \quad + \quad 2\ CuCl_2 \quad \xrightarrow[\text{CH}_3\text{COONa}]{\text{2.7\% PdCl}_2} \quad \textbf{12} \quad (50\text{-}84\%) $$

Formally this reaction corresponds to the addition of two different nucleophilic groups (Cl⁻ and OAc⁻) to an oxidized (withdraw of two electrons) norbornene double bond, and attachement to the positions 2 and 7! The palladium(II) catalyst is continuously regenerated by the copper chloride, which also furnishes the Cl⁻ nucleophile. Concerning the mechanism, the olefinic double bond of **11** is activated (coordination to Pd(II)) towards nucleophilic attack of the acetate ion. The intermediate O-acetyl palladium-σ-complex **14**, is now cleaved by CuCl₂ and rearranges to cation **15** giving the organic product **12** after substitution by Cl⁻.

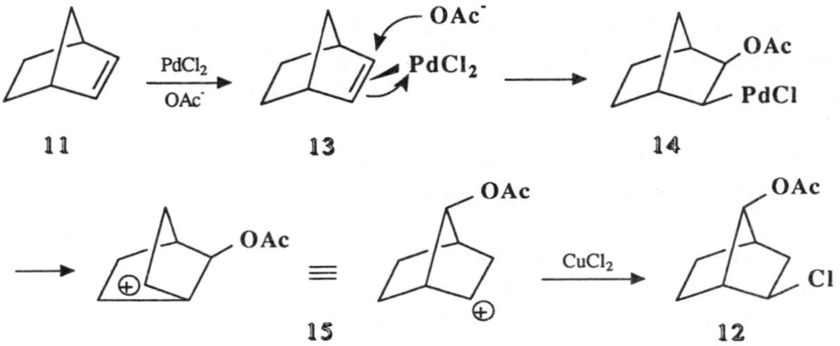

The stereochemistry of the different addition and substitution steps (exo attack) are substrate specific; the carbon-carbon bond migration has been shown to be the result of the copper chloride-promoted, oxidative cleavage of the C–Pd bond in σ-complex **14**.

Two experimental results support this statement:
i. the formation of a palladium-σ-complex **17** from norbornene and η³-Pd-complexes such as **16** [8, 9]:

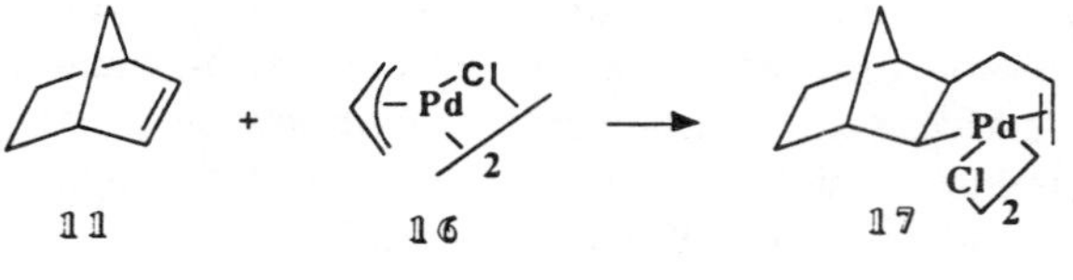

ii. the observation of a difference between the oxidative and non oxidative cleavage of the same Pd-σ-complex **18** and the formation of unrearranged epoxy-norbornane **19** and rearranged reaction products such as **20**, respectively [10, 11].

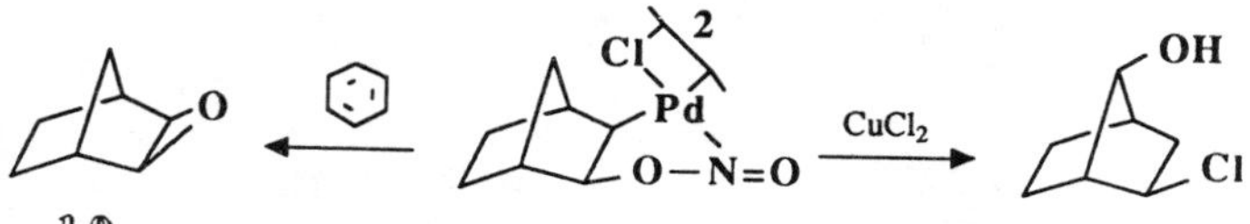

1 9 (quant.)   **1 8**   **2 0** (56%)

The formation of epoxide **19** is rationalized by a *syn*-elimination, whereas the relationship of the non-epoxidative rearrangement of **18** to Baird's reaction [7] (formation of **12**) is obvious.

Another interesting consequence of the heterolytic cleavage of C–Pd bonds is the possibility to induce variations at the carbon chain or at the ring size of carbocycles. The following examples of phenyl and methyl group migrations may be representative [12–16]:

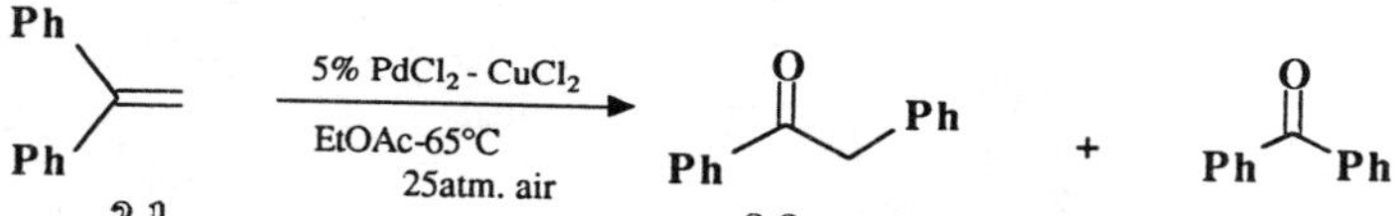

2 1   **2 2** (28.5%)   **2 3** (28.5%)

This reaction [12] proceeds via nucleophilic attack at the coordinated olefin and copper chloride induced rearrangement,whereas cationic complexes $[PdL_4]^{2+}$ have been shown to provoke rearrangements without the presence of $CuCl_2$ [13]:

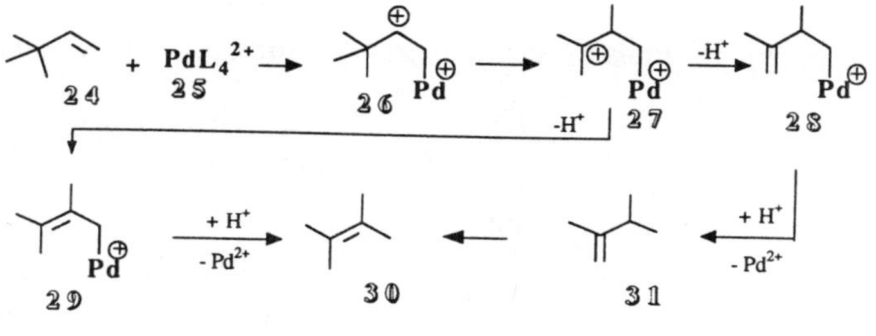

2 4   **2 5**   **2 6**   **2 7**   **2 8**

2 9   **3 0**   **3 1**

Metal exchange reactions, from organomercurials for example, are another possibility to selectively create carbon-palladium bonds, which may rearrange via phenonium ions [14] or methyl migration [15]:

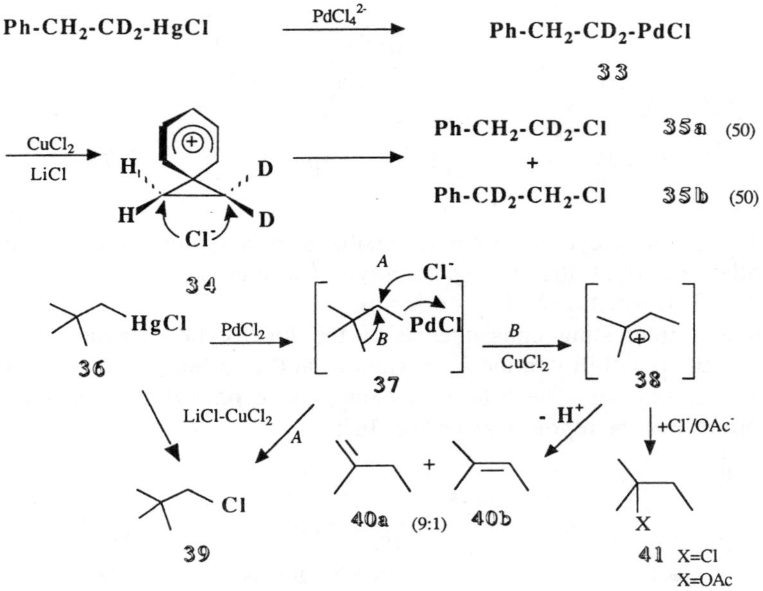

$$Ph\text{-}CH_2\text{-}CD_2\text{-}HgCl \xrightarrow{\ PdCl_4{}^{2-}\ } Ph\text{-}CH_2\text{-}CD_2\text{-}PdCl$$

33

Ph-CH$_2$-CD$_2$-Cl   **35a** (50)

+

Ph-CD$_2$-CH$_2$-Cl   **35b** (50)

A transposition without the auxiliary reoxidant CuCl$_2$ is the following nitrene transfer reaction [16]:

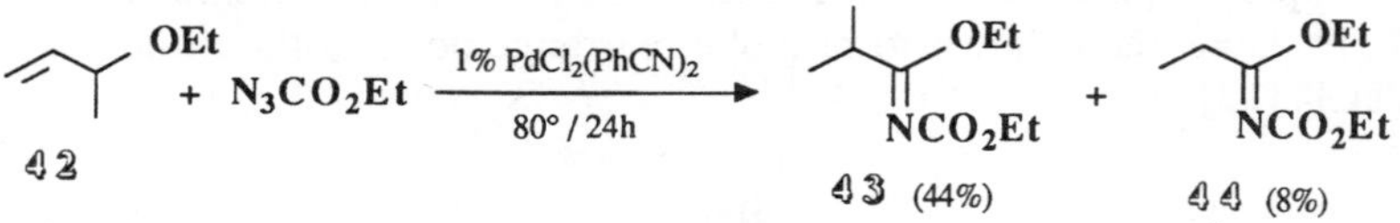

A general scheme resumes most of this transposition reactions (Scheme I):

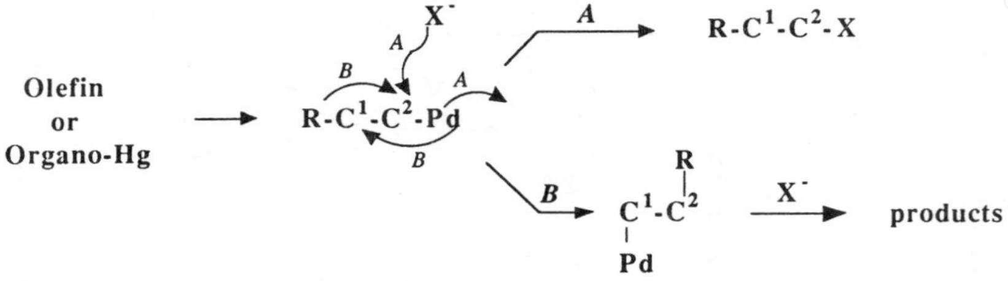

Olefins or organomercurials in the presence of Pd(II) form the σ-bonded R–C–C–Pd intermediate, and the cleavage of carbon-palladium bond via path B leads to rearranged products. It should be mentioned that added LiCl to the oxidation mixture sometimes suppresses the rearrangement and leads to substituted reaction products (path A) [15]. This particular behaviour of LiCl has already been observed during the oxidation of ethylene (vide infra) [17].

Alkyl groups which are part of a cyclic system also migrate easily, and the resultant modification of the ring size leads to ring-enlarged or ring-regressed mono- and bicyclic systems.

Partially strained (cyclobutane) [12]

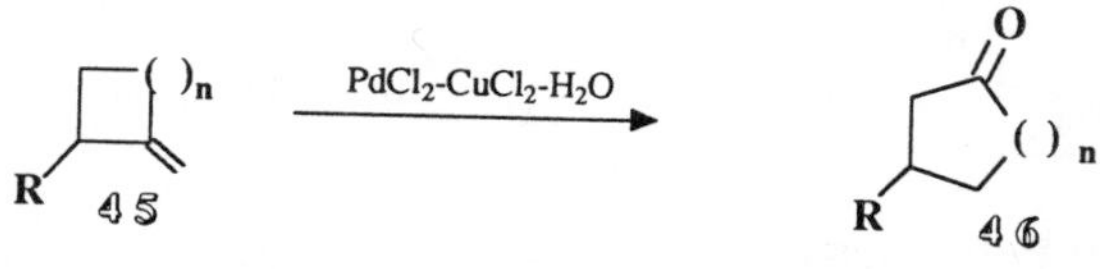

n=1	65°/different	65-82% (+ π-allyl complex)
n=2	solvents	20%
n=3		- (only π-allyl complexes)

or conformationaly favoured systems (cyclooctane) [18]

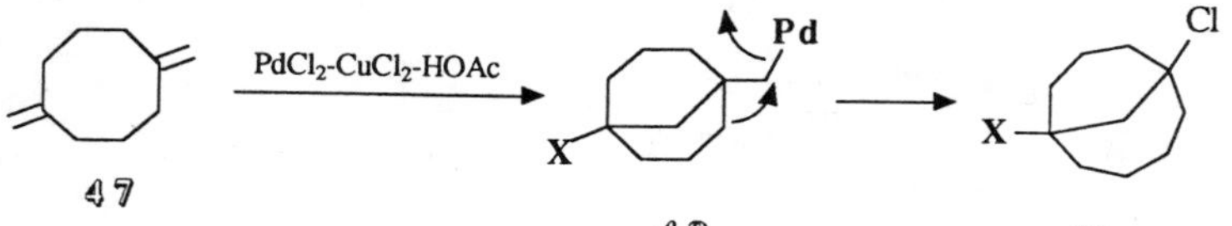

49a X=Cl
49b X=OAc

form intermediates, which rearrange selectively. This is also true when the unsaturated part of the molecule is an extracyclic vinyl [19] or acetylenic group [20]:

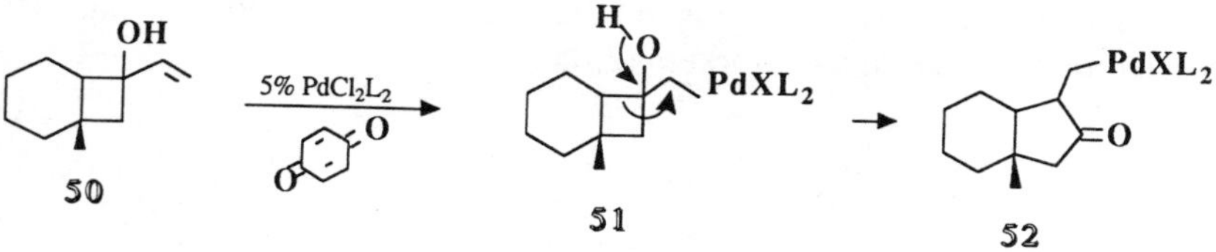

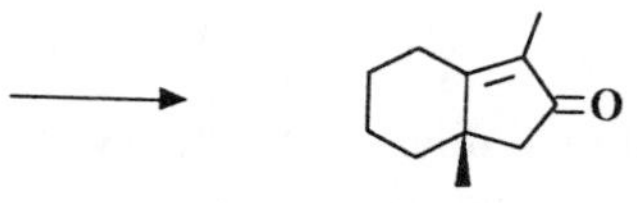

53 (67%)

In the latter case [20], 4-oxygenated-5-alkylidene cyclopentenones are formed which are structurally similar to pharmacologically interesting, marine natural products:

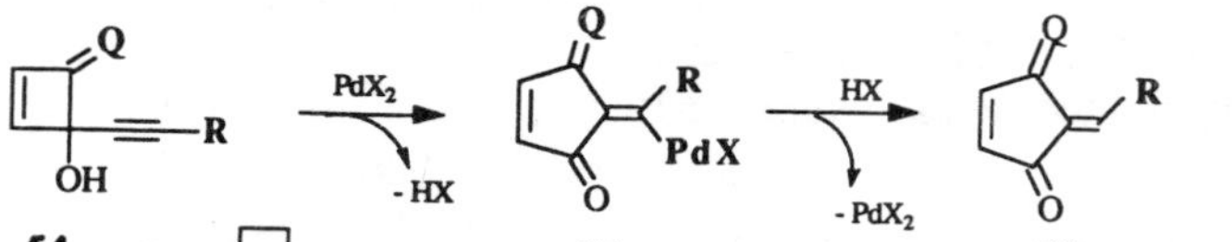

54 (Q=O, O̅ O̅)   55   56 (40-91%)

An interesting possibility to constitute the R–CH$_2$–Pd migratory unit **60** has been reported by Saegusa [21], a reaction sequence closely related to the oxidation of bis-1,5-methylene cyclooctane **47** [18]:

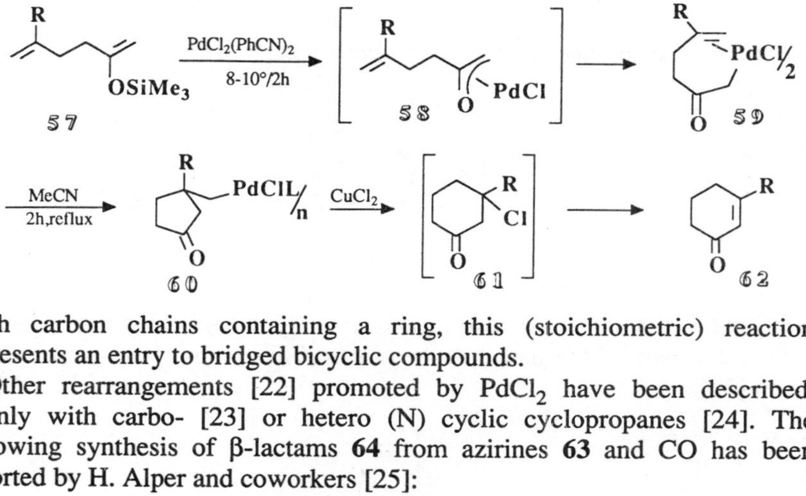

With carbon chains containing a ring, this (stoichiometric) reaction represents an entry to bridged bicyclic compounds.

Other rearrangements [22] promoted by PdCl$_2$ have been described, mainly with carbo- [23] or hetero (N) cyclic cyclopropanes [24]. The following synthesis of β-lactams **64** from azirines **63** and CO has been reported by H. Alper and coworkers [25]:

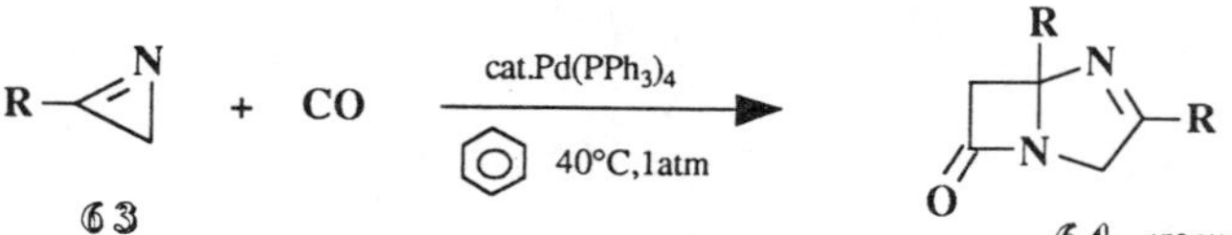

63   64 (50%)

Later, the same group found similar reactions, yet leading to monocyclic β-lactams **66** with rhodium [26] and palladium catalysis [27]:

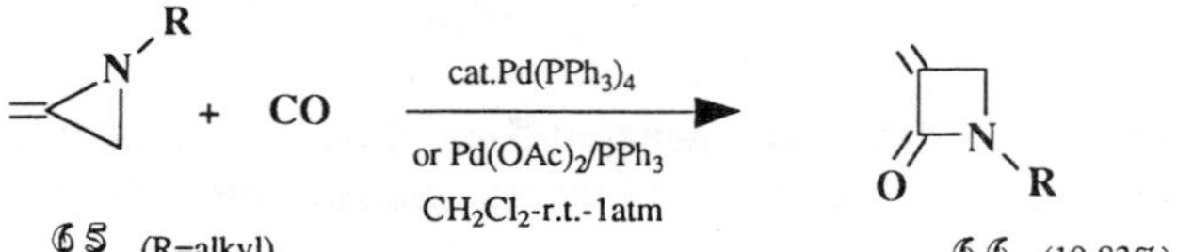

65 (R=alkyl)   66 (10-83%)

A particular case is the oxidation (using palladium acetate) of the tricyclic diene **67** [28]. First, this reaction corresponds to a ring enlargement (bicyclo[2.2.2]octane to bicyclo[3.2.2]nonane framework). However, as a consequence of cyclobutene and non conjugated diene structure elements, two other basic reactions take place simultaneously: the ring contraction of cyclobutane to cyclopropane and the carbomethoxy assisted cyclisation of the non conjugated diene to a five membered carbocyclic ring. Ring contractions of cyclobutene are not without precedents [29] and cyclobutanes – especially those integrated in a bicyclic system, have been found to be cleaved by Pd(II) catalysis [30, 31]:

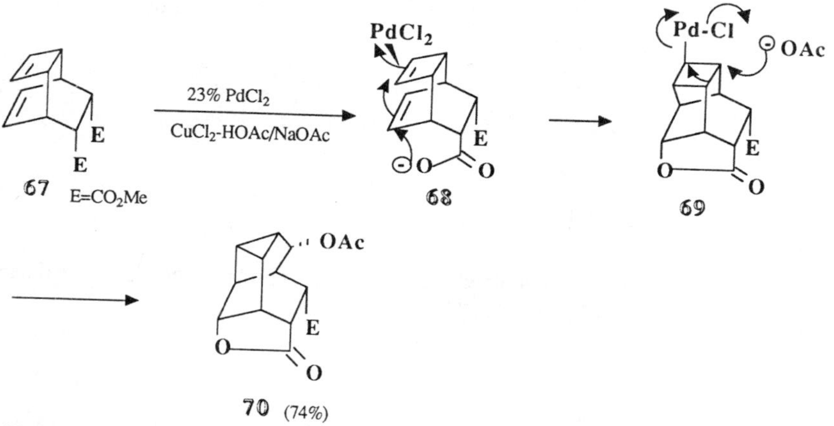

Until now, we have seen a number of quite selective, palladium catalyzed, alkyl migration reactions. Nevertheless, it should be emphazised that these reactions only occur when a hydrogen shift, the most common transposition encountered with this transition metal catalyst, does not occur. We will not discuss here in detail the well known «Wacker reaction» [32] *i.e.* the oxidation of ethene to acetaldehyde which proceeds *via* a 1.2-hydrogen shift, and which has been treated in numerous review articles and text books [33]:

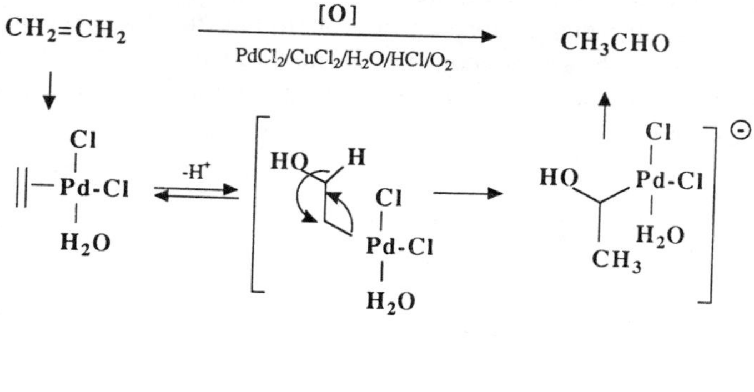

The fact that palladium(II) easily induces hydrogen shifts [2] (for example, the isomerization of higher olefins [34]) is considered to be rather an unwanted side reaction of this metal. Nevertheless regio- and stereoselective hydrogen migrations have been described:

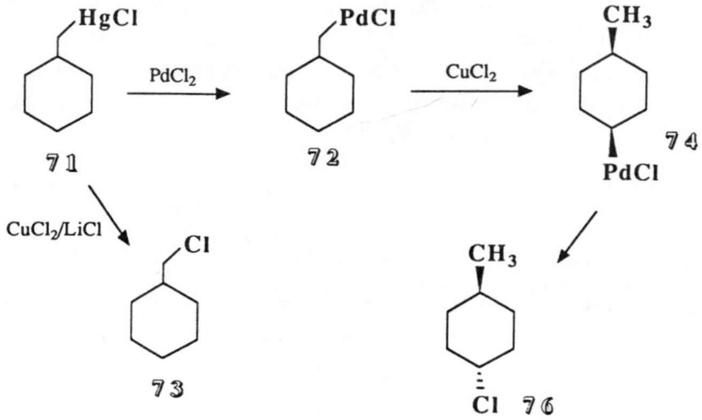

The mechanism most probably includes an (intramolecular) oxidative addition of a C–H bond to Pd(II) 77:

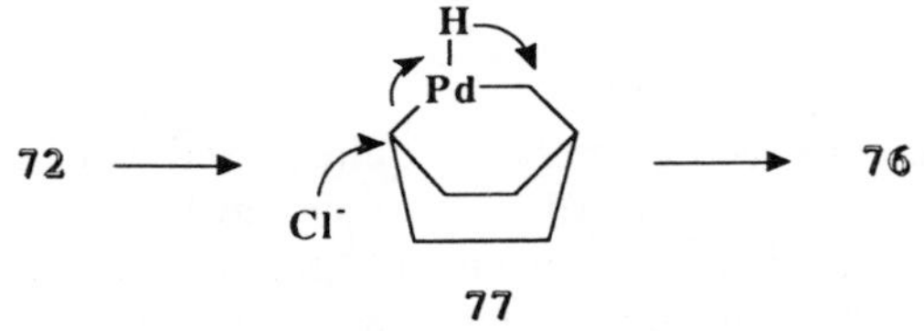

a principle that could be interesting for the intermolecular carbon-hydrogen activation [35].

Sigmatropic rearrangements of unsaturated systems are highly selective reactions without metal catalysis. Coordination to metallic electrophiles such as Pd^{2+} or Hg^{2+}, however reduces considerably the activation parameters and renders possible this type of rearrangement using extremely mild conditions [36].

A special case are 1,5-hexadienes. These dienes can react via [3.3]-sigmatropic rearrangements [36] termed «cyclization-induced rearrangement catalysis» (Scheme II):

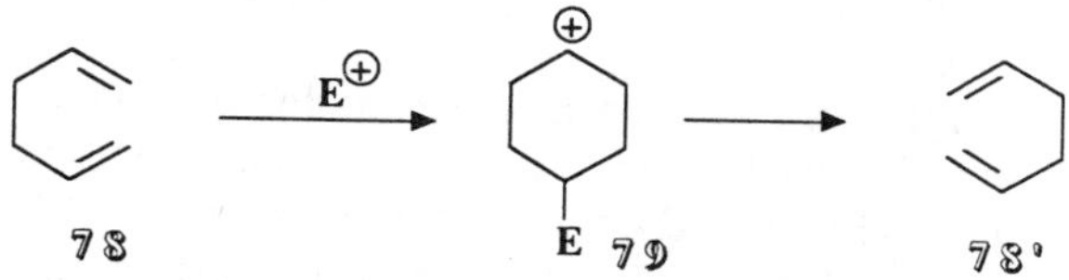

$$E^{\oplus} = Hg, Pd$$

As will be shown later, they also react selectively via oxidative cyclization [37].

Let us close this overview on palladium catalyzed transposition reactions with two further examples where carbon-oxygen bonds are the important factors for the selectivity of the rearrangements reactions. There is B. Trost's palladium catalyzed O to C migration [38]:

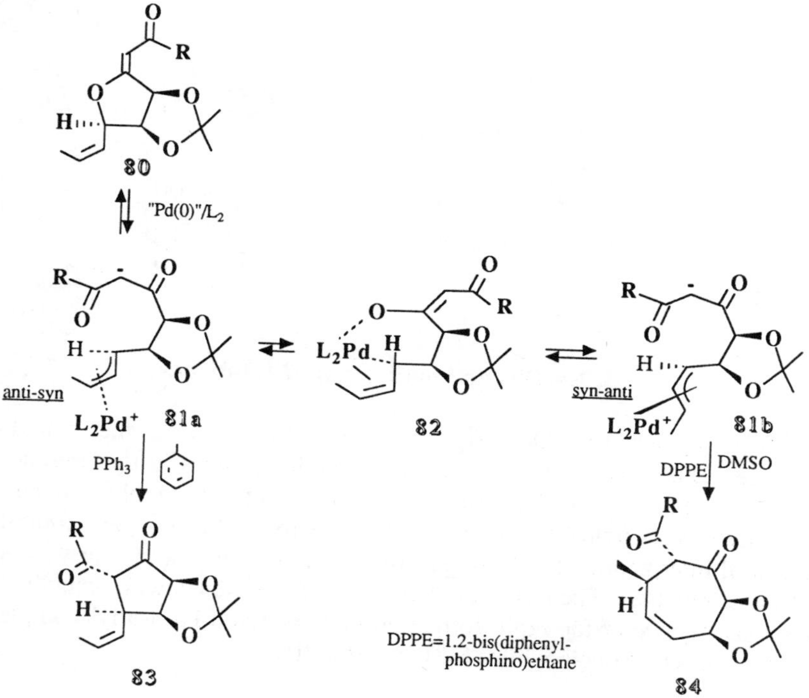

DPPE=1.2-bis(diphenyl-phosphino)ethane

The active species is a π-allyl complex, formed in situ from an intracylic allyl ether and palladium(O). The remarkable regio- and stereoselectivity is rationalized by the influence of ligands and solvent on the *syn-anti* interconversion of different π-allyl units and proceeding through σ-bonded intermediates.

The other example concerns the mechanism for the oxidation of olefins with $PdCl(NO_2)(CH_3CN)_2$. In this reaction, equal amounts of isomeric glycol monoacetates are formed under certain conditions [39]. It is possible to show, with deuterated terminal olefins **85** [40] that the following selective rearrangement takes place via acetoxonium intermediates **87**:

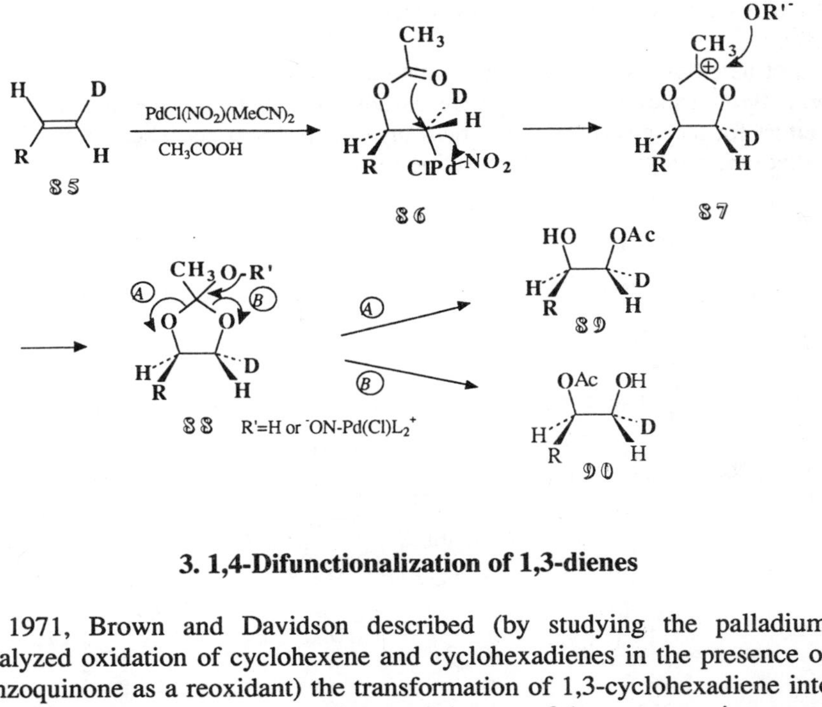

3. 1,4-Difunctionalization of 1,3-dienes

In 1971, Brown and Davidson described (by studying the palladium catalyzed oxidation of cyclohexene and cyclohexadienes in the presence of benzoquinone as a reoxidant) the transformation of 1,3-cyclohexadiene into 1,4-diacetoxy-2-cyclo-hexene [41]. 1,4-Acetoxy-1-butene, an important industrial intermediate for 1,4-butanediol, is accesible in a similar way from 1,3-butadiene [42]. The easy transformation of 1,3-dienes with palladium(II) is a consequence of the rapid formation of substituted Pd-π-allyl complexes in the presence of nucleophiles [43] (Scheme III):

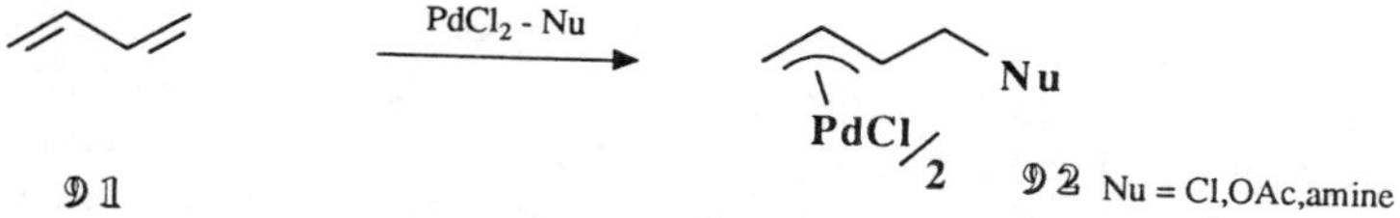

9 1   **9 2** Nu = Cl,OAc,amine

As has been studied extensively by J. Bäckvall and his group [44–49] these complexes may react with remarkable selectivities, giving 1,4- disubstituted products [45]:

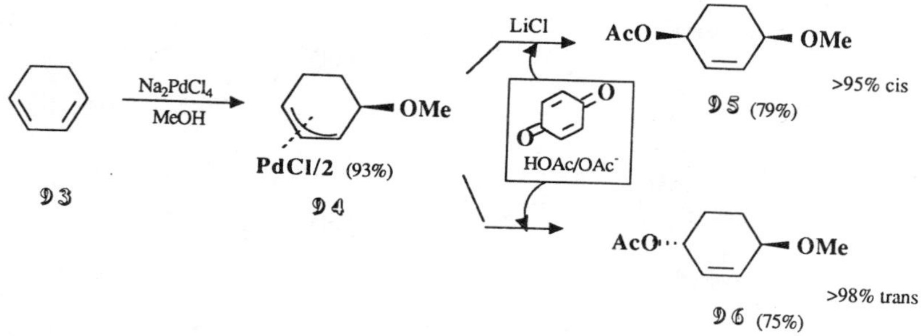

The acetate attack to the palladium-π-allyl complex **94** is induced by benzoquinone, and the sole presence (formation of **95**) or absence (formation of **96**) of LiCl determines the stereochemical mode of substitution. The formation off the *cis*-compound **95** is rationalized by a *trans*-attack of OAc to the π-allyl complex, whereas the formation of *trans*-methoxy ester **96** occurs, most probably, via the σ-alkyl palladium complex **98** by a *cis*-migration reaction:

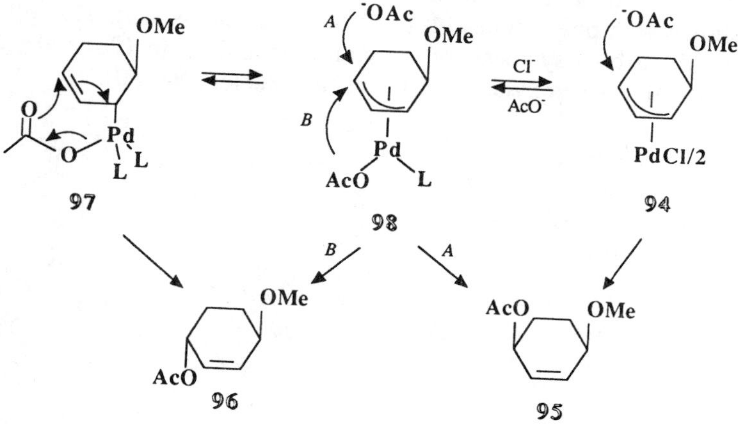

This **dual stereoselectivity** in the nucleophilic attack is not restricted to stoichiometric reactions on isolated π-allyl complexes. On modifying the reaction conditions and carefully avoiding Diels-Alder reactions between the 1,3-dienes and benzoquinone, a great number of catalytic 1,4-diacetoxylation [46] and 1,4-acetoxy-chlorination reactions [47] have been described by the same group. In addition to the high 1,4-selectivity the simple variation of the LiCl-concentration allows one to obtain different products, under otherwise identical conditions [46, 47]:

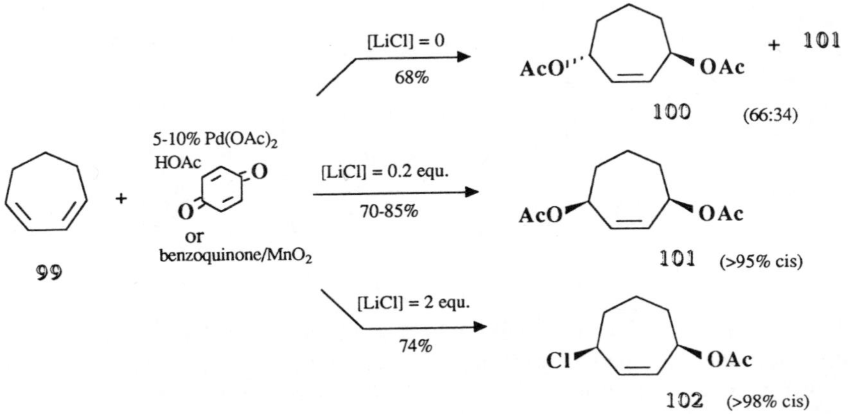

The high selectivities observed with cyclic systems also holds with acyclic dienes. Both nucleophilic groups in the 1,4-disubstituted olefins are allylic in nature and may be replaced sequentially with or without palladium catalysts. Thus, their use in natural product synthesis seems to be of great value, as has been shown with pheromones [48] or tropane alkaloids [49]. The following (recently described) synthesis of the antibiotic anticapsin **105** is also based on the 1,4-chloro-acetoxylation of 1,3-cyclohexadiene **93** [50]:

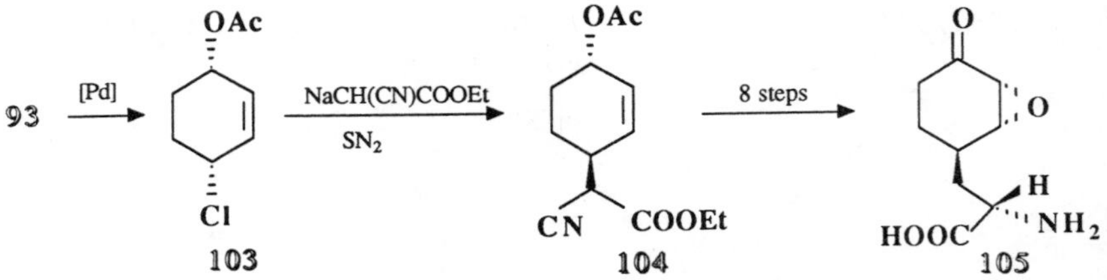

4. Cyclization of non-conjugated dienes

Most stable, and thus easily accessible palladium(II) complexes are formed from non-conjugated dienes and palladium salts such as $PdCl_2(PhCN)_2$ or Na_2PdCl_4 [51]). The double bonds in these complexes become *activated* and, as a consequence of coordination and a suitable size of the carbon chain, different kinds of cyclization reactions [52] become possible. This field has found an explosive expansion during the last years and we will brievely discuss the following reaction types: i. organopalladation of dienes; ii. ene reaction of dienes and olefin coupling.

4.1. ORGANOPALLADATION OFF NON-CONJUGATED DIENES

The most important feature of palladium (II) compounds, the coordination of olefins and the subsequent addition of nucleophiles [51] can easily be used for cyclization reactions provided a second double bond is present in the molecule. Nearly 25 years ago, an I.C.I. Research Group described an «*unusual cyclisation reaction*», the first example of the formation of five membered carbocycles from non conjugated dienes [53]:

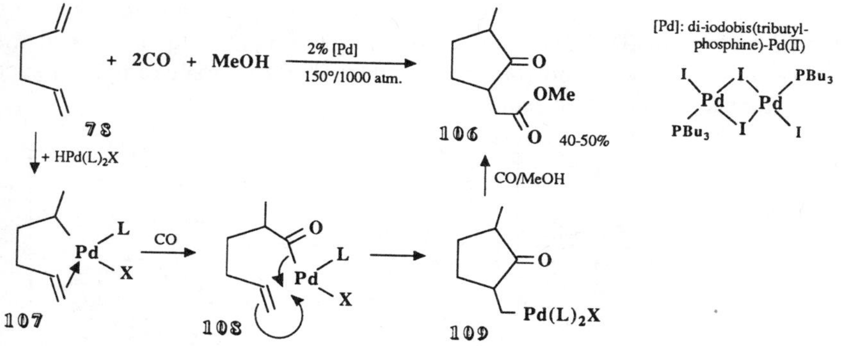

Despite the rough reaction conditions (150°/1000atm.), this constitutes a remarkable efficient and selective reaction: formation of 3 new carbon-carbon bonds and 2 different chemical functionalities during catalytic and (most probably) stepwise double carbonylation, cyclization and C–Pd cleavage. Another example is the transformation of 1,5-cyclooctadiene **110** to bicyclo[3.3.1]non-2-en-9-one **114** according to the following reaction sequence [54]:

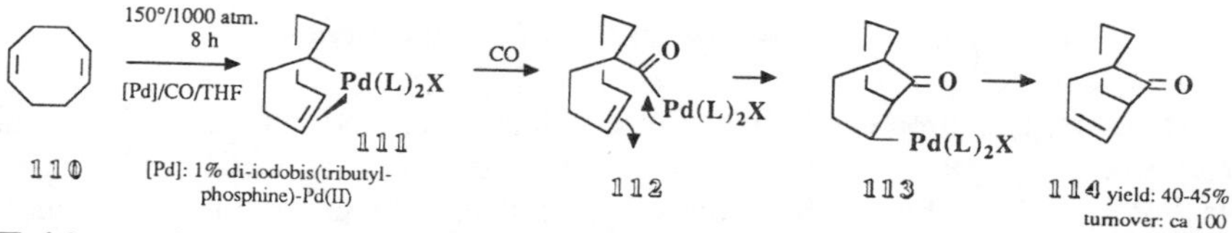

Evidence for this kind of mechanism has been obtained since stable palladium complexes **115** and **116a** are easily formed with 1,5-cyclooctadiene **110** [55]:

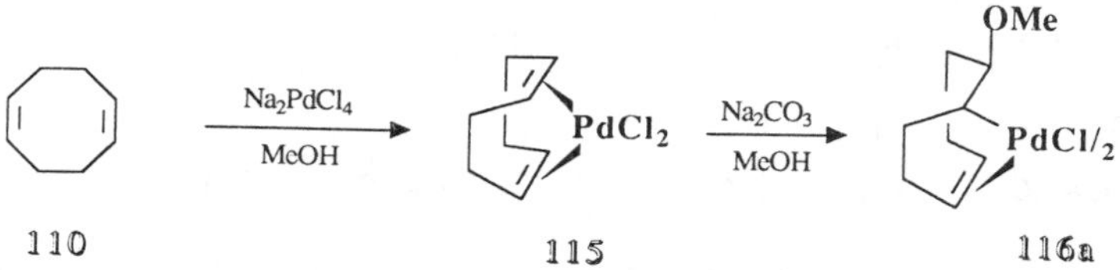

Complex **116a**, the product of nucleophilic addition to coordinated 1,5-cyclooctadiene **110** [51] contains a Pd carbon-σ-bond together with a Pd π-bond (σ-π-complex). Hence, another basic organometallic reaction – the insertion of an unsaturated «ligand» – becomes possible. The newly formed carbon-carbon bond is part of the carbocyclic system, and the overall process is a cyclization reaction. It should be emphasized that the insertion step usually requires a second oxidant, which on the other side recreates the original palladium(II), and allows reactions with catalytic quantities of the catalyst. P.M. Henry and coworkers have described the first example of such a catalytic carbocyclization sequence [56]:

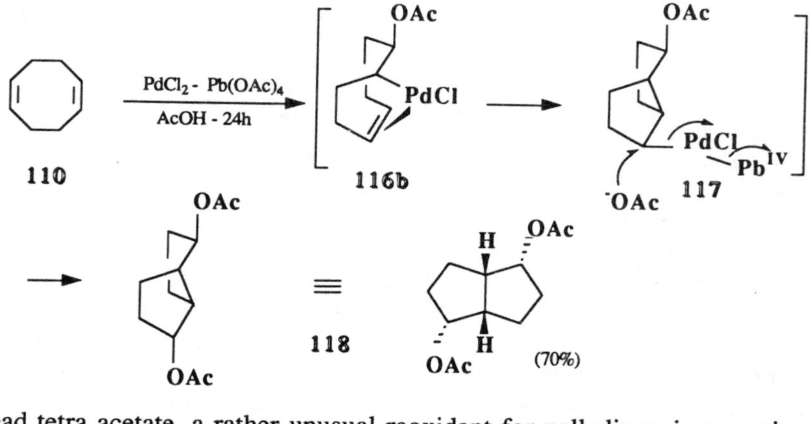

Lead tetra acetate, a rather unusual reoxidant for palladium, is operative in this reaction. We have already mentioned $PdCl_2$–$CuCl_2$-catalyzed cyclization reactions where the intermediary formed cyclic (or bicyclic) palladium σ-complex gave rise to a skeletal transposition, starting from dienes **47** [18] and **57** [21]. When other 1,5-dienic olefins are the substrates, $CuCl_2$ (the classical reagent in the Wacker Reaction) [32] is the efficient cooxidant as illustrated by the 1-step formation of the brendane derivative **120** [57]. In this reaction, molecular oxygen can be the ultimate oxidant [58].

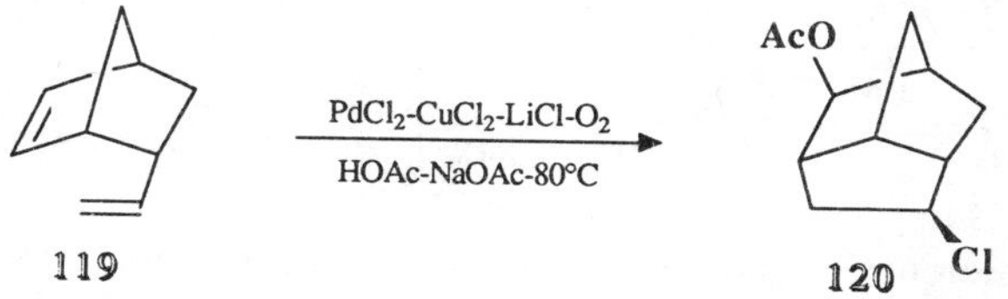

$$\xrightarrow[\text{HOAc-NaOAc-80°C}]{PdCl_2\text{-}CuCl_2\text{-}LiCl\text{-}O_2}$$

119

120 Cl (yield >75% selectivity >90%)

Although quite selective, these reactions are not very general and we could show that the palladium reoxidant combination is very important in the oxidative cylization of non conjugated dienes. With the three component system «palladium(II)/benzoquinone/MnO₂» it is possible to render the cyclization general *and* selective, at least for aliphatic divinyl starting olefins [37]:

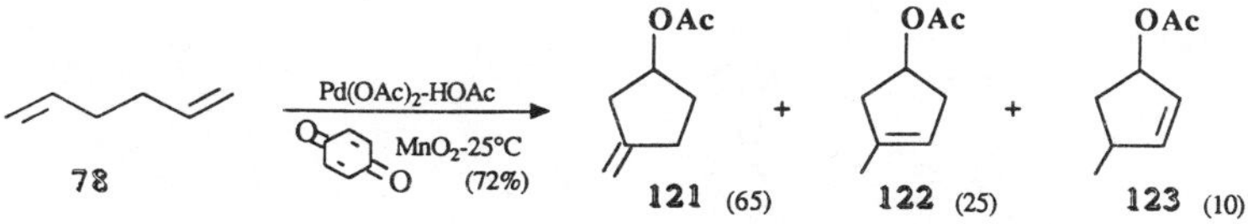

$$\xrightarrow[\text{MnO}_2\text{-25°C}]{Pd(OAc)_2\text{-HOAc}}$$

78   (72%)

121 (65) + **122** (25) + **123** (10)

This catalyst combination is superior to the stoichiometric reaction [59]. Depending on substitution pattern of the diene, high regio- and stereoselectivities are observed [37, 60].

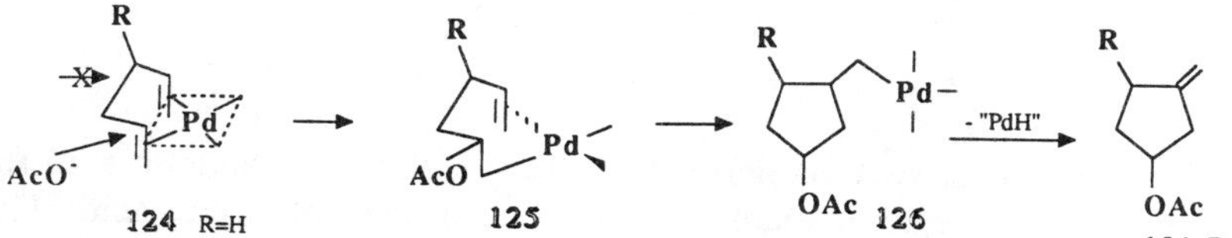

124 R=H
R=Ph
R=C₅H₈OAc

125

126 OAc

- "PdH"

121-R OAc

In monosubstituted dienes, only the regio-control is high. However in the disubstituted diene, *cis*-1,5-divinylcyclohexane **128**, the chemoselectivity ($PdCl_2$–$CuCl_2$ vs. $Pd(OAc)_2$-benzoquinone-MnO_2) as well the regio- and stereo control during the oxidative cyclization, is very high [37, 60]:

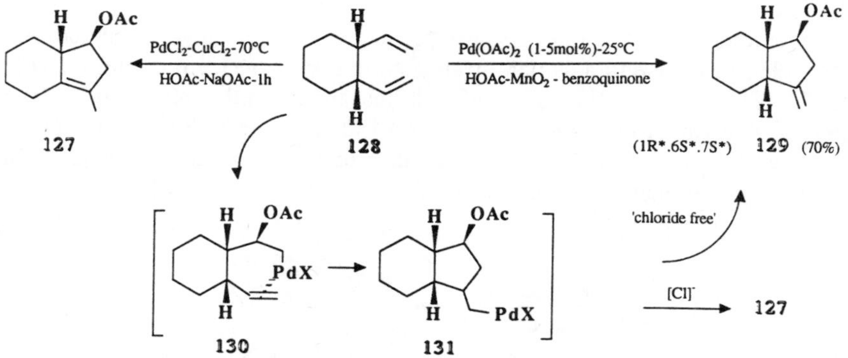

127   128   (1R*.6S*.7S*) 129 (70%)

130   131   'chloride free'   [Cl]⁻ → 127

The complete diastereoselectivity encountered in this system makes this reaction a powerful model for stereochemical studies in catalytic oxidation reactions.

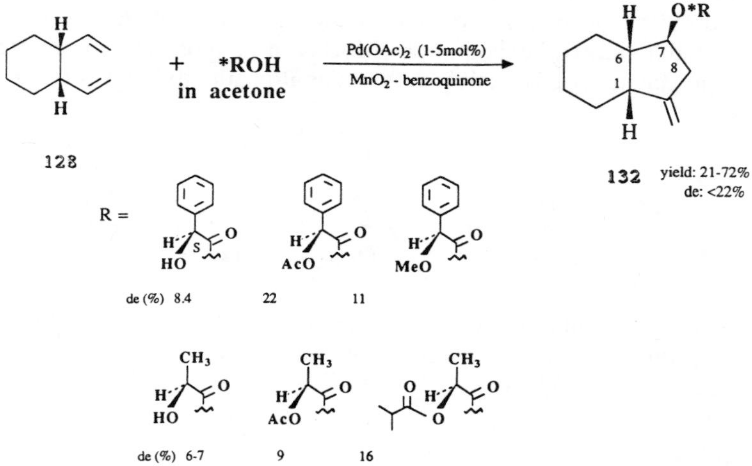

128   + *ROH in acetone $\xrightarrow[\text{MnO}_2 \text{ - benzoquinone}]{\text{Pd(OAc)}_2 \text{ (1-5mol\%)}}$ 132 yield: 21-72% de: <22%

R =

de (%) 8.4   22   11

de (%) 6-7   9   16

Chiral nucleophiles, such as derivatives from lactic- or mandelic acid may easily be used in organopalladation reactions instead of acetic acid. Their size and steric arrangement influences the addition of the nucleophile to the coordinated olefin in a way that gives rise to asymmetric discrimination with variable but significant diastereomeric excess.(de : < 22%) [61].

4.2. ENE REACTION OF DIENES AND OLEFIN COUPLING REACTIONS

It has been shown in numerous examples that diolefines, other than 1,5-

hexadienes, are good candidates for cyclization. Grigg and his coworkers [62–65] studied the cylization of a wide range of substituted olefines. Depending on the substituents of the double bonds different reaction types may be observed. The reaction of diene **134** [62] corresponds to a metal catalysed ene-reaction. Two different hydride shifts are operative and the reaction requires some specific reaction conditions (traces of HCl, ethanol free chloroform).

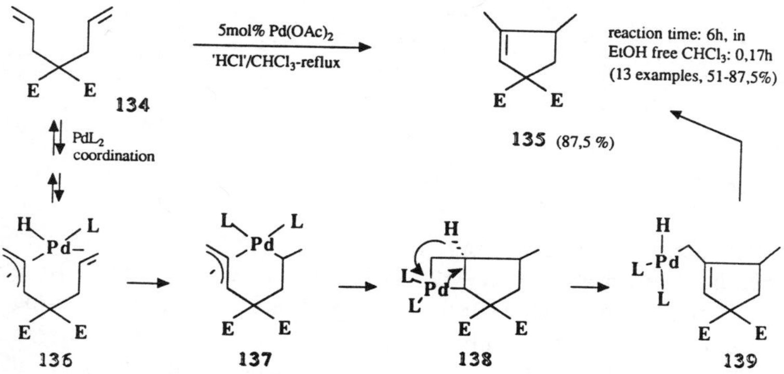

E = "CO" containing groups,e.g. COOMe

The mono- or di substitution of the double bonds permits to observe an intramolecular Heck type reaction [63, 66]

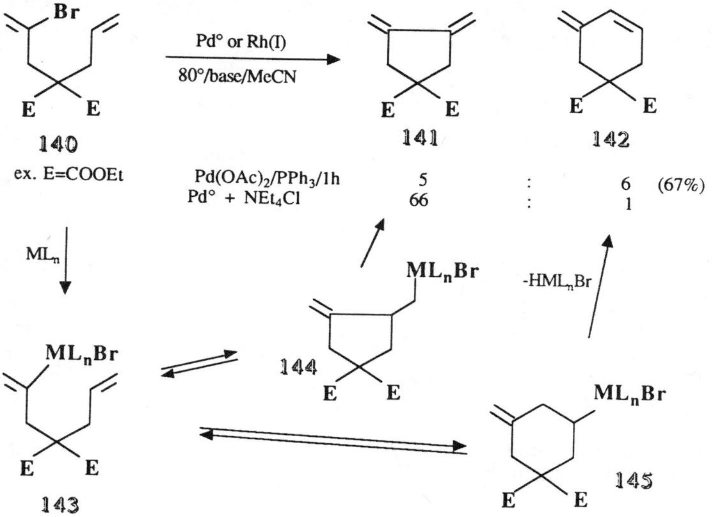

or a divinylbromide coupling [64].

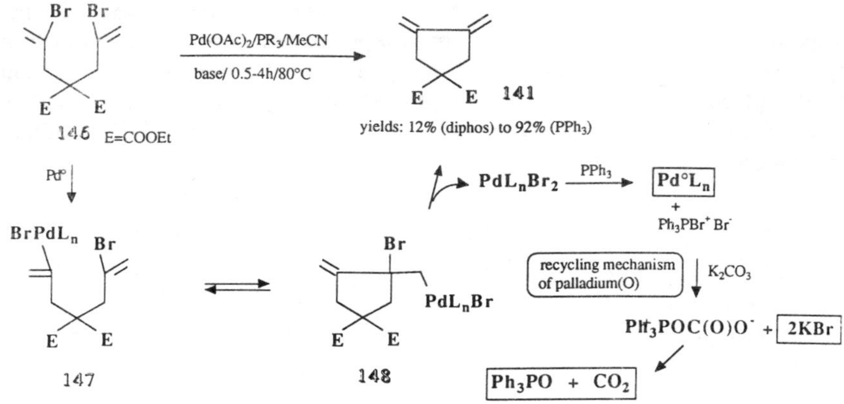

yields: 12% (diphos) to 92% (PPh₃)

In both reactions, additives are necessary in order to ensure higher reactivity and elevated selectivity. In the case of bromo diene **140**, the role of tetraethyl ammonium chloride is supposed to convert the intermediate vinyl palladium bromide into the corresponding chloride which is more effective in the cyclisation and elimination steps. Vinylbromide **146** leads to identical products, however the catalyst palladium(0) is oxidized to $PdBr_2$. An interesting recycling procedure is added, in order to maintain the catalytic cycle.

Another kind of reaction control can be operative when hydride sources, in form of formic acid derivatives are added to the system. An example of this «tandem cyclisation-anion capture process» [65] is the following reaction:

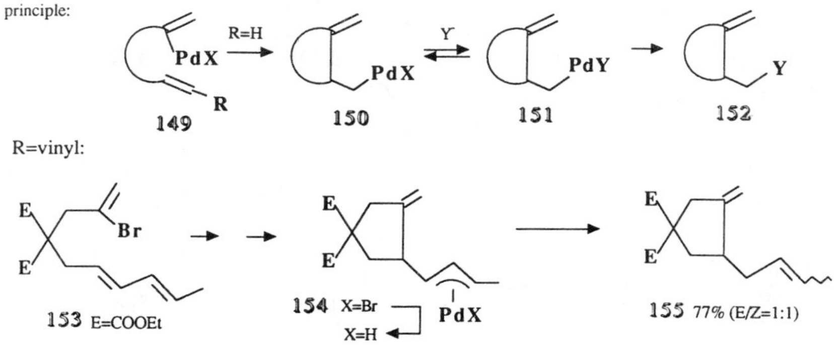

principle:

R=vinyl:

cond.: Pd(OAc)₂/PPh₃/MeCN
hydride source: pyrrolidine/HCOOH

The intermediate (cyclised) palladium-σ complex **150**, which «normally» eliminates PdHX to form a diene, is trapped by the hydride and leads via **151** (Y=H) to the monoolefin **152**. This capture process is also possible with π-allyl intermediates (**153→154→155**).

J.K. Stille and his group [67] have developed crosscoupling reactions with vinyl triflates as electrophiles. An intramolecular version of this reaction is a remarkably powerful method to create large ring lactones **157** under high dilution conditions (56–57% isolated yield of 12–15 membered rings).

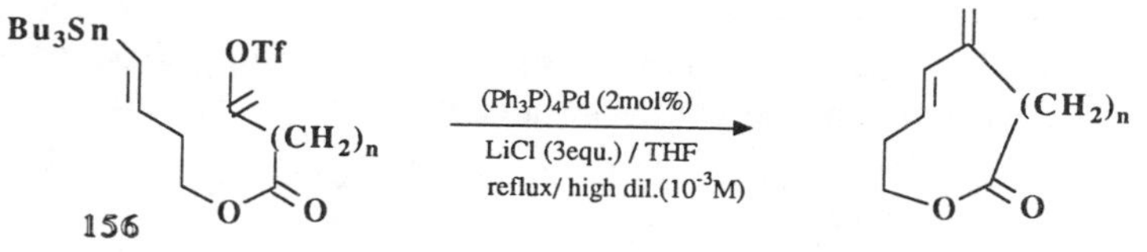

157 n=5-8; y: 56-57%

5. Asymmetric alkylation via π-allyl Pd complexes

Allylic alkylation via palladium-π-allyl complexes is probably one of the most useful and best known reaction in organic palladium chemistry [68]. It is generally accepted now that, in the presence of palladium-phosphine complexes, the following catalytic cycle is operative:

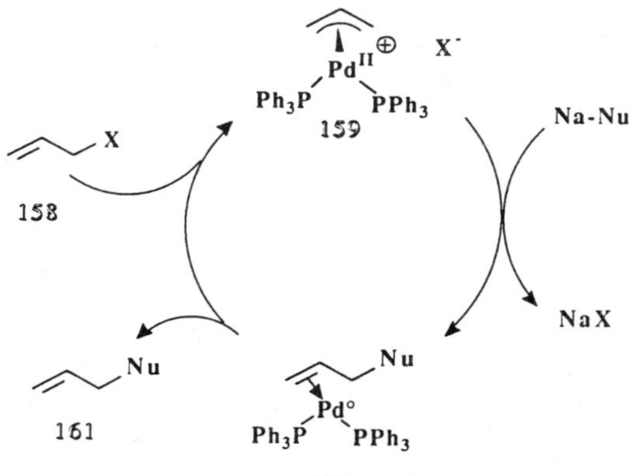

The replacement of PPh$_3$ by chiral phosphines introduces chirality into the palladium complexes. Consequently the transition states become diastereomeric and the formation of the organic (allylic) products may proceed with asymmetric induction [69].

The first examples of chiral allylic alkylation, according to this principle, have been reported by Trost and Strege [70] with phosphines such as (+)DIOP, (+)CAMPHOS and (-)DIPAMP. The optical yields (16–46%) varied considerably depending on the substrate, the chiral ligands, the nucleophile and the reaction conditions. New chiral phosphines, containing a «chiral functional group remote from the phosphine groups» [71] could slightly raise the enantiomeric excess to 52%.

Some years later, a careful mechanistic study, the examination of diastereomeric equilibria of a series of [palladium-(chiral diphosphine)(chiral π-allyl)]$^+$ complexes permitted, to B. Bosnich [72] to establish some major features for the asymmetric allylation of substituted allylic acetates:

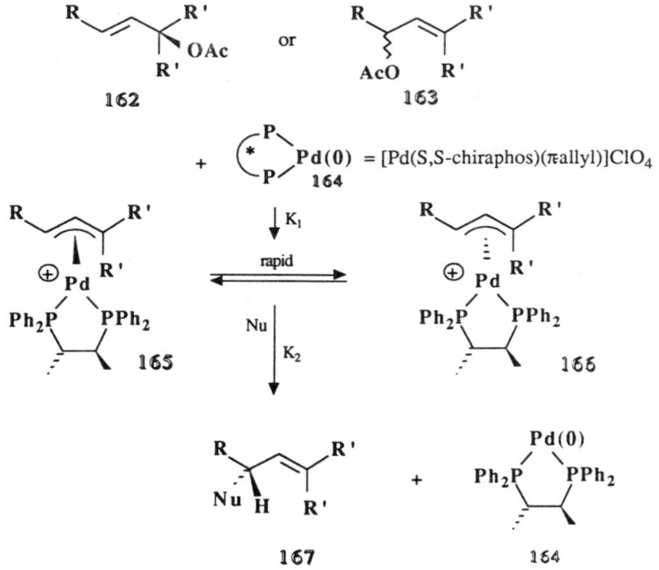

With the anti displaced π-allyl substituents as the major source of discrimination, optical yields up to 86% are attained in these reactions – sensitive to the chiral phosphine (CHIRAPHOS) and insensitive to the nature of the newly introduced nucleophile. Kinetically, two primary steps are distinguished, both occurring with inversion of configuration: the oxidative addition of Pd(0) to the allylic acetate (K_1) and the nucleophilic attack (stabilized carbon nucleophile) to the rapidly epimerizing, chiral π-allyl intermediate (K_2). This last and slow step is the turnover limiting as well as the enantioselective step.

The products of this most efficient catalytic asymmetric induction are of practical value since they are readily transformed into «useful» material (ex.: phenylsuccinic acid **172**).

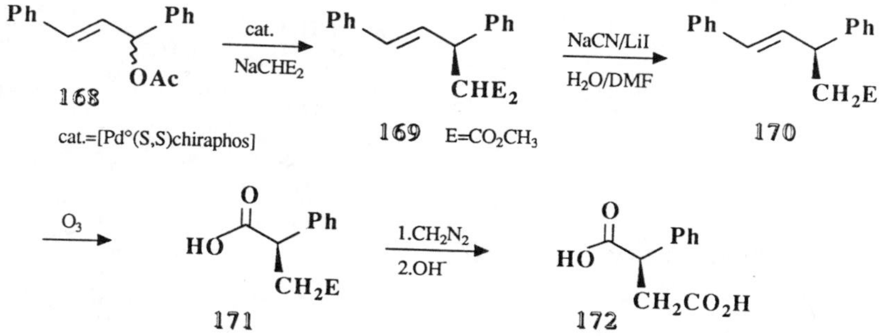

Chiral platinum complexes may also be used for chiral allylation reactions, however the optical inductions did not exceed 23% [73]. Apparently, palladium seems to be the *ideal* metal for these reactions, as has been demonstrated before by a japanese group who developed real «tailor made *ligands»: chiral ferrocenyl phosphines containing aminoalcohol side chain functional groups [74, 75]:

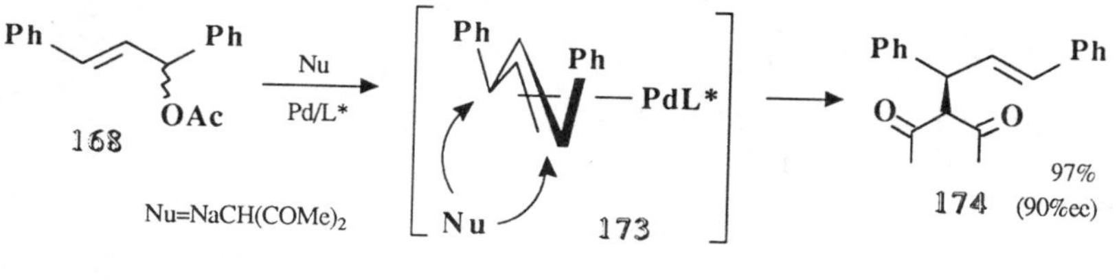

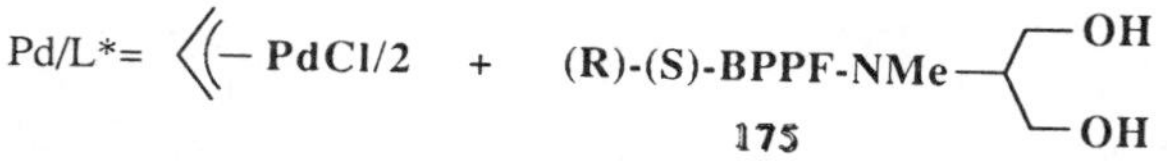

Finally, the best ee value (95%) is reported on 2-propenyl acetates, substituted with two different aryl groups at 1 and 3 position [75a].

6. Asymmetric hydroformylation

Let us conclude this promising, and rapidly developing field of metal catalyzed chiral induction reactions with a reaction, catalyzed by platinum, the metal which in 1828 opened the era of organometallic chemistry [76].

Often much less efficient in organic reactions than palladium, platinum is one of the best catalysts in hydroformylation reactions, and the asymmetric variation of this important «CO/H$_2$ activation» is of great potential for the synthesis of a variety of chiral products.

176 + H$_2$ + CO $\xrightarrow{179}$ 177 + 178

catalyst: [(-)BPPM]PtCl$_2$-SnCl$_2$
2700psi, 60°C, 40h

70% conversion
branched:normal = 0.5

- AcOH

179

Parinello and Stille [77] have shown that this metal, in combination with SnCl$_2$ and chiral amino-phosphine ligands is able to catalyse the transformation of a large variety of prochiral olefins into asymmetric aldehydes with high optical inductions (60–85%), which, in the presence of orthoformiates, may reach 98% (ee).

Université d'Aix-Marseille, URA 1410,
ENSSPICAM, Faculté de St.-Jérôme,
F 13013 Marseille (France)

References

1. *Modern Synthetic Methods 1983 Transition Metals in Organic Synthesis*; Scheffold, R., Salle and Sauerländer, Frankfurt am Main, 1983.
2. (a) Segnitz, A. in Houben-Weyl *Methoden der Organischen Chemie*; Bd. 13, Teil 9b, Thieme, 1983; (b) Heck, R.F. *Palladium Reagents in Organic Synthesis*; Academic Press, London, 1985.
3. Harrington, P.J.; Hegedus, L.S. and McDaniel, K.F. *J.Am.Chem.Soc.* **1987**, *109*, 4335; Harrington, P.J. and Hegedus, L.S. *J. Org. Chem.* **1984**, *49*, 2657.
4. (a) Heck, R.F. *Org. React. (N.Y.)* **1982**, *27*, 345; (b) Reissig, H.U. *Nachr. Chem. Tech. Lab.* **1986**, *34*, 1066.
5. Colquhoun, H.M. *Chem. Ind. (London)* **1987**, 612.
6. For a discussion of synthetical useful rearrangements: Mundy, B.P. *Concepts of Organic Synthesis*; M. Dekker, New York, 1979; p.119.

7. Baird, W.C. Jr. *J. Org. Chem.* **1966**, *31*, 2411.
8. Hughes, R.P. and Powell, J. *J. Organomet. Chem.* **1973**, *60*, 387.
9. Horino, H.; Arai, M. and Inoe, N. *Tetrahedron Lett.* **1974**, 647; Catellani, M.; Chiusoli, G.P. and Peloso, C. *Tetrahedron Lett.* **1983**, *24*, 813.
10. Andrews, M.A.; Chang, T.C.T.; Cheng, C.W.F., Emge, T.J.; Kelly, K.P. and Koetzle, T.F. *J. Am. Chem. Soc.* **1984**, *106*, 5913; Andrews, M.A.; Chang, T.C.T.; Cheng, C.W.F. and Kelly, K.P. *Organometallics* **1984**, *3*, 1777.
11. Chauvet, F.; Heumann, A. and Waegell, B. *J. Org. Chem.* **1987**, *52*, 1916.
12. Boontanonda, P. and Grigg, R. *Chem. Commun.* **1977**, 583
13. Sen, A. and Lai, T.W. *J. Am. Chem. Soc.* **1981**, *103*, 4627.
14. Bäckvall, J.E. and Nordberg, R. *J. Am. Chem. Soc.* **1980**, *102*, 393.
15. Heumann, A. and Bäckvall, J.E. *Angew. Chem. Int. Ed. Engl.* **1985**, *24*, 207.
16. Migita, T.; Saitoh, N.; Iizuka, H.; Ogyu, C.; Kosugi, M. and Nakaido, S. *Chem. Lett.* **1982**, 1015; Migita, T.; Hongoh, K.; Naka, H.; Nakaido, S. and Kosugi, M. *Bull. Chem. Soc. Japan* **1988**, *61*, 931.
17. Stangl, H. and Jira, R. *Tetrahedron Lett.* **1970**, 3589.
18. Heumann, A.; Reglier, M. and Waegell, B. *Tetrahedron Lett.* **1983**, *24*, 1971.
19. Clark, G.R. and Thiensathit *Tetrahedron Lett.* **1985**, *26*, 2503.
20. Liebeskind, L.S.; Mitchell, D. and Foster, B.S. *J. Am. Chem. Soc.* **1987**, *109*, 7908.
21. Ito, Y.; Aoyama, H. and Saegusa, T. *J. Am. Chem. Soc.* **1980**, *102*, 4519.
22. Tsuji, J. *Organic Synthesis with Palladium Compounds*; Springer Verlag Berlin, 1980; pp. 173.
23. Ouelette, R.J. and Levin, C. *J. Am. Chem. Soc.* **1971**, *93*, 471; Albelo, G. and Rettig, M.F. *J. Organomet. Chem.* **1972**, *42*, 183; Balme, G.; Fournet, G. and Goré, J. *Tetrahedron Lett.* **1986**, *27*, 3855; Rautenstrauch, V.; Burger, U. and Wirthner, P. *Chimia* **1985**, *39*, 225; Rautenstrauch, V. *Chimia* **1985**, *39*, 227.
24. Isomura, K.; Uto, K. and Taniguchi, J.C.S. *Chem. Comm.* **1977**, 664; Wiger, G. and Rettig, M.F. *J. Am. Chem. Soc.* **1976**, *98*, 4168.
25. Alper, H. and Perera, C.P. *J. Am. Chem. Soc.* **1981**, *103*, 1289.
26. Alper, H.; Urso, F. and Smith, D.J.H. *J. Am. Chem. Soc.* **1983**, *105*, 6737.
27. Alper, H. and Hamel, N. *Tetrahedron Lett.* **1987**, *28*, 3237.
28. Sasaki, T.; Kanematsu, K. and Kondo, A. *J. Chem. Soc. Perkin Trans. 1* **1976**, 2516.
29. Byrd, J.E.; Cassar, L.; Eaton, P.E. and Halpern, J. *Chem. Commun.* **1971**, 40.
30. Giddings, R.M. and Whittaker, D. *Tetrahedron Lett.* **1978**, 4077.
31. Loubinoux, B. and Caubère, P. *J. Organomet. Chem.* **1974**, *67*, C48.
32. Smidt, J.; Hafner, W.; Jira, R.; Sedlmeier, J.; Sieber, R.; Rüttinger, R. and Kojer, H. *Angew. Chem.* **1961**, *71*, 176.
33. March, J. *Advanced Organic Chemistry*, Third Edition; Wiley, 1985; pp. 1084; Hegedus, L.S. *Tetrahedron* **1984**, *40*, 2415 (Tetrahedron Report n°166).
34. Henry, P.M. *Palladium Catalyzed Oxidation of Hydrocarbons*; Academic Press, New York, 1980.
35. Gretz, E.; Oliver, T.F. and Sen, A. *J. Am. Chem. Soc.* **1987**, *109*, 8109.
36. Overman, L.E. *Angew. Chem. Int. Ed. Engl.* **1984**, *23*, 579
37. Antonsson, T.; Heumann, A. and Moberg, C. *J. Chem. Soc., Chem. Comm.* **1986**, 518.
38. Trost, B.M. and Runge, T.A. *J. Am. Chem. Soc.* **1981**, *103*, 2485.
39. Mares, F.; Diamond, S.E.; Regina, F.J. and Solar, J. P. *J. Am. Chem. Soc.* **1985**, *107*, 3545.
40. Bäckvall, J.E. and Heumann, A. *J. Am. Chem. Soc.* **1987**, *108*, 7107

41. Brown, R.G. and Davidson, J.M. *J. Chem. Soc.(A)* **1971**, 1321.
42. Numerous patents, c.f. ref. 34, p. 243.
43. Robinson, S.D. and Shaw, B.L. *J. Chem. Soc.* **1963**, 4806 & **1964**, 5002.
44. Bäckvall, J.E. *Acc. Chem. Res.* **1983**, *16*, 335; *Pure Appl. Chem.* **1983**, *55*, 1669; *Advances in Metal-Organic Chemistry*, Vol.1; JAI Press Inc., 1989; p. 135.
45. Bäckvall, J.E.; Nordberg, R.E. and Wilhelm, D. *J. Am. Chem. Soc.* **1985**, *107*, 6892.
46. Bäckvall, J.E.; Byström, S.E. and Nordberg, R.E. *J. Org. Chem.* **1984**, *49*, 4619.
47. Bäckvall, J.E.; Nyström, J.E. and Nordberg, R.E. *J. Am. Chem. Soc.* **1985**, *107*, 3676.
48. Bäckvall, J.E.; Byström, S.E. and Nyström, J.E. *Tetrahedron* **1985**, *42*, 5761.
49. Bäckvall, J.E.; Renko, Z.D. and Byström, S.E. *Tetrahedron Lett.* **1987**, *28*, 4199
50. Souchet, M.; Baillargé, M. and Le Goffic, F. *Tetrahedron Lett.* **1988**, *29*, 191.
51. Omae, I. *Organometallic Intramolecular-coordination Compounds*; Elsevier, Amsterdam & New York, 1986.
52. Hegedus, L.S. *Tetrahedron* **1984**, *40*, 2415 (Tetrahedron Report n°166); *Angew. Chem. Int. Ed. Engl.* **1988**, *27*, 1113; Oppolzer, W. *Angew. Chem. Int. Ed. Engl.* **1989**, *28*, 38; Trost, B. M. *Angew. Chem. Int. Ed. Engl.* **1989**, *28*, 1173.
53. Brewis, S. and Hughes, R.P. *J. Chem. Soc. Chem. Comm.* **1965**, 489.
54. Brewis, S. and Hughes R.P. *J. Chem. Soc., Chem. Comm.* **1966**, 6 and **1967**, 71; Brewis, S. Brit. 1, 144, 166, 1969, CA 75: 11804r, 1971.
55. Chatt, J.; Vallarino, L.M. and Venanzi, L.M. *J. Chem. Soc.* **1957**, 3413; Schultz, R. G. *J. Organomet. Chem.* **1966**, *6*, 435.
56. Henry, P.M.; Davies, M.; Ferguson, G.; Phillips, S. and Restivo, R. *J. Chem. Soc., Chem. Comm.* **1974**, 112.
57. Heumann, A; Reglier, M. and Waegell *Angew. Chem. Int. Ed. Engl.* **1979**, *18*, 866 and 867.
58. Heumann, A., unpublished results.
59. Adachi, N.; Kikukawa, K.; Takagi, M. and Matsuda, T. *Bull. Chem. Soc. Japan* **1975**, *48*, 521.
60. Antonsson, T.; Moberg, C.; Tottie, L. and Heumann, A. *J. Org. Chem.* **1989**, *54*, 4914.
61. Heumann, A. and Moberg, C. *J. Chem. Soc., Chem. Comm.* **1988**, 1516.
62. Grigg, R.; Malone, J.F.; Mitchell, T.R.B.; Ramasubbu, A. and Scott, R.M. *J. Chem. Soc. Perkin Trans. 1* **1984**, 1745.
63. Grigg, R.; Stevenson, P. and Worakun, T. *Tetrahedron* **1988**, *44*, 2033.
64. Grigg, R.; Stevenson, P. and Worakun, T. *Tetrahedron* **1988**, *44*, 2049.
65. Burns, B.; Grigg, R.; Sridharan, V. and Worakun, T. *Tetrahedron Lett.* **1988**, *29*, 4325; Burns, B.; Grigg, R.; Ratananukul, P.; Sridharan, V.; Stevenson, P. and Worakun, T. *Tetrahedron Lett.* **1988**, *29*, 4329.
66. Narula, C.K.; Mak, K.T. and Heck, R.F. *J. Org. Chem.* **1983**, *48*, 2792; Shi, L.; Narula, C.K.; Mak, K.T.; Kao, L.; Xu, Y. and Heck, R.F. *J. Org. Chem.* **1983**, *48*, 3894.
67. Stille, J.K. and Tanaka, M. *J. Am. Chem. Soc.* **1987**, *109*, 3785.
68. Trost, B.M. *Acc. Chem. Res.* **1980**, *13*, 385; Tsuji, J. *Tetrahedron* **1986**, *42*, 4361
69. Bosnich, B. *Asymmetric Catalysis*; NATO ASI Series E, Applied Sciences 103, Martinus Nijhoff, Dordrecht, 1986.
70. Trost, B.M. and Strege, P.E. *J. Am. Chem. Soc.* **1977**, *99*, 1649; Trost, B.M. and Murphy, D.J. *Organometallics* **1985**, *4*, 1143.
71. Hayashi, T.; Kanehira, K.; Tsuchiya, H. and Kumada, M. *J. Chem. Soc., Chem.*

Comm. **1982**, 1162.

72. Auburn, P.R.; Mackenzie, P.B. and Bosnich, B. *J. Am. Chem. Soc.* **1985**, *107*, 2033; Mackenzie, P.B.; Whelan, J. and Bosnich, B. *J. Am. Chem. Soc.* **1985**, *107*, 2046.
73. Brown, J.M. and MacIntyre, J.E. *J. Chem. Soc. Perkin Trans. 2* **1985**, 961.
74. Hayashi, T.; Yamamoto, A.; Hagihara, T. and Ito, Y. *Tetrahedron Lett.* **1986**, *27*, 191.
75. (a) Hayashi, T.; Yamamoto, A. and Ito, Y. *Chem. Lett.* **1987**, 177; (b) Hayashi, T.; Kanahira, K.; Hagihara, T. and Kumada, M. *J. Org. Chem.* **1988**, *53*, 113.
76. Zeise, W.C. *Pogg. Ann. Phys.* **1827**, *9*, 632.
77. Parrinello, G. and Stille, J.K. *J. Am. Chem. Soc.* **1987**, *109*, 7122.

DIETRICH ARNTZ AND ADOLF SCHÄFER

ASYMMETRIC HYDROGENATION

1. Introduction

In the last few years the demand for enantiomerically pure compounds has grown rapidly. The main reason for this development is, that a lot of pharmaceutical and chemical compounds have to be sold in an optically active form. In nature often only the one enantiomer of a substance has any biological activity. Sometimes the other one even causes toxic side effects. Figure 1 shows some examples.

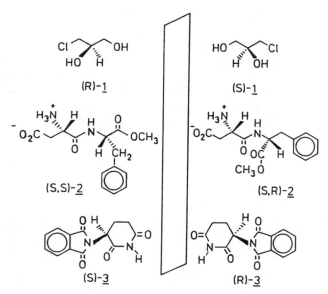

Figure 1. Chiral compounds having different biological effects.

A. F. Noels et al. (Eds.), Metal promoted selectivity in organic synthesis, 161–189.

In the case of chloropropandiole 1 the (S)-enantiomer is used as a rodent chemosterilant, while the (R)-form is toxic. The new sweetener Aspartam («Nutrasweet™») 2 tastes bitter in its (S,R)-form. Another spectacular example is thalidomide 3 («Contergan™»). One enantiomer causes fetal abnormalities, while the other one was used as a sedative and hypnotic. Therefore the racemic form had to be withdrawn from the market [1]. In principle there are several different routes to achieve enantiomerically pure compounds in organic synthesis:

— resolution of racemic mixtures
— enzymatic methods
— stereoselective synthesis
— asymmetric catalysis
— chiral pool

Asymmetric hydrogenation is one of the most powerfull tools for the synthesis of chiral compounds [2]. Chiral hydrogenation catalysts are able to distinguish enantiotopic faces of a molecule via diastereomeric transition states. These catalysts are able to multiply optical information during the course of the catalytic reaction. Therefore they provide an interesting completion of biological transformations, classical chemical resolution and preparative methods.

The development of this methods started in the late 60's with a chiral modified Wilkinson complex, a catalyst which is well known from homogenous hydrogenation of olefins [3]. Today chiral catalysts are used in commercial processes for asymmetric hydrogenation, for example to produce (S)-Dopa by Monsanto Co./USA and VEB Isis-Chemie/GDR [4a–c] and for (S)-phenylalanine [4d] by ANIC S.P.A. in Italy.

2. General principles

2.1. CHIRAL PHOSPHINE LIGANDS

In 1968 Horner and Knowles showed that asymmetric hydrogenation is possible with Wilkinsons complex $RhCl(P(C_6H_5)_3$ modified with chiral ligands [5].

They tried to hydrogenate simple prochiral olefins but the optical yields were low. Later acetamido acrylic acids were tested as substrates and the results improved. But still the optical yields remained insufficient. Figure 2 demonstrates this development for the synthesis of (S)-Dopa [6].

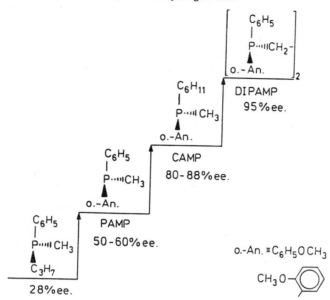

Figure 2. Increase of optical yields in synthesis of (S)-Dopa.

Introducing bulky substituents into the phosphine ligands increased the ee.'s, but still the values did not exceed 80% ee. Important progress was made through the investigations of Kagan in France and Knowles, Sabacky and coworkers at Monsanto/USA. They prepared chiral bidentate phosphines and introduced them into the synthesis of Schrock/Osborn type, square planar 16 electron rhodium complexes [7].

Problems arise in synthesis of suitable ruthenium compounds with chiral bidentate phosphine ligands, because Ru(II) prefers an octahedral coordination sphere and therefore often complex mixtures are found in the synthesis of the catalyst precursors. Literature data indicate that complexes like HRuCl(P–P)$_2$ and [RuCl$_2$(P–P)]$_2$ (P–P = CHIRAPHOS, DIOP) exhibit sufficient activity [8]. In the case of BINAP [Ru$_2$Cl$_4$(P–P)$_2$]NEt$_3$, Ru(P–P)(O$_2$CR)$_2$, HRuCl(P–P)$_2$ and [RuH(P–P)$_2$]BF$_4$ have been used successfully [9, 10].

The synthesis of chiral bidentate phosphines mostly starts with cheap materials from the «chiral pool». Good examples for that route are DIOP, PYRPHOS, BPPM and NMDPP. Details for their preparation are given in literature [11]. Figure 3 shows examples of commonly used chiral bidentate phosphines.

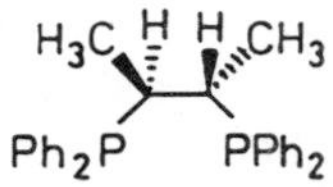

CHIRAPHOS
5

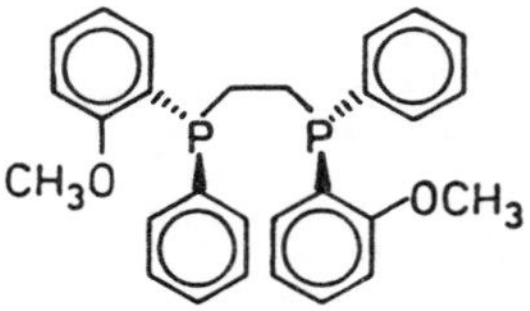

DIPAMP
6

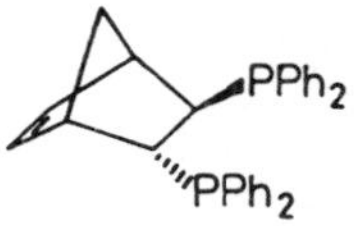

NORPHOS
7

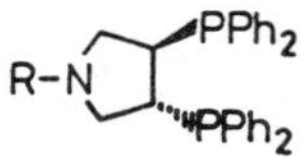

PYRPHOS
8

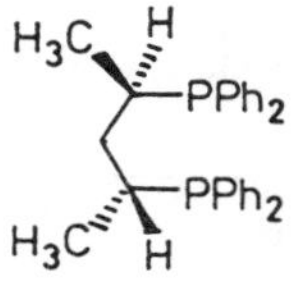

SKEWPHOS
9

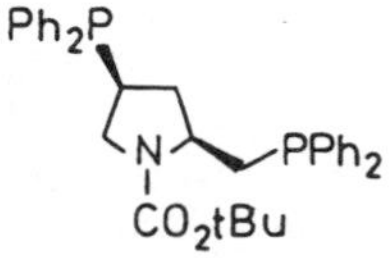

BPPM
10

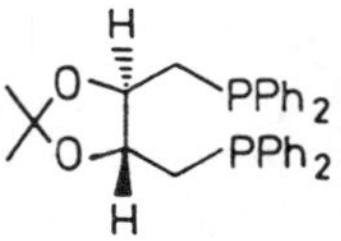

DIOP
11

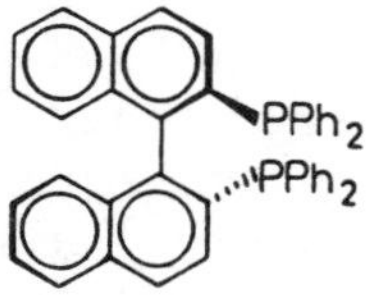

BINAP
12

Figure 3. Chiral bidentate phosphines.

They differ in size and positions of the chiral information. The chirality can either be located on the phosphorus atoms or in the carbon skeleton of the phosphine. Or they are atropisomers like the BINAP ligands **12**.

On coordination to the metal (Ru, Rh, Ir) the chiral phosphine ligands form puckered chelate ring systems of different sizes. For example 1,2-

diphosphines like CHIRAPHOS 5 and DIPAMP 6 lead to fivemembered rings, 1,3-diphosphines like SKEWPHOS 8 form sixmembered chelates and sevenmembered systems will be found with 1,4-diphosphines DIOP 11 [12].

The chelate rings of 1,2-diphosphines have two enantiomeric conformations λ and δ with C_2-symmetry. Substituents at the carbon atoms C(1) and/or C(2) arrange preferably in equatorial positions. Thus the chiral carbon centers induce the chelate ring to adopt one prefered chiral conformation. The phenyl groups at the phosphorus atoms will then be found in axial and equatorial positions of the chelate ring. Figure 4 shows the Rh-(S,S)-CHIRAPHOS complex as an example.

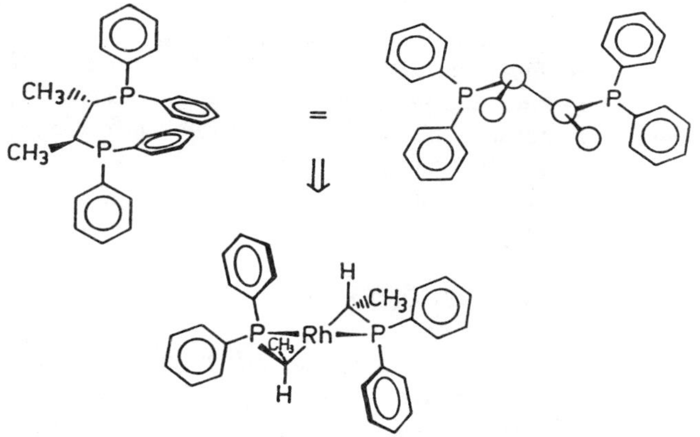

(S,S)−CHIRAPHOS / Rh (I)

Figure 4. Structure of the Rh-(S,S)-CHIRAPHOS complex.

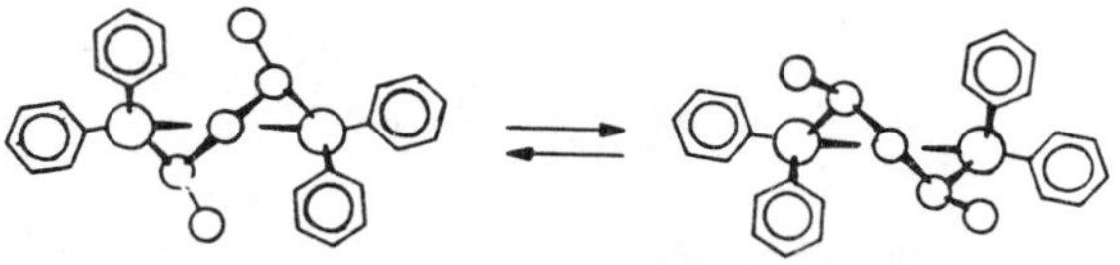

Figure 5. Chiral skew conformations of (S,S)-BDPP 9.

All chiral 1,2-diphosphines lead two enantiomeric complexes with the same type of chiral array of phenyl groups.

Figure 5 shows some conformations of (S,S)-2,4-bis(diphenyl-phosphino)pentane-rhodium ((S,S)-SKEWPHOS–Rh or (S,S)–BDPP–Rh). The chelate rings are in the skew conformation. In this conformation the methyl substituents are in the axial or in the equatorial positions. Conformational studies of the corresponding diamine complexes suggest that the most stable conformation is skew with two methyl groups in the equatorial position [13, 14]. Both skew conformations provide a chiral array of phenyl groups with C_2-symmetry around the rhodium atom.

In the case of (3S)-1,3-Bis(diphenylphosphino)butane ((S)-CHAIRPHOS) a chair and a skew conformation exist both having an equatorial exposed methyl group. The chair conformation which is intrinsically more stable should be preferred [15].

DIOP 11, which forms heptagonal chelate rings, exists in various conformations in solution. Spectroscopic data suggest that two are related to the chair and the twist boat conformations of cycloheptane. They both have C_2-symmetry with a helical arrangement of the phenyl rings. Structural data of a Pt(DIOP)$_2$ complex confirm these considerations [16].

^{31}P-NMR studies imply that there may be an equilibrium between both stereoisomers. Therefore reinforcement of the carbon skeleton is necessary for 1,4-diphosphines because the number of possible conformations will then be restricted. Ligands without any reinforcement posses a high flexibility and lead to less effective catalysts.

2.2. STRUCTURE OF THE SUBSTRATES

In synthesis of amino acids it was found, that high enantiomeric excesses were only achieved only with the (Z)-isomers of the amino acid precursors. Structural data indicate that the aryl oxygen atom of the acetamido acrylic acid coordinates to the rhodium thus forming a chelate ring.

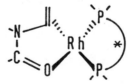

Figure 6. Chelation of Rh-catalysts by acetamido acrylic acid (detail).

This secondary coordination provides a rigid system with a special phosphine/substrate arrangement at the rhodium atom which leads to high optical yields [17]. In the case of ethyl-(Z)-α-acetamidocinnamate Halpern and coworkers were able to isolate such a complex and characterized it by x-ray cristallography [18].

The electronic properties of the substituents have hardly any influence on the optical yields of the reactions. It is more important that the substrate has a functional group next to the olefinic or carbonyl double bond, because hydroxy, carbonyl, amino groups etc. in α-position to the double bond are able to chelate providing such rigid catalyst-substrate arrangements [19].

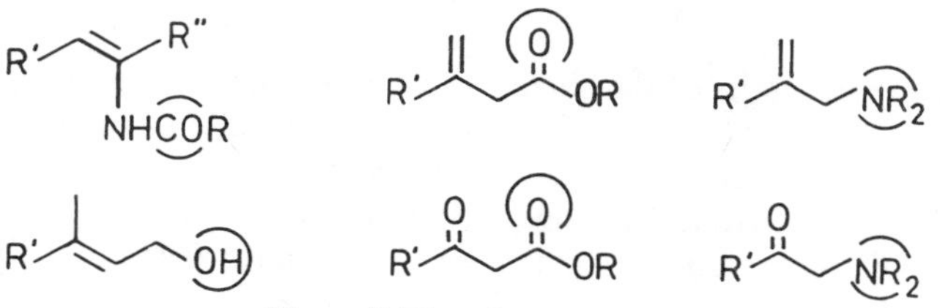

Figure 7. Directing groups.

Simple ketones like acetophenone or olefins like α-ethylstyrene which do not provide such a directing group lead to lower optical yields [20].

2.3. STEREOCHEMISTRY OF THE HYDROGENATIONS

When a prochiral substrate, for example a (Z)-enamide, is coordinated to the catalyst two diastereomeric complexes are formed. According to steric interaction between the substituents at the double bond and the phenyl rings of the ligands these two diastereomers have different stabilities [18, 21], therefore one diastereomer (major diastereomer) is found in higher concentrations than the less stable catalyst-substrate complex (minor diastereomer) in solution. These diastereomeric equilibria have been examined by NMR-spectroscopy. The ratios between major and minor diastereoisomer are sometimes greater than 20 to 1. In the case of enamides interconversion between these two species occurs by dissociation of the substrate followed by recombination. And a second intramolecular process was observed where the carbon-carbon double bond dissociates and the amide oxygen remains coordinated to the metal [22]. The direction of the asymmetric induction is determined by the conformation of the metal chelates. In the case of (Z)-enamides the λ-conformation leads to the (S)-enantiomer, the δ-form generates the (R)-enantiomer of the product.

Other substrates which also provide the possibility for chelation, should follow the similar stereochemical course.

For substrates like simple olefins, ketones or imines without any directing groups only sterical interaction between the catalyst and the substrate determines the stereochemical course of the reaction.

2.4. MECHANISM OF HYDROGENATIONS WITH RHODIUM(I)- AND IRIDIUM(I)-CATALYSTS

2.4.1. Hydrogenation of substrates with directing groups

The mechanistic details of the rhodium catalysed reaction were examined by Halpern and Brown intensively [18, 21–24]. In the case of the hydrogenation of (Z)-enamides the minor diastereomer controls the stereochemistry of the product. Figure 8 illustrates the mechanism of the asymmetric hydrogenation of methyl-(Z)-α-acetamido cinnamate by Rh(1,5-Cyclooctadien)((R,R)-DIPAMP)]BF$_4$.

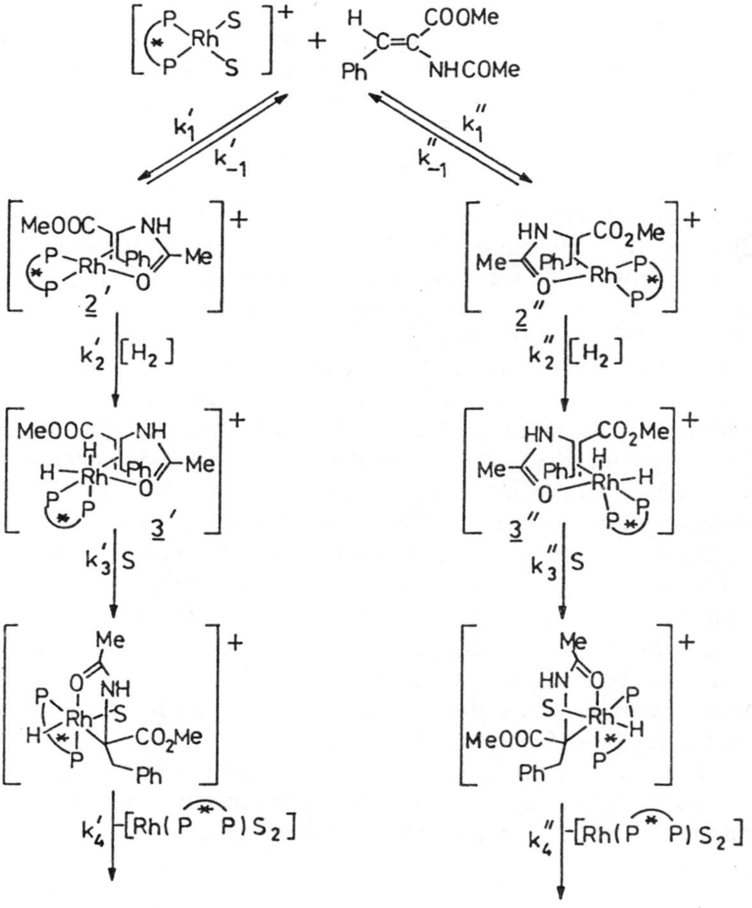

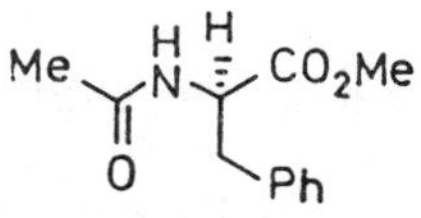

S = Solvent

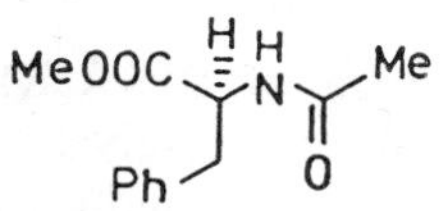

Figure 8. Mechanism of asymmetric hydrogenation of methyl-(Z)-acetamido cinnamate by Rh-complexes.

This mechanism seems to be the general mechanism for all asymmetric hydrogenations carried out with square planar d^8 rhodium(I)- and iridium(I)-complexes. The same kind of reaction pathway will be found also for hydrogenations of other (Z)-enamides, β-keto carbonyl compounds, allylic alcohols, α,β-unsaturated carboxylic acids and all systems which provide directing in the molecule.

The first irreversible step in the course of the hydrogenation is the oxidative addition of hydrogen (Fig. 8). According to the relative rates of this step the (R)- or the (S)-enantiomer is formed.

In the case of acetamido acrylic acids the minor diastereomer of the catalyst-substrate adduct activates hydrogen faster and leads to the (S)-product.

Both rates and enantioselection exhibit a complex dependence on hydrogen partial pressure and temperature. Details are given in literature [24].

Summarised briefly one can point out that in the limit of low temperatures and/or high hydrogen partial pressures, interconversion between the major and the minor diastereomer is frozen out and the relative rates of the formation of the diastereomeric adducts determine the enantioselection.

(1) $$\frac{[S]}{[R]} = \frac{k_1{}'}{k_1{}''} .$$

Here the enantioselection is independent from the relative stabilities and equilibrium concentrations of the substrate-catalyst complex. Therefore in this case more of the (R)-enantiomer is formed and the optical yield drops.

In the limit of high temperatures and/or low hydrogen partial pressures the major and the minor diastereomeric substrate-catalyst complexes interconvert rapidly. Then the relative rates of the oxidative addition of hydrogen, first irreversible step, determine the enantioselection.

(2) $\dfrac{[S]}{[R]} = \dfrac{k_2''[2'']}{k_2'[2']}$.

In the case of Rh/(R,R)-DIPAMP predominantly the (S)-enantiomer is formed and the catalyst has its maximum performance. In the intermediate regimes the rates and enantioselection shows a complex dependence on the hydrogen pressure.

2.4.2. Hydrogenation of substrates without directing groups

In the case of substrates like imines and α-ketoesters like ketopantolactone it is necessary to block one coordination site in the catalyst to obtain higher optical yields. Therefore mostly neutral in situ generated catalysts are used.

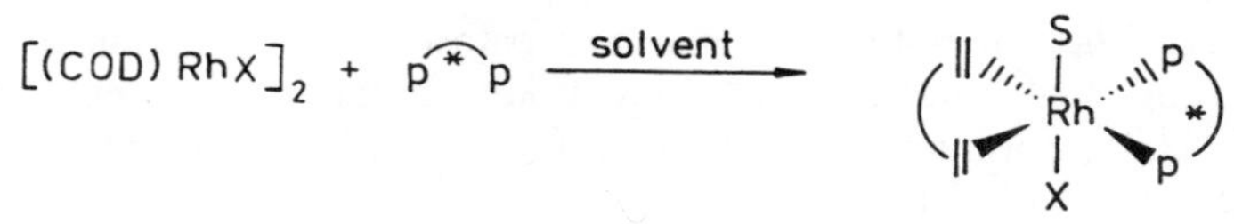

COD = cyclooctadiene

X   = halogen   .

S   = solvent

p͡*p = chiral bidentate phosphine

Figure 9. Synthesis and proposed structure for neutral Rh-BPPM-complexes 13.

In this case data derived from an x-ray analysis [25a] and spectroscopic studies of the BPPM–Rh catalysts show that the lower vacant site could be occupied by a solvent molecule or a chlorine ligand.

Ojima and coworkers have proposed a mechanism for hydrogenation of ketopantolactone [25b] with neutral rhodium complexes.

The mechanism involves oxidative addition of hydrogen to a rhodium(I) specie, formation of the substrate-rhodium(III)-dihydride intermediate, hydrogen migration giving an alkoxy-rhodium(III) intermediate and reductive elimination of the alcohol to regenerate the catalyst.

Sometimes tertiary amines are added as a cocatalyst. They may increase the activity of the catalyst, but mostly their function is unclear.

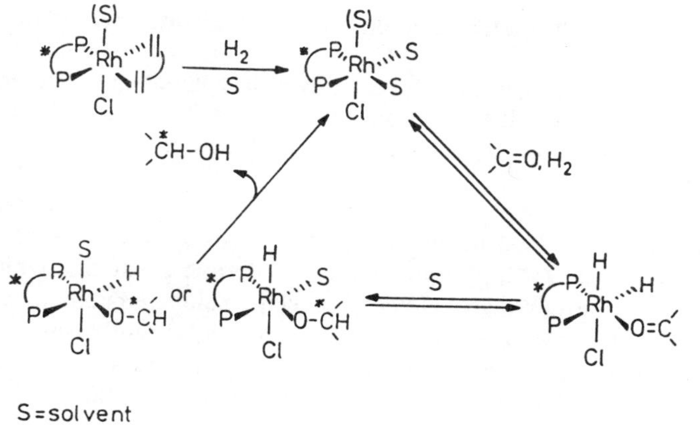

Figure 10. Proposed mechanism for the asymmetric hydrogenation of carbonyl double-bonds.

S = solvent

2.5. HYDROGENATION WITH RUTHENIUM-CATALYSTS

Asymmetric hydrogenations by a d^6-ruthenium(II)-center are much more complicated. Despite of the high catalytic activity of ruthenium catalysts [26] only few information is given about the mechanism of these hydrogenations. Only for trans-HRuCl((R,R)-DIOP)$_2$ a proposal was made based on experimental data (Fig. 11). This study showed that the catalyst precursor loses a phosphine ligand at the beginning of the catalytic cycle to provide a vacant coordination site for the substrate. The reaction proceeds via a σ-alkyl complex to the saturated product and a compound like HRuCl(DIOP) which rapidly coordinates another DIOP ligand to form the starting compound [27].

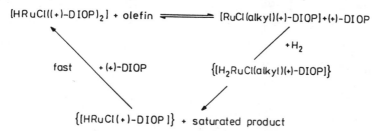

Figure 11. Mechanism of the hydrogenation with HRuCl((R,R)-DIOP)$_2$ **14**.

But still it is not quite clear if the reaction proceeds via Ru-cluster compounds. Compounds like [HRuCl(P–P)]$_3$ (P–P=CHIRAPHOS) have been isolated and their structur was characterized [8].

3. Asymmetric hydrogenation of substituted olefins

The following examples shall give an overview about the different kinds of substrates which have been hydrogenated with different chiral catalysts. The number of applications is still growing and therefore this synopsis cannot be complete.

As mentioned above functional groups next to the double-bond in the substrate molecule lead to chelation because of coordination to the metal. This is favourable for high optical yields. Olefinic substrates which provide such directing groups are N-acetyl acrylic acids, enamides, allylic and homoallylic alcohols, α,β- and β,γ-unsaturated carboxylic acids.

Other interesting target molecules are α,β-unsaturated aldehydes like pulegone and neral as precursors for menthol and citronellol.

3.1. AMINO ACID PRECURSORS

The problem of the hydrogenation of aminoacid precursors is solved [28, 29]. Table I gives an overview about the results for different chiral catalysts.

$$R\diagup\!\!\!\diagdown\!\!\!\overset{\displaystyle COOH}{\underset{\displaystyle NHCOCH_3}{|}} \xrightarrow{\ [Rh(COD)(P\text{-}P)]X\ } R\diagup\!\!\!\diagdown\!\!\!\overset{\displaystyle H}{\underset{\displaystyle NHCOCH_3}{\overset{\displaystyle |}{\text{—}COOH}}}$$

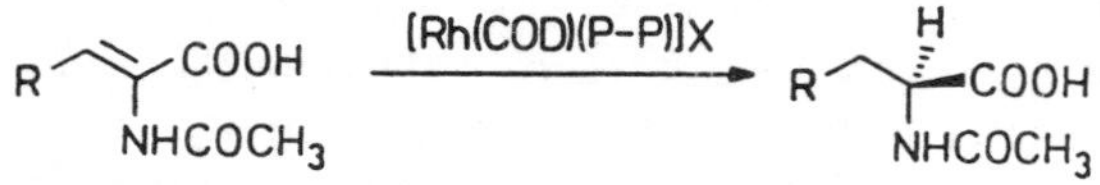

P–P = DIPAMP,...

X = BF_4^-. ClO_4^-,...

TABLE I
Chiral diphosphines, optical purity (%)

Substrates	DIPAMP	DIOP	CHIRAPHOS	BPPM	BINAP
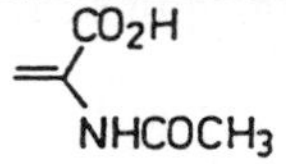 CO₂H / =< / NHCOCH₃	94	73	91	98,5	67
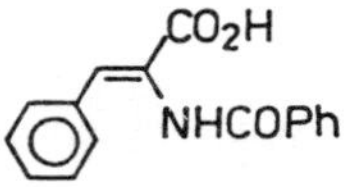 CO₂H / NHCOPh	96	64	99	–	96
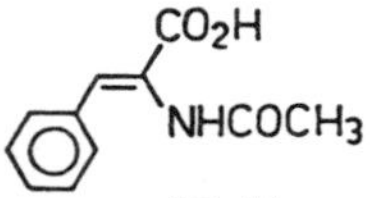 CO₂H / NHCOCH₃	95	81	89	91	84
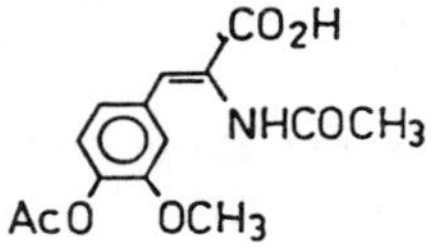 CO₂H / NHCOCH₃ (AcO, OCH₃)	94	84	83	86	70

The optical yields for acetamido acrylic acids are very high and the hydrogenation conditions are moderate. But still new ligands are under investigation.

Recently Gladysz and coworkers have published their work on a mixed diphenylphosphin-diphenylphosphido-ligand which contains a chiral rhenium center in its skeleton [30]. They used this chiral metal complex as ligand for the hydrogenation of acetamido acrylic acids and were able to achieve an enantiomeric excess of 98 % for acetamido acrylic acid.

cat.:

98% ee.

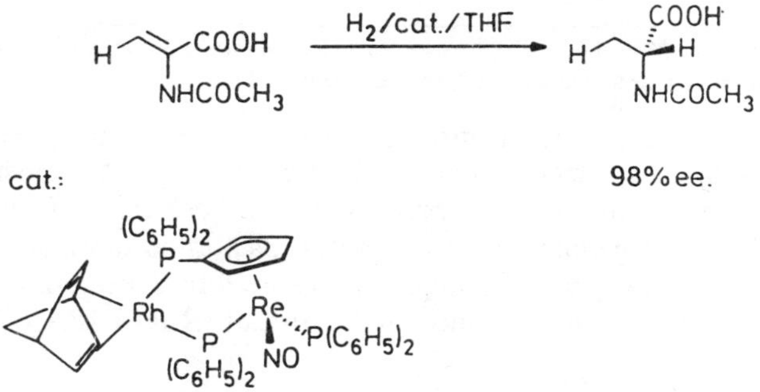

Figure 12. Hydrogenation of acetamido acrylic acid.

Summarizing literature data about asymmetric hydrogenations of acetamido acrylic acids, the ideal chiral phosphines should fulfil the following requirements [31]:

1. bidentate (1,2-diphosphine) ligands,
2. formation of fivemembered chelate rings,
3. rigid carbon backbone,
4. aryl substituents at the phosphorous atoms,
5. cheap chiral starting material and a short high yield synthesis.

The PYRPHOS ligands recently developed by Beck, Nagel and Degussa satisfy these requirements very well and the hydrogenation results of the corresponding catalyst (Deguphos®) are very good [32].

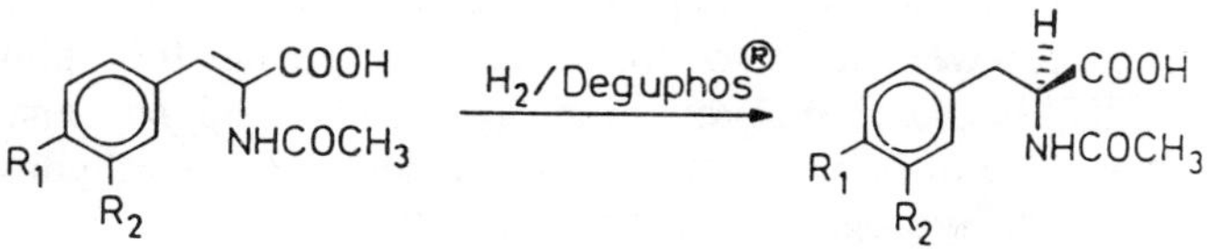

Figure 13. Hydrogenations with Deguphos®.

TABLE II

Typical experiments

Entry	[s]/[c]	P_{H_2}(bar)	T (°C)	Time (h)	Yield (%)	O.Y. (%)[a]
$R_1,R_2 = H$	10000[b]	10–15	50	3–4	100	99,5
$R_1,R_2 = H$	13000	40–50	50	2	100	98,5
$R_1,R_2 = H$	16000	40–50	60	7	100	97,0
R_1•AcO,R_2•MeO	10000	40–50	50	3	100	97,0

[a] based on $\alpha_0 = 47,5°$ and $\alpha_0 = 40,8°$ (c=1,95%, EtOH) for the pure products.
[b] 16,6g catalyst, 40 kg substrate, 25% in methanol.

Another interesting application is given in literature [33]. Dehydro-phosphinothricin derivatives could be hydrogenated to (S)-phosphinothricin derivatives which are used as precursors for herbicides. Chiral rhodium catalysts also hydrogenate dehydropeptides [34], important intermediates for the synthesis of enkephalin analogs etc.. Generally diastereomeric excesses between 95 and 98 % are obtained by the use of (R,R)-DIPAMP 6 or (S,S)-BPPM 10.

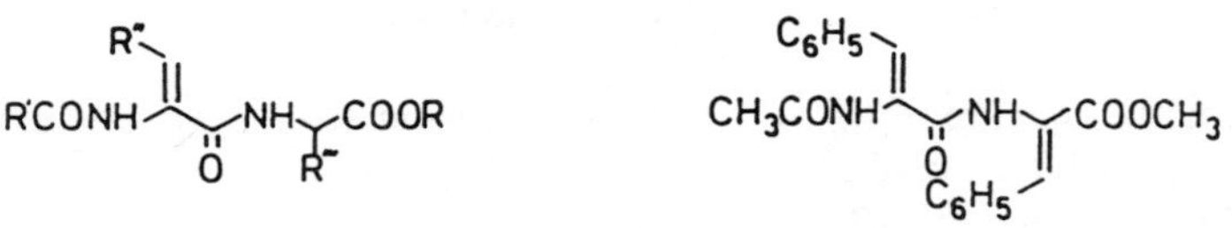

Figure 14. Examples for dehydropeptides.

The N-protecting group in dehydrodipeptides or tripeptides has some influence on the outcome of the reaction. Bulky substituents at the amino nitrogen may lead to lower enantiomeric excesses.

3.2. (Z)-ENAMIDES

A lot of interesting data have been published recently on the hydrogenation of (Z)-enamides like isoquinoline precursors [35] (Figure 15). First rhodium-BPPM complexes were used for this reaction [36], but the application of ruthenium-BINAP dicarboxylate complexes lead to higher enantiomeric excesses. A variety of N-acyl-(Z)-1-benzylidene-1,2,3,4-tetrahydroisoquino-isoquinolines were hydrogenated in 95 to 100 % ee. [37].

Figure 15. Hydrogenation of isoquinolin precursors.

An important example is shown in Figure 16. Using (R)-BINAP catalysts predominantly (1R)-products are formed. This hydrogenation is a key step in the synthesis of Dextromorphan, a morphinan which is used as an analgetic and bronchodilator in pharmacy [38].

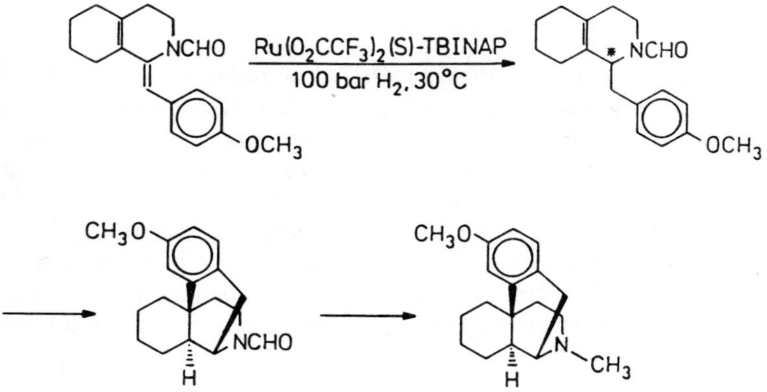

Figure 16. Synthesis of Dextromorphan 15.

3.3. UNSATURATED CARBOXYLIC ACID DERIVATIVES

There are a lot of examples for the hydrogenation of terminal olefins. One of the best investigated examples for this type of substrate is itaconic acid 16 [39].

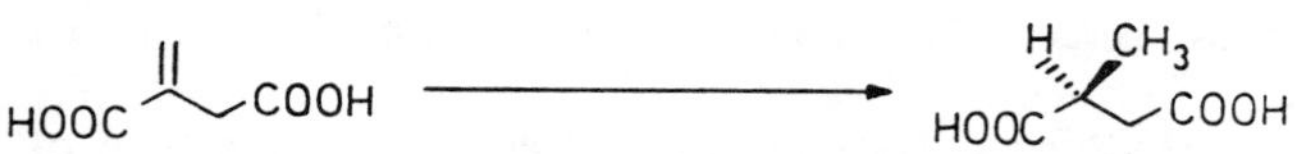

Figure 17. Hydrogenation of itaconic acid 16 to methyl succinic acid 17.

A summary of the results for itaconic acid is given in Table III.

TABLE III

Catalyst	Pressure [atm]	Temperature [°C]	Solvent	Opt. yield [% (S)]
Rh/BPPM*	50	20	methanol	93.6
Rh/BPPM*	1	20	methanol	94.8
Rh/BCPM*	50	20	methanol	58.0
Rh/BCPM*	1	20	methanol	92.0
Rh/FPPM*	50	20	methanol	94.2
$[Ru_2Cl_4(S)\text{-}BINAP_2]NEt_3$	3	35	toluene/THF	90.0

* NEt₃-added.

Mostly in situ generated rhodium-BPPM complexes were used [39a–c], but cationic rhodium-compounds and ruthenium-BINAP catalysts also show good results [39d]. The addition of triethylamine increases the optical yield under given hydrogen pressures and enhances the activity of the catalysts. The amine is added in stoichiometric amounts compared to the substrate.

It is well known from 2-arene propionic acids like Naproxen, Ibuprofen and Fluorbiprofen etc. that only the (S)-enantiomer posses any antiinflammatory effects. (S)-Naproxen is still produced by resolution of the racemate. The next example shows two new and interesting approaches for this class of compounds.

When 2-(4-methoxy-5-phenyl-3-thienyl)acrylic acid 18 is hydrogenated with an in situ rhodium-DIOP complex, (S)-2-(4-methoxy-5-phenyl-3-thienyl)propionic acid 19 is obtained with 97 % yield and an enantiomeric excess of 88 % ee. [40].

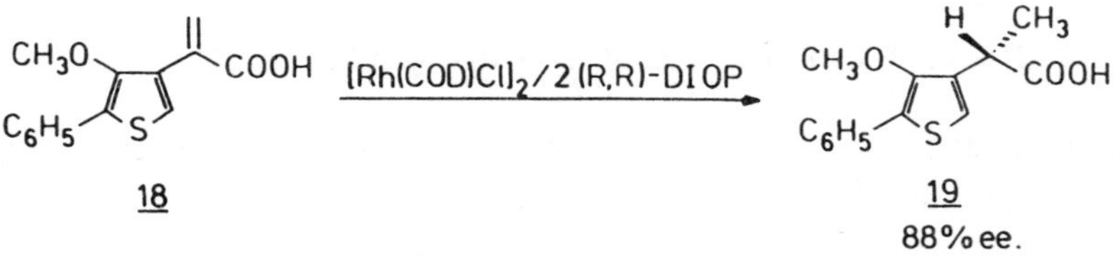

Figure 18. Synthesis of (S)-2-(4-methoxy-5-phenyl-3-thienyl)-propionic acid 19.

Ruthenium-BINAP catalysts are also effective for this type of substrate [41]. Enantiomeric excesses of 97% ee. were obtained for (S)-Naproxen 20. Also precursors for β-lactam antibiotics can be prepared [41].

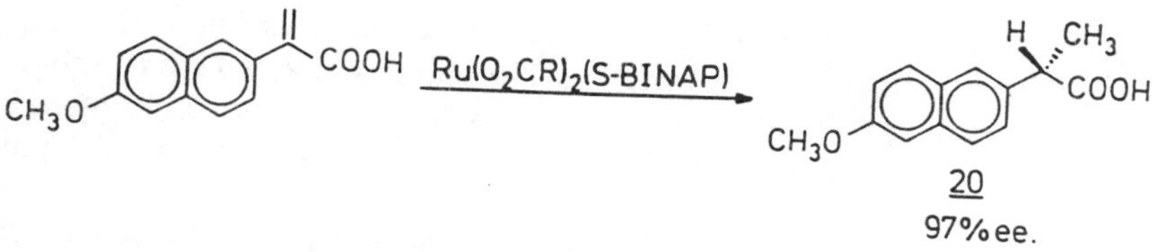

Figure 19. Synthesis of (S)-Naproxen 20.

Figure 20. β-Lactam antibiotic precursors.

For a long time asymmetric hydrogenation of trisubstituted acrylic acids (tetrasubstituted olefins) had been a challenging problem and optical purities of the hydrogenation products remained low. Among this class of compounds there are some important insecticides. one example is Fenverelate. Only the (S)-α-isopropyl acrylic esters are biological active components. At beginning 80's Paxson of Shell Oil Company applicated cationic rhodium-BPPM and rhodium-PPFA complexes (PPFA=ferrocenylphosphine derivative) in the synthesis of these compounds [42] and enantiomeric excess could be increased up to 71 and 95% ee. Later Hayashi and coworkers improved this reaction and ee.'s from 97.4 to 98.4% (S) were obtained by using (aminoalkyl)ferrocenylphosphine-rhodium catalysts. Good results were also found with β-disubsituted α-phenylacrylic acids [43, 44].

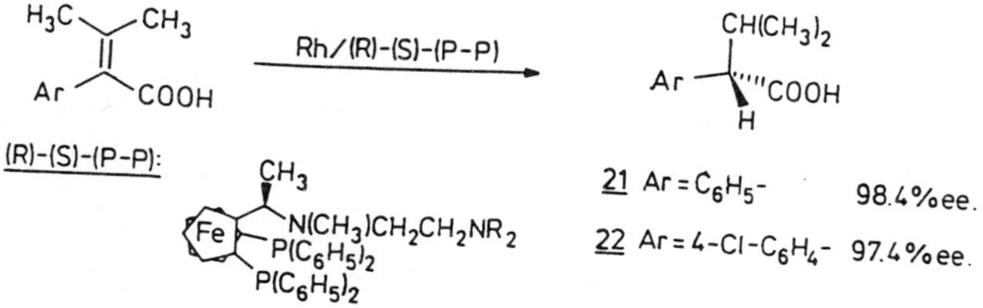

Figure 21. Hydrogenation of trisubstituted acrylic acids.

The substrates as shown in Figure 22 represents an other interesting target molecule for asymmetric hydrogenation. The chiral moiety which is found in the hydrogenation product is also present in side chains of several important natural products like α-tocopherol (vitamin E) and phylloquinone (vitamin K1).

Therefore Valentine and coworkers tried to prepare citronellic acid 23 by asymmetric hydrogenation with cationic rhodium catalysts. The best optical yield of 79 % was reached with the monodentate NMDPP ligand [45].

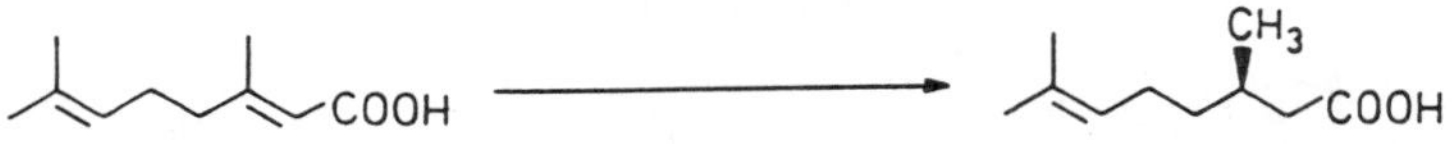

Figure 22. Asymmetric hydrogenation of citronellic acid 23.

In this case the asymmetric hydrogenation of homoallyl alcohols as precursors is more promising (Chapt. 3.5).

3.4. α,β-UNSATURATED ALDEHYDES AND KETONES

First attempts to hydrogenate α,β-unsaturated aldehydes and ketones were made by Solodar and Dang. Piperitenone was converted into pulegone with 61 to 92% yield and 38% optical purity, using cationic rhodium-diolefin-CAMP as catalyst [46].

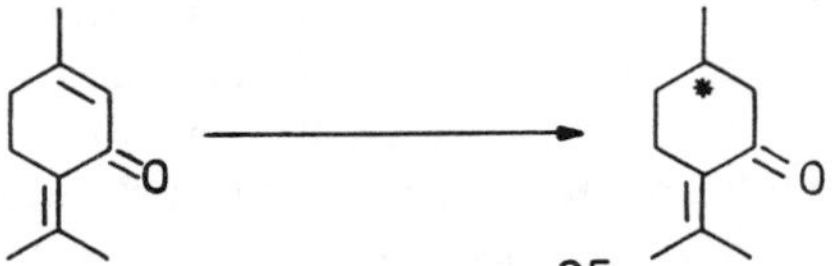

Figure 23. Synthesis of pulegone 25.

Neral 26 was hydrogenated to (R)-citronellal 27 with 70 to 79% ee. In this case rhodium complexes with 1,4 diphosphines (DIOP, and (+)-TBPC) were used [47].

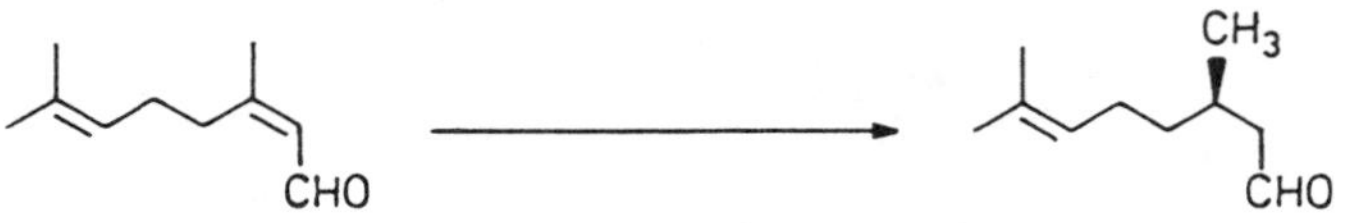

Figure 24. Asymmetric hydrogenation of neral 26 to (R)-citronellal 27.

Unfortunately the enantiomeric excess was too low for a commercial process.

3.5. ALLYLIC AND HOMOALLYLIC ALCOHOLS

In the hydrogenation of nerol and geraniol ruthenium-BINAP dicarboxylate catalysts exhibit a very high efficiency. Using reaction pressures greater than 30 atmospheres good enantioselectivities were obtained and substrate to catalyst ratios up to 50000 to 1 were reached [48]. The values of the enantiomeric excess range from 96 to 99% ee. and the regioselectivity is very high. Only the C(2)–C(3) double bond is hydrogenated. The C(6)–C(7) double bond remaines untouched. Figure 25 shows the reaction scheme for the conversion of geraniol 28 and nerol 29 to either (R)- or (S)-citronellol 30.

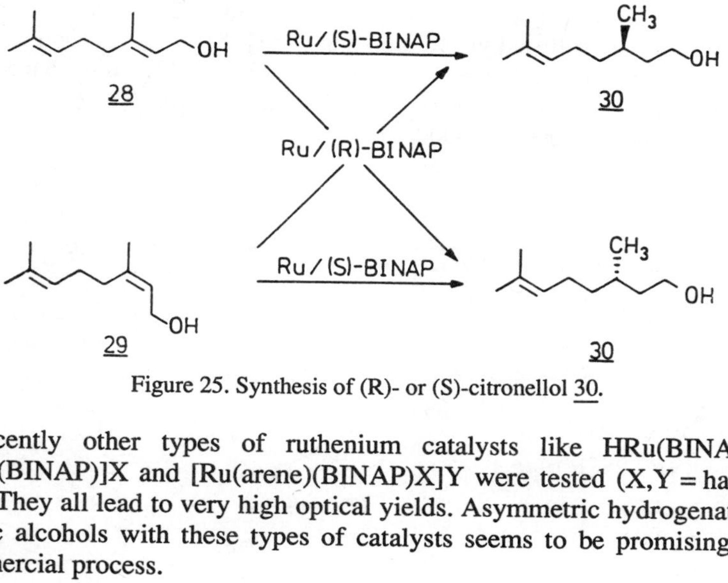

Figure 25. Synthesis of (R)- or (S)-citronellol 30.

Recently other types of ruthenium catalysts like $HRu(BINAP)_2X$, $[HRu(BINAP)]X$ and $[Ru(arene)(BINAP)X]Y$ were tested $(X,Y = halogen)$ [49]. They all lead to very high optical yields. Asymmetric hydrogenation of allylic alcohols with these types of catalysts seems to be promising for a commercial process.

4. Asymmetric hydrogenation of carbon-oxygen double-bounds

4.1. α-KETOACID DERIVATIVES

(R)-(–)-Pantolactone 32 is an important precursor for several biological active compounds like pantothenic acid, its calcium salt (vitamin B5) and N-pantoyl-propanolamine (Dexpanthenol, Panthenol) [50]. Therefore a lot of research groups started investigations on hydrogenation of ketopantolactone 31, the precursor of pantolactone (Table IV). In 1978 Achiwa and coworkers published the hydrogenation with an in situ rhodium-BPPM complex [51].

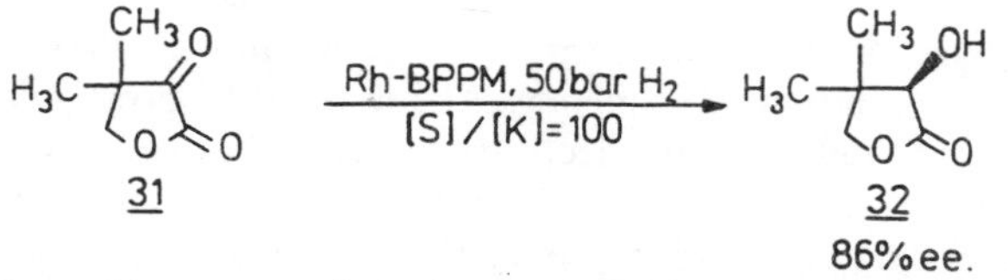

Figure 26. Asymmetric hydrogenation of ketopantolactone 31.

Asymmetric hydrogenation was also carried out with a supported catalyst [52]. But optical yields reached only 73.4 to 75.7% (R). Later the activity of the rhodium complexes was enhanced by replacing the aryl substituents at the phosphorus by fully alkylated analogues [53]. Symmetrical and unsymmetrical substituted diphosphines like BCPM, MCPM and MCCPM, derivatives of BPPM, gave optical yields of 90 to 92% (R). The catalyst-substrate ratios reached 1000 to 10000/1.

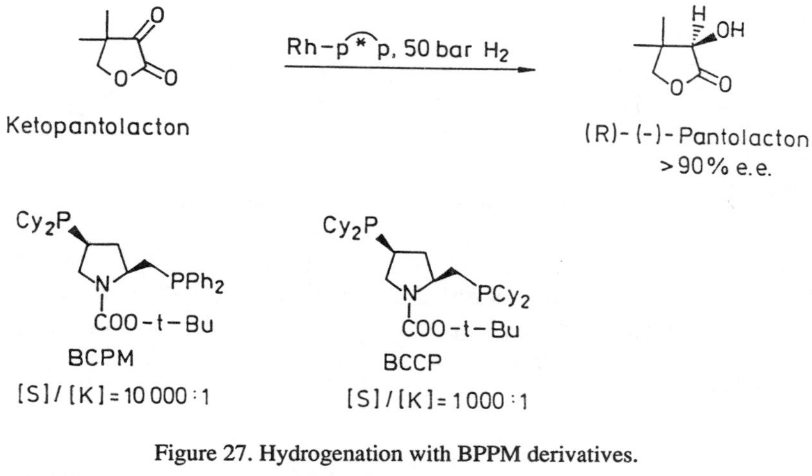

Figure 27. Hydrogenation with BPPM derivatives.

A further increase in activity and enantioselectivity was observed on changing the anion of the neutral in situ complexes [54]. Researchers at

Hoffmann-La Roche in Basle demonstrated, that fluorinated acetic acid derivatives accelerate the reaction. Thus substrate to catalyst ratios of 15000 to 65000/1 were possible, accompanied by a small increase in enantiomeric excess up to 95% ee. (R) [55].

TABLE IV

Catalyst	[S]/[C]	Pressure [bar]	Temp. [°C]	Solvent	React. time [h]	Opt. yield [% (R)]
Rh-BPPM	100	50	50	benzene	45	54.6–59.2
Rh-PPM-derivative	100	50	50	benzene	45	67.5–80.5
Rh-PPM-polymer	100	50	50	benzene	45	73.4–75.7
Rh-BPPM	100	50	30	benzene	48	86.0
Rh-BPPM	200	50	30	toluene	48	35.0–83.9
Rh-BPPM	200	50	30–50	toluene	45–48	70.0–81.2
Rh-DIOP or BPPM	100	50	30	benzene	48	18.7–86.7
Rh-PPM-polymer	100	50	50	benzene	45	76.0
Rh-PPM-derivative	12000	130	60	toluene	70	56.9–77.3
Rh-PPM-derivative	3000	40	40	toluene	25	81.8–84.3
Rh-BCCP	200	1	35	THF	1	41.0–66 (S)
Rh-Cy-DIOP	50	20	20	benzene/E+OH	1	45.0
Rh-BPPM	100	50	30	benzene	48	86.7
Rh-BPPM	100	50	30–50	benzene	48	78.0–84.0
Rh-BPPM/cokat.	12500	40	35	toluene	5	89.1–93.8
Rh-BCPM	100–1000	50	50	THF	45	90.5–92.0
Rh-BCPM	10000	50	50	THF	45	90.0
Rh-DIOCP	1000	15	50	THF	70	75.0
Rh-BPPM/cokat.	8000–10000	40	40	toluene	1–2	88.5–92.5
Pt (5% on activated carbon, 1% Cinchonidine)				benzene		36.0
Rh-BCPM	1000–10000	50	50	THF	45	90.0–91.0
Rh-AMPP	10000	1	20	toluene	3	77.0

Various α-keto esters have been tested. Achiwa and coworkers reached up to 87% optical yield for methyl pyruvate with rhodium/MCCPM and MCPM catalysts [56a].

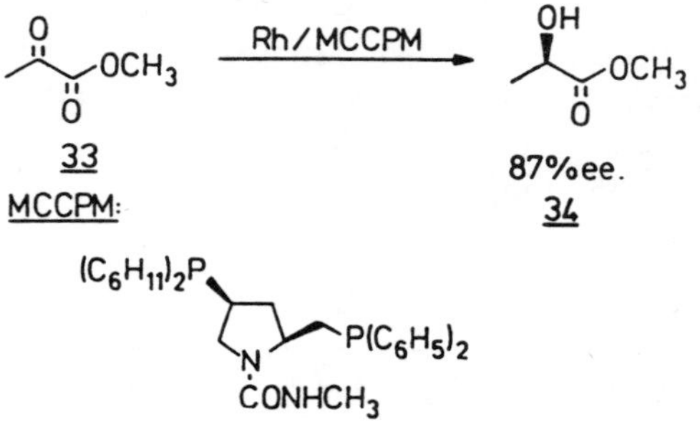

Figure 28. Hydrogenation of methylpyruvate 33 to methyl lactic acid 34.

Similar results are obtained with rhodium-BPPM (79–80% ee. (R)) complexes by changing its anionic ligand [54]. And pyruvic acid was hydrogenated in presence of a small amount of triethylamine with 83% optical yield [56b].

4.2. HYDROGENATION OF β-KETO COMPOUNDS

Recently there has been growing interest in the chemical industry for (L)-carnitine chloride 35, which is a derivative of the naturally occurring hydroxide compound, responsible for the transport and metabolism of longchain fatty acids in mitochondrial membranes. Today (L)-carnitine is produced by resolution of the racemic mixture. But recently Noyori and coworkers have published a synthesis with ruthenium-BINAP as hydrogenation catalyst [57].

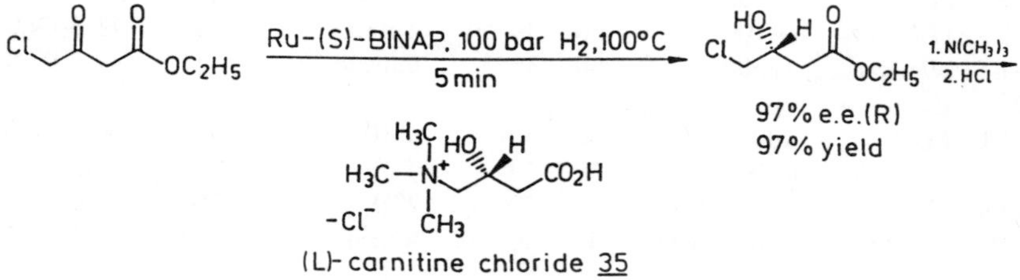

Figure 29. Synthesis of (L)-carnitine chloride 35.

Under standard conditions for asymmetric hydrogenation for β-ketoacids with ruthenium catalysts normally enantiomeric excess does not exceed 70% ee. But an increase in pressure up to 100 bar and temperature up to 100°C leads to 97% ee. (L) and completion of the reaction within five minutes.

It is also possible to hydrogenate other β-keto carboxylic esters like ethyl acetoacetate 36 in high enantiomeric excess [58]. Catalysts like HRuX(BINAP)$_2$ (X = Cl, Br, I) are also effective for this reaction.

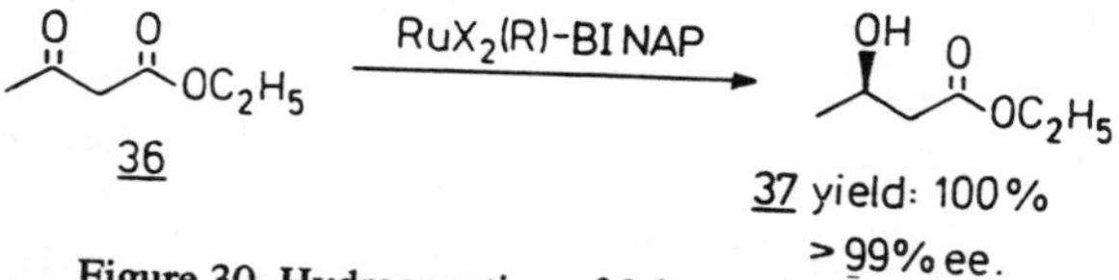

Figure 30. Hydrogenation of β-keto compounds.

The enantiomeric excesses range from 98 to 100 %ee. Noyori et al. also developed a highly efficient synthesis for precursors of statine 38 and its analogues [59].

The results for other β-keto compounds are summarized in [60].

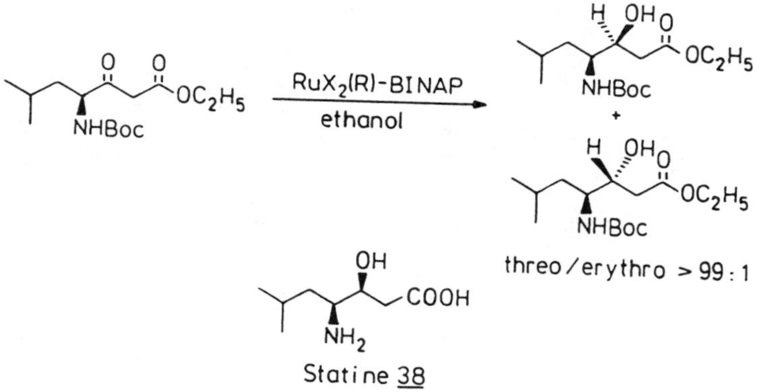

Figure 31. Synthesis of statine precursors.

4.3. ALKYLAMINO ARYL KETONES

Asymmetric hydrogenation of alkylamino aryl ketones were reported with rhodium-BPPFOH and -DIOP catalysts [61]. Either cationic or neutral in situ complexes can be used.

Ruthenium complexes are also effective for this reaction. Optical yields range from 93 to 96 % for simple alkyl dimethylamino ketones [62].

Alkylamino aryl ketones are important intermediates for pharmaceuticals products. Figure 33 shows a synthesis for (S)-(–)-Levamisol a remedy for treatment of rheumatism and infections [63].

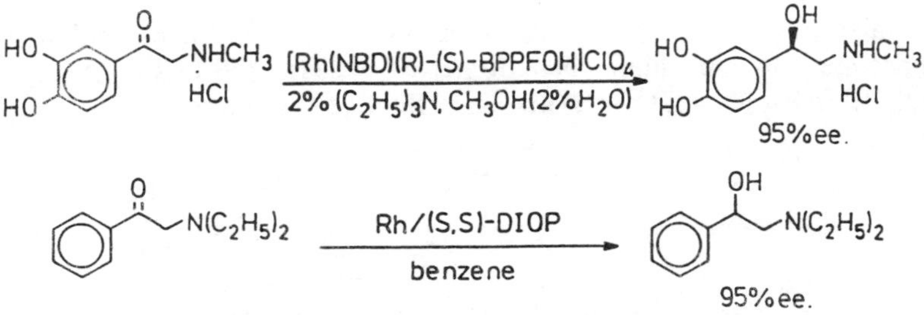

Figure 32. Hydrogenation of alkylamino aryl ketones.

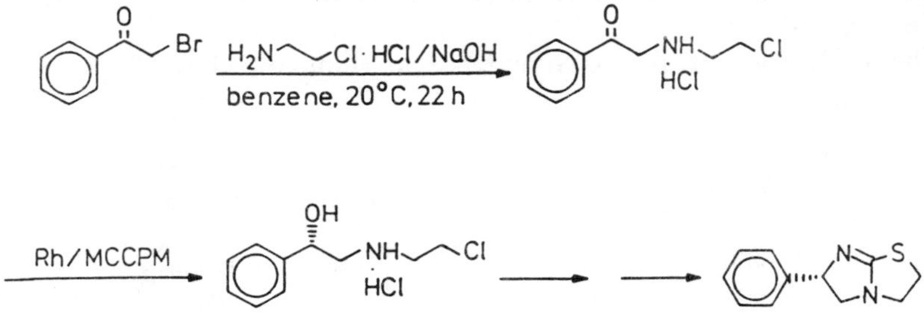

Figure 33. Synthesis of (S)-Levamisol 39.

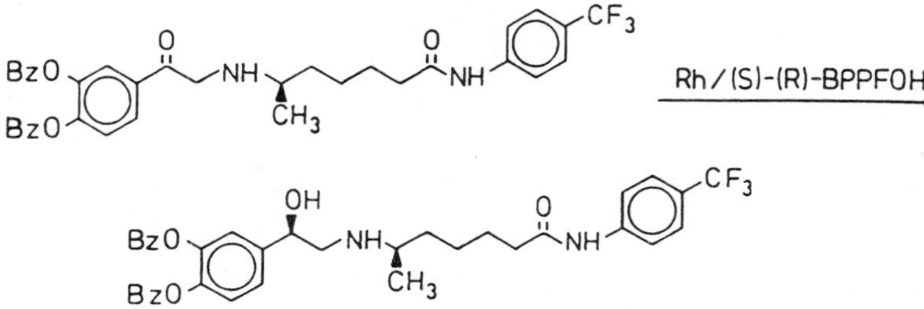

yield: 87.4%
98% de.

Figure 34. Hydrogenation of alkylamino aryl ketone in a synthesis of an β-adrenoreceptor antagonist.

Another interesting application of asymmetric hydrogenation is shown in Figure 34. Here this method is used in the synthesis of a β-adrenoreceptor antagonist [64].

Diastereomeric excesses are about 98 % with an overall yield of 87.4 %.

4.4. KETONES

Asymmetric hydrogenation of simple ketones to chiral alcohols is a well known problem. For the hydrogenation of acetophenone $\underline{40}$ to phenylethanol $\underline{41}$ the best results did not exceed 82% ee. [65].

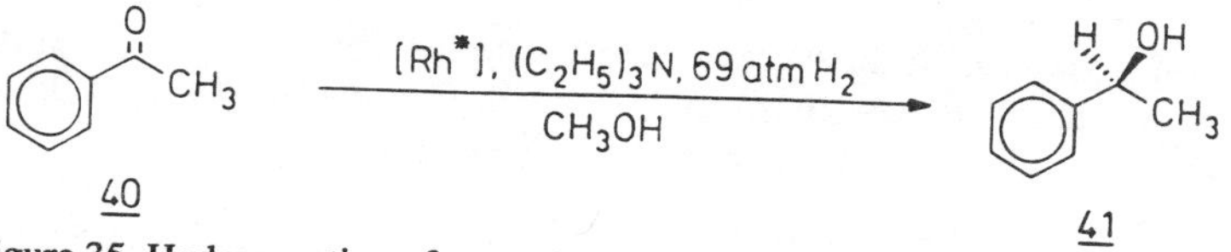

Figure 35. Hydrogenation of acetophenone $\underline{40}$ with chiral in situ Rh-complexes.

Bakos et al. used a rhodium-SKEWPHOS catalyst prepared in situ for this reaction. Recently Chan and Landis reported promoter effects for a rhodium-DIOP-chloride catalyst system [66]. Triethanolamine leads to an optical yield of 79% ee. (S) but at low conversion. At higher conversion optical yields drop to 74% (S).

5. Asymmetric hydrogenation of carbon-nitrogen double-bonds

The asymmetric hydrogenation of prochiral imines has received only little attention. First attempts to hydrogenate carbon-nitrogen double bonds were made with rhodium-DIPAMP and -DIOP catalysts [67]. As mentioned above optical yields improve when neutral in situ catalysts are used. In situ catalysts of PHEPHOS and VALPHOS were also investigated as ligands [68].

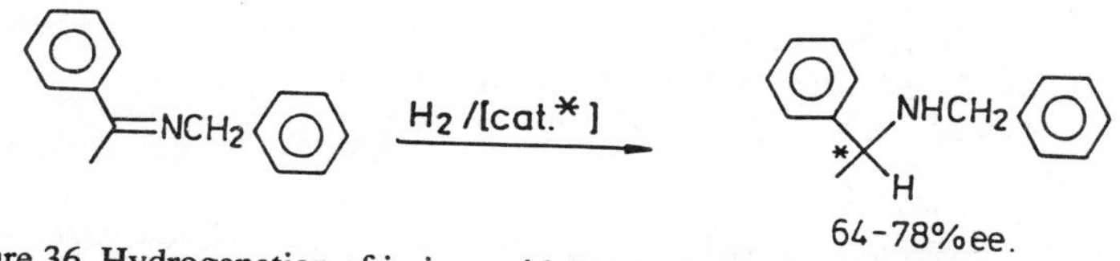

Figure 36. Hydrogenation of imines with PHEPHOS and VALPHOS-Rh-catalysts.

The enantiomeric excesses of the corresponding secondary amines are moderate and vary remarkably. Temperature and solvent have a strong influence on the outcome of the hydrogenation.

Recently James and coworkers could enhance the results by adding cocatalysts like potassium iodide to the reaction mixture [69]. Maximum values of 91% ee. are achieved at temperatures of minus 25°C.

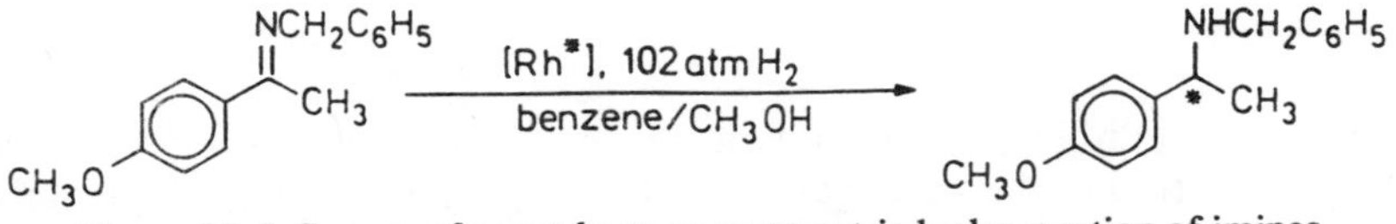

Figure 37. Influence of cocatalysts on asymmetric hydrogenation of imines.

Research groups in industry also show interest for this type of substrate [70]. They were able to obtain up to 75% ee. with novel iridium catalysts and tetrabutyl ammonium iodide as cocatalyst (TBAI).

Degussa AG
Department of Chemical and Biological Research
P.O. Box 1345
D-6450 Hanau 1 (Germany)

References

1. *The Merck Index*, Tenth Edition, page 1324, No. 9076, Merck & Co., Inc. Rahway New Jersey, USA, 1983.
2. a) B. Heil, Wiss. Zeitschr. *THLM* **1985**, *27* 6; b) H.B. Kagan, *Bull. Chem. Soc. France*, **1988**, *5* 846.
3. a) J.A. Osborn, F.H. Jardine, J.F. Young and G. Wilkinson, *J. Chem. Soc. A*, **1966** 1711; b) Schrock-Osborn type catalysts: R.R. Schrock and J.A. Osborn, *J. Am. Chem. Soc.*, **1971**, *93*, 2397.
4. a) DD 240 372, R. Selke, W. Vocke (VEB Isis-Chemie, Zwickau), 29.10.86; b) DD 248028, H. Foken, H. Grüner, H.-W. Krause, R. Selke (VEB Isis-Chemie, Zwickau), 29.7.87; c) W. Vocke, R. Haenel, F.-U. Floether, *Chem. Techn.*, **39** (1987) 123; d) EP 0077099, M. Fiorini, M. Riocci, M. Giongo (ANIC S.p.A.), 6.10.82.
5. a) L. Horner, H. Siegel, H. Buethe, *Angew. Chem.*, **1968**, *80*, 168; b) L. Horner, H. Siegel, H. Buethe, *Angew. Chem., Int. Ed. Engl.* **1968**, *7*, 942; c) W.S. Knowles and M.J. Sabacky, *J. Chem. Soc., Chem. Commun.*, **1968**, 1445.
6. W.S. Knowles, M.J. Sabacky and B.D. Vineyard, *J. Chem. Soc.*, **1972**, 10.
7. a) R.R. Schrock and J.A. Osborn, *J. Am. Chem. Soc.*, **1971**, *93*, 2397; b) W.S. Knowles, *Acc. Chem. Res.*, **1983**, *16*, 106.
8. B.R. James, A. Pacheco, S.R. Rettig, I.S. Thorburn, R.G. Ball and James Ibers, *J. Mol. Catal.*, **1987**, *41*, 147.
9. a) T. Ikariya, Y. Ishii, H. Kawano, T. Arai, M. Saburi, S. Yoshikawa and S. Akutagawa, *J. Chem. Soc., Chem. Commun.*, **1985**, 922; b) R. Noyori and M. Kitamura, «Enantioselective Catalysis with Metal Complexes. An Overview», page 128-138, in R. Scheffold, «Modern Synthetic Methods» Vol. 5, Springer publishers, 1989.
10. T. Ohta, H. Takaya and R. Noyori, *Inorg. Chem.*, **1988**, *27*, 566.
11. H.B. Kagan, «Chiral Ligands for Asymmetric Synthesis»,in J.D. Morrison «Asymmetric Synthesis», Vol. 5, pages 1–35, Academic Press, Inc., 1985.
12. a) M.D. Fryzuk and B. Bosnich, *J. Am. Chem. Soc.*, **1977**, *99*, 6262; b) R. Glaser,

Asymmetric hydrogenation 187

Sh. Geresh and M. Twaik, *Isr. J. Chem.*, **1980**, *20*, 102; c) P.A. MacNeill, N.K. Roberts and B. Bosnich, *J. Am. Chem. Soc.*, **1981**, *103*, 2273.
13. H. Boucher and B. Bosnich, *Inorg. Chem.*, **1976**, *15*, 1471.
14. M. Kojima, M. Fujita and J. Fujita, *Bull. Chem. Soc. Jpn.*, **1977**, *50*, 898.
15. a) L.J. DeHayes and D.H. Busch, *Inorg. Chem.*, **1973**, *12*, 2620; b) M. Kojima and J. Fujita, *Bull. Chem. Soc. Jpn.*, **1977**, *50*, 3237.
16. J.M. Brown and P.A. Chaloner, *J. Am. Chem. Soc.*, **1978**, *100*, 4321.
17. T.P. Dang, J.C. Poulin and H.B. Kagan, *J. Organomet. Chem.*, **1975**, *91*, 105.
18. A.S.C. Chan, J.J. Pluth and J. Halpern, *J. Am. Chem. Soc.*, **1980**, *102*, 5952.
19. J.M. Brown, I. Cutting and A.P. James, *Bull. Chem. Soc. France*, **1988**, *2*, 211.
20. Y. Kawabata, M. Tanaka and I. Ogata, *Chem. Lett.*, **1976**, 1213.
21. J. Halpern, *Pure & Appl. Chem.*, **1983**, *55*, 99.
22. J.M. Brown, P.A. Chaloner and G.A. Morris, *J. Chem. Soc., Perkin Trans. II*, **1987**, 1583.
23. J. Halpern, *Inorg. Chim. Acta*, **1981**, *50*, 11.
24. C.R. Landis and J. Halpern, *J. Am. Chem. Soc.*, **1987**, *109*, 1746.
25. a) Y. Ohga, Y. Yitaka, K. Achiwa, T. Kogure and I. Ojima, *25th Symposium on Organometallic Chemistry*, Osaka/Japn, 1978, Abstract 3A15; b) I. Ojima and T. Kogure, *J. Organomet. Chem.*, **1980**, *195*, 239.
26. T.W. Deklava, I.S. Thorburn and B.R. James, *Inorg. Chim. Acta*, **1985**, *49*, 49 and references therein.
27. a) B.R. James, R.S. McMillan, R.H. Morris and D.K. Wang, *Adv. Chem. Ser.*, **1978**, *167*, 122; b) B.R. James and D.K. Wang, *Can. J. Chem.*, **1980**, *58*, 245.
28. K.E. Koenig, «*The Applicability of Asymmetric Homogenous Catalytic Hydrogenation*»,in J.D. Morrison «Asymmetric Synthesis, Vol. 5, page 71-79, Academic Press, Inc., 1985 and references therein.
29. H. Takahashi and K. Achiwa, *Chem. Lett.*, **1989**, 305.
30. a) B.D. Zwick, A.M. Arif, A.T. Patton and J.A. Gladysz, *Angew. Chem.*, **1987**, *99*, 921; b) B.D. Zwick, A.M. Arif, A.T. Patton and J.A. Gladysz, *Angew. Chem. Int. Ed. Engl.*, **1987**, *26*, 910.
31. H. Brunner, *Angew. Chem.*, **1983**, *95*, 425.
32. a) U. Nagel, *Angew. Chem.*, **1984**, *96*, 425; b) EP 0151282, W. Beck, U. Nagel (Degussa AG), 14.12.84; c) EP 0185882, J. Andrade, U. Nagel, G. Prescher (Degussa AG), 1984. d) U. Nagel, E. Kinzel, J. Andrade and G. Prescher, *Chem. Ber.*, **1986**, *119*, 3326.
33. DE 3609818, H.-J. Zeiß (Hoechst AG, Frankfurt), 22.3.86.
34. a) I. Ojima, T. Kogure, N. Yoda, T. Suzuki, M. Yatabe and T. Tanaka, *J. Org. Chem.*, **1982**, *47*, 1329; b) I.Ojima, N. Yoda, M. Yatabe, T. Tanaka and T. Kogure, *Tetrahedron*, **1984**, *40*, 1255; c) T. Yamagishi, M. Yatagai, H. Hatakeyama and M. Hida, *Bull. Chem. Soc. Jpn.*, **1984**, *57*, 1897 and references therein; d) J.M. Nuzillard, J.C. Poulin and H.B. Kagan, *Tetrahedron Lett.*, **1986**, *27*, 2993; e) S. El-Baba, J.M. Nuzillard, J. Poulin and H.B. Kagan, *Tetrahedron*, **1986**, *42*, 3851; f) T. Yamagishi, S. Ikeda, M. Yatagai, M. Yamaguchi and M. Hida, *J. Chem. Soc. Perkin Trans. I*, **1988**, 1787.
35. R. Noyori, M. Ohta, Y. Hsiao, M. Kitamura, T. Ohta and H. Takaya, *J. Am. Chem. Soc.*, **1986**, *108*, 7117.
36. K. Achiwa, *Heterocycles*, **1977**, *8*, 247.
37. EP 0245960, R. Noyori, M. Kitamura, H. Takaya, H. Kumobayashi and S. Akutagawa (Takasago Perfumery Co., Ltd., Tokyo/Japan), 13.4.1987.

38. M. Kitamura, Y. Hsiao, R. Noyori and H. Takaya, *Tetrahedron Lett.*, **1987**, *28*, 4829.
39. a) K. Achiwa, *Chem. Lett.*, **1978**, 561; b) I. Ojima, T. Kogure and N. Yoda, *J. Org. Chem.*, **1980**, *45*, 4728; c) H. Takahashi and K. Achiwa, *Chem. Lett.*, **1987**, 1921; d) H. Kawano, Y. Ishii, T. Ikariya, M. Saburi, S. Yoshikawa, Y. Uchida and H. Kumobayashi, *Tetrahedron Lett.*, **1987**, *28*, 1905.
40. A.P. Stoll and R. Suess, *Helv. Chim. Acta*, **1974**, *57*, 2487.
41. T. Ohta, H. Takaya, M. Kitamura, K. Nagai and R. Noyori, *J. Org. Chem.*, **1987**, *52*, 3176.
42. US 4409397, T.E. Paxson (Shell Oil Comp., Houston/USA), 11.10.1983.
43. T. Hayashi, N. Kawamura and Y. Ito, *J. Am. Chem. Soc.*, **1987**, *109*, 7876.
44. T. Hayashi, N. Kawamura and Y. Ito, *Tetrahedron Lett.*, **1988**, *29*, 5969.
45. D. Valentine, K.K. Johnson, W. Priester, R.C. Sun, K. Toth and G. Saucy, *J. Org. Chem.*, **1980**, *45*, 3698.
46. J. Solodar, *J. Org. Chem.*, **1978**, *43*, 1787.
47. T.P. Dang, P. Aviron-Violet, Y. Colleuille and J. Varagnat, *J. Mol. Catal.*, **1982**, *16*, 51.
48. H. Takaya, T. Ohta, N. Sayo, H. Kumobayashi, A. Akutagawa, Sh.-I. Inoue, I. Kasahara and R. Noyori, *J. Am. Chem. Soc.*, **1987**, *109*, 1596.
49. a) EP 0256634, T. Hidemasa, T. Ohta, R. Noyori, S. Naburo, H. Kumobayashi and S. Akutagawa (Takasago Perfumery Co., Ltd., Tokio/Japan), 15.6.87; b) T. Tsukahara, H. Kawano, Y. Ishii, T. Takahashi, M. Saburi, Y. Uchida and S. Akutagawa, *Chem. Lett.*, **1988**, 2955; c) K. Mashima, K.-H. Kusano, T. Ohta, R. Noyori and H. Takaya, *J. Chem. Soc., Chem. Commun.*, **1989**, 1208.
50. a) M. Purko, W.O. Nelson and W.A. Wood, *J. Biol. Chem.*, **1954**, *207*, 51; b) G.M. Brown and J.J. Reynolds, *Ann. Rev. Biochem.*, **1963**, *32*, 419.
51. a) K. Achiwa, T. Kogure and I. Ojima, *Tetrahedron Lett.*, **1977**, 4431; b) K. Achiwa, T. Kogure and I. Ojima, *Chem. Lett.*, **1978**, 297; c) GB 1592536 (Sagami Research Center/Japan), 27.2.1978.
52. K. Achiwa, *Heterocycles*, **1978**, *9*, 1539.
53. a) H. Takahashi, M. Hattori, M. Chiba, T. Morimoto and K. Achiwa, *Tetrahedron Lett.*, **1986**, *27*, 4477; b) T. Morimoto, H. Takahashi, K. Fujii, M. Chiba. and K. Achiwa, *Chem. Lett.*, **1986**, *12*, 2061.
54. EP 0158875, E. Broger and Y. Crameri (Hoffmann-La Roche), 26.10.1985.
55. EP 0218970, E. Broger and Y. Crameri (Hoffmann-La Roche), 22.4.1987.
56. a) H. Takahashi, T. Morimoto amd K. Achiwa, *Chem. Lett.*, **1987**, 855; b) T. Hayashi, T. Mise and M. Kumada, *Tetrahedron Lett.*, **1976**, *48*, 4351.
57. M. Kitamura, T. Ohkuma, H. Takaya and R. Noyori, *Tetrahedron Lett.*, **1988**, *29*, 1555.
58. R. Noyori, T. Okhuma, M. Kitamura, H. Takaya, N. Sayo, H. Kumobayashi and S. Akutagawa, *J. Am. Chem. Soc.*, **1987**, *109*, 5856.
59. T. Nishi, M. Kitamura, T. Okhuma and R. Noyori, *Tetrahedron Lett.*, **1988**, *29*, 1555.
60. T. Nishi, M. Kitamura, T. Okhuma and R. Noyori, *Tetrahedron Lett.*, **1988**, *29*, 6327.
61. a) T. Hayashi, A. Katsumara, M. Konishi and M. Kumada, *Tetrahedron Lett.*, **1979**, *20*, 425; b) S. Törös,, L. Kollar, B. Heil and L. Marko, *J. Organomet. Chem.*, **1982**, *232*, C17.
62. M. Kitamura T.Ohkuma, S. Inoue, N. Sayo, H. Kumobayashi, S. Akutagawa, T.

Ohta, H. Takaya and R. Noyori, *J. Am. Chem. Soc.*, **1988**, *110*, 629.

63. H. Takeshi, T. Tachinami, M. Aburatani H. Takahashi, T. Morimoto and K. Achiwa, *Tetrahedron Lett.*, **1989**, *30*, 363.

64. H.P. Märki, Y. Crameri, R. Eigermann, A. Krasso, H. Ramuz, K. Bernauer, M. Goodman and K.L. Melmon, *Helv. Chim. Acta*, **1988**, *71*, 320.

65. J. Bakos, I. Toth, B. Heil and L. Marko, *J. Organomet. Chem.*, **1985**, *279*, 23.

66. A.S.C. Chan and C.R. Landis, *J. Mol. Catal.*, **1989**, *49*, 165.

67. a) H.B. Kagan, N. Langlois, and T.P. Dang, *J. Organomet. Chem.*, **1975**, *279*, 283; b) A. Levi, G. Modena and G. Scorrano, *J. Chem. Soc., Chem. Commun.*, **1975**, 6; c) S. Vastag, B. Heil, S. Toros and L. Marko, *Transition Met. Chem.*, **1977**, *2*, 58.

68. a) S. Vastag, J. Bakos, S. Toros, N.E. Takach, R.B. King, B. Heil and L. Marko, *J. Mol. Catal.*, **1984**, *22*, 283. b) J. Bakos, I. Toth, B. Heil and L. Marko, *J. Organomet. Chem.*, **1985**, *279*, 23.

69. a) G.-J. Kang, W.R. Cullen, M.D. Fryzuk, B.R. James and J.P. Kutney, *J. Chem. Soc., Chem. Commun.*, **1988**, 1466; b) EP 0302021, W.R. Cullen, M.D. Cullen, B.R. James, G.-J. Kang, J.P. Kutney, R. Spogliarich and I.S. Thorburn (The University of British Columbia), 25.7.88.

70. a) EP 0301457, B. Pugin, G. Ramos and F. Spindler (Ciba-Geigy AG, Basle), 25.7.88; b) H.-U.-Blaser, B. Pugin and F. Spindler «Enantioselective Hydrogenation of Imines by Means of Iridium Diphosphine Catalysts», Sixth International Symposium on Homogenous Catalysis, 21–26.8.88, Vancouver/Canada.

E. FARNETTI AND M. GRAZIANI

THE DESIGN OF A CHEMOSELECTIVE REDUCTION CATALYST

1. Introduction

The capability of a reduction catalyst to affect only one reducible group in a polyunsaturated organic molecule is of considerable interest in organic synthesis. In particular, reduction of a carbonyl group in the presence of C=C bonds proves to be a difficult task to achieve, and is at the same time a reaction which is often required for the synthesis of natural products employed as pharmaceuticals or pesticides. Whereas some catalysts are known to promote the reduction of α,β-unsaturated aldehydes to the corresponding allylic alcohols [1–5], the selective reduction of a keto group cannot be obtained in α,β-unsaturated ketones, where the olefinic double bond is invariably affected. Only recently has this reaction been reported to be promoted by Cp_2ZrH_2 and Cp_2HfH_2, to give selective carbonyl group reduction in hydrogen transfer reaction using propan-2-ol as source of hydrogen [6].

2. Reduction of α, β-unsaturated ketones

In the last years we have been interested in iridium/phosphine reduction catalysts, and gradually gained enough knowledge of the systems under investigation, as to allow us to design a catalyst possessing the desired selectivity. We will discuss here the properties of a number of homogeneous iridium catalysts with phosphine ligands, which promote the reduction of α, β-unsaturated ketones to the corresponding allylic alcohols. The factors affecting the chemoselectivity will be examined, both in hydrogenation and in hydrogen transfer reactions.

A. F. Noels et al. (Eds.), Metal promoted selectivity in organic synthesis, 191–205.

2.1. HYDROGENATION REACTIONS

2.1.1. Hydrogenation with iridium/monophosphine systems

The hydrogenation of benzylideneacetone which we used as model substrate can be performed in the presence of a catalyst prepared in situ from $[Ir(cod)(OMe)]_2$ and a monodentate phosphine in toluene.

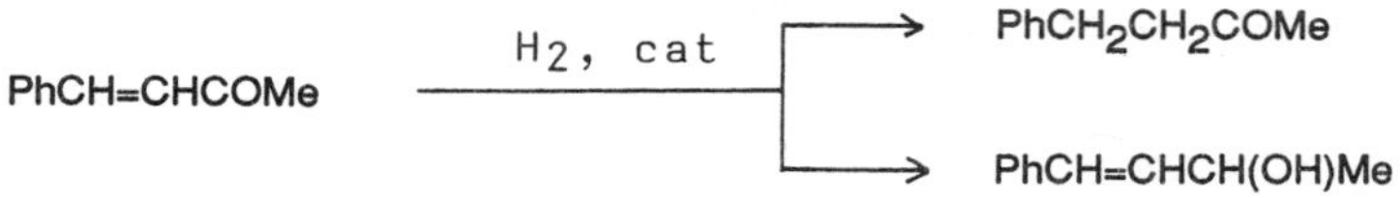

$$PhCH=CHCOMe \xrightarrow{\quad H_2, \ cat \quad} \begin{array}{l} PhCH_2CH_2COMe \\ PhCH=CHCH(OH)Me \end{array}$$

Scheme 1.

The chemoselectivity of the reaction is markedly dependent on the choice of the ligand, and on the amount of phosphine employed [7, 8]. With a doublefold excess of PEt_2Ph (or $PMePh_2$, $PEtPh_2$, PPh_2Pr^i), fast reduction of the C=C bond is observed; the saturated ketone so formed is subsequently reduced to the saturated alcohol, the rate of the latter reaction being one order of magnitude lower than the former. When a higher excess of one of these phosphines is employed (P/Ir > 4) hydrogenation of the C=C bond is suppressed, and only the slower carbonyl group reduction occurs, giving the unsaturated alcohol in yields up to 100%; this product is not further hydrogenated to the saturated alcohol (see Table I). In the presence of bulkier phosphines such as PBu^t_2Ph only C=C bond reduction is observed, whatever the excess of ligand employed (P/Ir=2–80).

When at variance a phosphine with a smaller cone angle than PEt_2Ph is used, the catalytic activity and the selectivity of the system obtained drop dramatically.

TABLE I

Hydrogenation of PhCH=CHCOMe catalyzed by $[Ir(cod)(OMe)]_2$ + P

P	(%)PhCH$_2$CH$_2$COMe	(%)PhCH$_2$CH$_2$CH(OH)Me	(%)PhCH=CHCH(OH)Me
PPh$_2$Pr^i	6	2	85
PEtPh$_2$	2	1	96
PEt$_2$Ph	5	2	90
PMePh$_2$*	3	2	85

[Ir] = 4×10^{-4} M; [P]/[Ir] = 10; [sub]/[cat] = 500; p(H$_2$) = 30 atm;
T = 100°C; solvent = toluene; reaction time 24 hours.
* Reaction time 48 hours.

In Figure 1 is reported the chemoselectivity of the system as a function of the cone angle [9] of the phosphine employed, at P/Ir=10: the volcano-shaped curve so obtained has a maximum in the cone angle range 135–155°. Moreover, within this range, the excess of ligand required to observe selective carbonyl group reduction increases with increasing the value of cone angle.

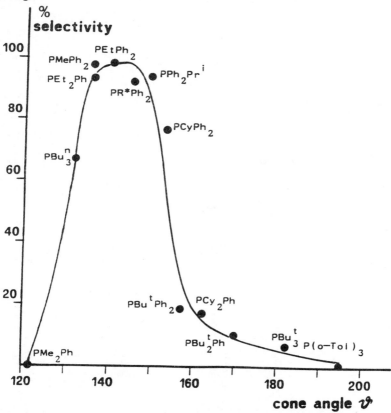

Figure 1. Hydrogenation of PhCH=CHCOMe to PhCH=CHCH(OH)Me catalyzed by [Ir(cod)(OMe)]$_2$ + P (P/Ir = 10): selectivity as a function of the cone angle of the phosphine.

Spectroscopic investigations on the catalytic system were crucial to understand the reasons determining the selectivity. In the experimental conditions employed for the catalytic reactions, the only species detectable with PEt$_2$Ph, at P/Ir=2, is the bisphosphino complex H$_5$Ir(PEt$_2$Ph)$_2$ (see in Figure 2 the ^{1}H NMR spectra at high field of TMS). When a higher excess of this ligand is employed, formation of another hydridic species is observed: the latter, which is the only hydride formed in the presence of added phosphine, is identified as *mer*-$H_3Ir(PEt_2Ph)_3$. A similar behaviour is observed when $PMePh_2$ or $PEtPh_2$ are used: with the former *fac*-H_3IrP_3 is obtained, whereas the latter gives a mixture of *mer* and *fac* isomers. However the selectivity does not appear to be dependent on the isomer

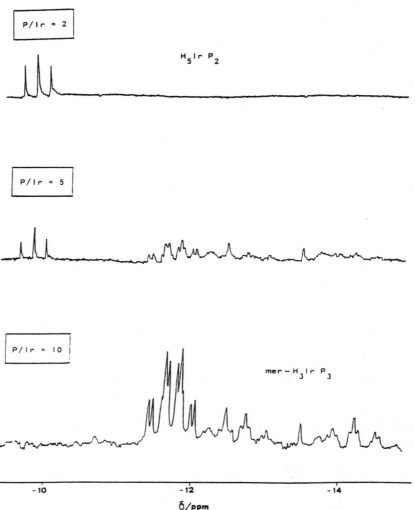

Figure 2. ^1H NMR (high field of TMS, solvent C_6D_6) of the system $[Ir(cod)(OMe)]_2$ + PEt_2Ph, after treatment with H_2.

formed. Therefore, it appears that a bisphosphino complex is responsible for C=C bond reduction, whereas the catalyst which gives carbonyl group reduction is likely to be a species with three coordinated phosphines.

If the same NMR experiment is made using a bulky phosphine (e.g. PBu^t_2Ph), the pentahydrido species is formed even at very large values of P/Ir, showing that a third phosphine cannot coordinate to iridium and H_5IrP_2 is the stable species. This is the reason why phosphines with a large cone angle are not suitable for carbonyl group reduction: coordination of three bulky phosphines to the same iridium atom is very unlikely [10] even in the presence of a large excess of phosphine. As a matter of fact, in the case of phosphines having a cone angle in the range 135-150° an excess of ligand is required to prevent dissociation of a phosphine from H_3IrP_3, forming in the presence of H_2 the species H_5IrP_2.

On the other hand, when phosphines with very small cone angle are used (*e.g.* PMe_2Ph), the prevalent species formed with excess of phosphine is probably a tetraphosphino complex which, having a saturated coordination sphere, will not allow substrate coordination, hence no catalytic activity is expected in these conditions.

Let us now examine more in detail the two catalysts H_5IrP_2 and H_3IrP_3. The former has the two phosphines coordinated in apical position, and therefore the approach of the substrate should not involve unfavorable steric interactions. Coordination of the unsaturated ketone is therefore likely to occur via C=C bond. Such a coordination, which is thermodynamically more favoured in comparison to coordination of C=O group, becomes impossible when the steric situation around the iridium atom is crowded, as is the case of trisphosphino complexes. With such species however the substrate can still approach the iridium centre via the C=O group, which is less sterically demanding if we assume that it coordinates in an end-on rather than in a side-on fashion, the latter being generally unfavourable for ketones, as proved in the comparison between $CpRe(NO)(PPh_3)(PhCOMe)$ and the analogous aldehydo complex [11, 12].

According to these hypotheses, the control of the selectivity in the catalytic systems under investigation stands on steric factors: different steric properties of the catalyst determine the coordination mode of the substrate, and subsequently the chemoselectivity of the reaction.

At variance, the electronic properties of the phosphines employed do not seem to have a great influence on the selectivity of the catalytic system, as it can be seen in the three-dimensional curve shown in Figure 3, where the selectivity is reported as a function of both the cone angle and the electronic parameter ν of the phosphine [9].

We have so far made the assumption that the function which coordinates is necessarily the same one which is reduced. Although this seems to be reasonable, some examples are known in the literature where coordination of a certain group does not guarantee its reduction. In particular, it has been found that coordination of unsaturated aldehydes to the species $[Ir(L)(CO)P_2]^+$ (L = unsaturated nitrile) occurs via carbonyl group, but it is the C=C bond which is hydrogenated [13].

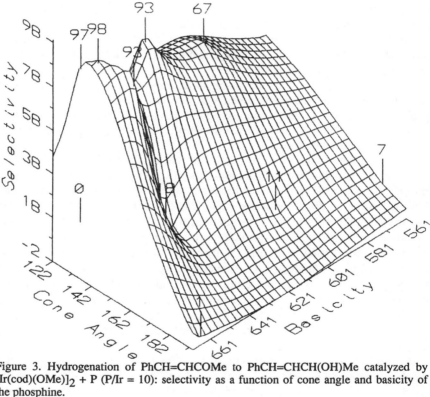

Figure 3. Hydrogenation of PhCH=CHCOMe to PhCH=CHCH(OH)Me catalyzed by $[Ir(cod)(OMe)]_2$ + P (P/Ir = 10): selectivity as a function of cone angle and basicity of the phosphine.

2.1.2. Hydrogenation with iridium/diphosphine systems

An interesting modification of the catalytic systems described above could be obtained by employing chiral phosphines, as in the presence of such ligands one step chemo and enantioselective hydrogenation of unsaturated ketones is in principle feasible. The catalytic system $[Ir(cod)(OMe)]_2$ + S-(–)-$P(CH_2CH(Me)Et)Ph_2$ proves to be highly chemoselective in carbonyl group hydrogenation (96% yield of unsaturated alcohol) but gives only poor optical yields (< 10%). This result is not surprising, as catalysts employing chiral monodentate ligands, owing to their lack of rigidity, often prove to be poorly enantioselective. From that point of view, chiral bidentate phosphines appear to be more promising, therefore a series of iridium complexes employing such ligands were tested as catalysts for the hydrogenation of benzylideneacetone.

TABLE II

Hydrogenation of PhCH=CHCOMe catalyzed by $[Ir(P–P)_2]+$

P–P	(%) conversion (hours)	(%) PhCH=CHCH(OH)Me	(%) e.e.
chiraphos	98(24)	0	–
prophos	91(48)	2	–
diphos-2	100(24)	0	–
diphos-3	81(48)	1	–
diphos-4	40(120)	35	–
diop	96(120)	89	18(S)–(–)
prolophos	93(24)	85	31(S)–(–)

$[Ir] = 4 \times 10^{-4}$ M; [sub]/[cat] = 500; $p(H_2)$ = 30 atm; T = 100°C; solvent = toluene.

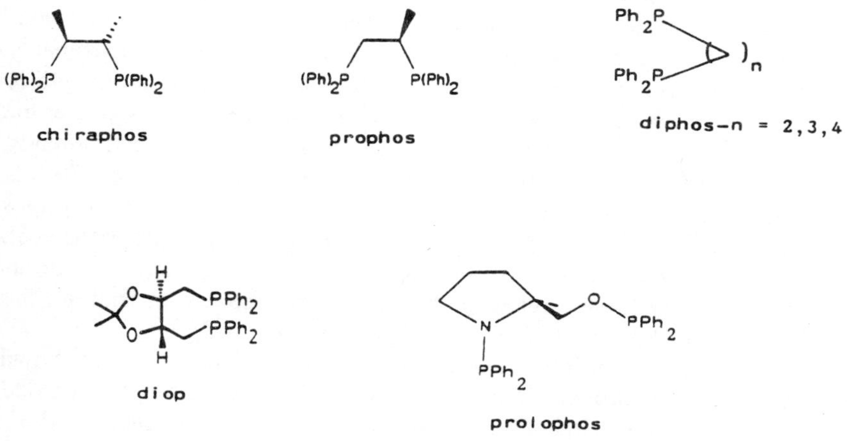

chiraphos

prophos

diphos-n = 2,3,4

diop

prolophos

Figure 4.

When employing the cationic complex $[Ir(P-P)_2]^+$ as the catalyst precursor, use of phosphines which give a five or six-membered ring on chelation leads to C=C bond hydrogenation. At variance, when ligands with a larger bite such as diop are chosen, the system displays a good chemoselectivity in carbonyl group reduction (see Table II and Figure 4). Unfortunately, values of enantiomeric excess are never satisfactory, reaching a maximum of 31% in the case of (S)-(−)-prolophos. Moreover, all the systems employing bidentate phosphines are generally less active than the previously discussed Ir/monophosphine catalysts.

A reasonable explanation of these results could be the following. The starting cationic complex reacts with molecular hydrogen to give [14] the Ir(III) species $[H_2Ir(P-P)_2]^+$, which can behave as a catalyst only by making a coordination site free, in order to allow substrate coordination: this can happen in two ways, as shown in Scheme 2.

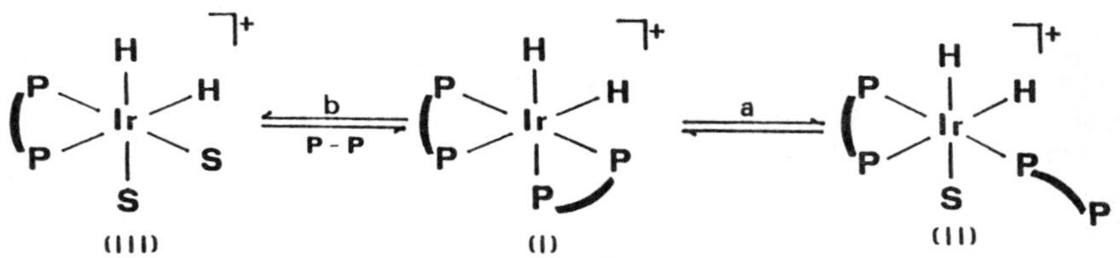

Scheme 2.

Partial dissociation of one phosphine leads to a species which possesses three coordinated phosphorous atoms, and it can be expected to behave as $H_3Ir(PEt_2Ph)_3$, and to catalyze carbonyl group reduction. With phosphines with a large bite, species (II) so obtained should be formed in reasonably high concentration, however any attempt to identify them in the NMR spectrum of the mixture failed. Such a behaviour for these ligands, due to the flexibility of the seven membered chelate ring [15], has been already proposed for the reaction of $[Rh(diop)_2]^+$ with molecular hydrogen. In this case a mechanism whose kinetics suggest the involvement of intermediates with a diphosphine bound in a monodentate fashion has been proposed [16]. At variance, ligands such as chiraphos after dissociation of one phosphorous atom will probably leave altogether the coordination sphere of iridium: the complex so formed promotes C=C bond hydrogenation [17].

In order to have direct information on the presence of the equilibria shown in Scheme 2, the hydrogenation reactions of PhCH=CHCOMe was carried out using as catalyst $[Ir(P-P)_2]^+$ in the presence of an excess of P-P. Depending on the diphosphine used, two different behaviours were found: (i) when P-P is a ligand forming a five membered ring, for example chiraphos, an excess of P–P kills completely the catalytic activity, showing that equilibrium (b) of Scheme 2 is operative in this case; however (ii) with diphosphines such as diop, the excess of ligand added does not affect the activity or the selectivity of the catalyst showing that the equilibrium (a) is very likely to be present in the system.

2.1.3. Hydrogenation with iridium / P–N systems

Hybrid P–N ligands could in principle offer some advantages in comparison to bidentate phosphines. These ligands should possess a good chelating ability owing to the presence of the amino group, which at the same time is

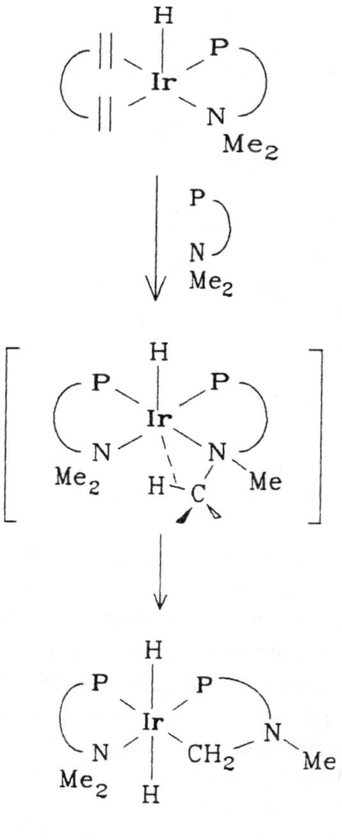

Scheme 3.

known [18, 19] to give ready dissociation in the presence of π-acceptor molecules. Catalytic systems employing such phosphines should combine the advantages of a satisfactory steric control of the coordination mode of the substrate together with a good catalytic activity. For this purpose we have employed the aminophosphines P–NH$_2$ (= P(o–C$_6$H$_4$NH$_2$)Ph$_2$) [20] and P–NMe$_2$ (= P(o–C$_6$H$_4$NMe$_2$)Ph$_2$) [21].

Hydrogenation of our model substrate benzylideneacetone in the presence of the system [Ir(cod)(OMe)]$_2$ + P–N (P/Ir=5) actually occurs at a rate which is at least one order of magnitude higher than the systems employing diphosphines. Moreover, a different chemoselectivity is obtained changing from P–NH$_2$ to P–NMe$_2$, the yield in unsaturated alcohol being near 90% with the former, and only 50% with the latter. These results have been found to be related to the different species which are formed in the catalytic conditions with these two ligands. When P–NMe$_2$ is employed, reaction of [Ir(cod)(OMe)]$_2$ with the ligand initially gives Ir(OMe)(cod)(P–NMe$_2$), which subsequently undergoes β-hydrogen elimination to form HIr(cod)(P–NMe$_2$) [22]. Such hydridic species in the presence of excess P–NMe$_2$ further reacts according to the reaction path shown in Scheme 3, to give H$_2$Ir(P–NCH$_2$Me) (P–NMe$_2$) as final product in the catalytic conditions [23]. A completely different behaviour is observed using P–NH$_2$, as in this case formation of the trisphosphino complex HIr(P–NH)$_2$(P–NH$_2$) takes place [24]. Once again, a high chemoselectivity in carbonyl group reduction appears to be related to the formation of a species with three coordinated phosphines, whereas when only two phosphines are coordinated the selectivity of the system is poor.

2.2. HYDROGEN TRANSFER REACTIONS:
CATALYSIS BY IRIDIUM / P–N SYSTEMS

From the results we have so far discussed it appears that, with the hydrogenation catalysts we have employed, steric factors have a great importance in determining the selectivity, whereas variations in the electronic properties of the ligands do not seem to have a marked influence. This observation might seem to be surprising, as carbonyl group reduction is generally thought to be affected by electronic factors, and particularly to be promoted by hydrides which possess a strongly hydridic character [6]. The anionic phosphine-hydrido complex [RuH$_2$(PPh$_3$)$_2$(PPh$_2$C$_6$H$_4$)]$^-$ was used as catalyst for hydrogenation of organic compounds possessing polar reducible groups. The strongly hydridic character of the hydride obtained by the combination of the anionic character and the presence of relatively poor π-acceptor phosphine ligands was expected to promote nucleophilic addition of the hydride to the positive carbon atom of the keto group [25, 26]. $[RuH_2(PPh_3)_2(PPh_2C_6H_4)]^-$ in hydrogen atmosphere is rapidly converted to $[RuH_3(PPh_3)_3]^-$ [27], but this species was not found to be an effective catalyst for hydrogenation of cyclohexanone. During the catalytic reaction the complex is slowly converted to the more active species $RuH_4(PPh_3)_3$ [28], whose high catalytic activity may be related to the formulation as non-classical hydride $Ru(H_2)H_2(PPh_3)_3$ [29] and to dihydrogen dissociation to form the coordinatively unsaturated species. In this case then the extent of negative charge on the metal does not appear to play an important role in polar group hydrogenation. James argued that, in alternative to hydride attack on the carbonyl carbon, the reduction of carbonylic compounds might proceed via electrophilic attack of H^+ to the oxygen, and proposed this to be the case for the hydrogen transfer reduction of unsaturated aldehydes catalyzed by $HIrCl_2(Me_2SO)_3$; however, this system was not effective for the reduction of unsaturated ketones [4].

The importance of electronic factors was suggested in the reduction of unsaturated aldehydes in hydrogen transfer from propan-2-ol, catalyzed by a system prepared in situ from $[Ir(cod)Cl]_2$ and an excess of phosphine. In a large variety of phosphines tested, only those which possess an ortho anisyl group give rise to a catalyst which reduces the aldehyde function to the unsaturated alcohol with a yield close to 100%. All the other ligands give poorly active and selective catalysts. These results were interpreted suggesting a chelation by the o-methoxy group, which enhances the electronic density on the iridium and subsequently on the coordinated hydride [3]. Such catalytic system however is not selective in the reduction of unsaturated ketones; if our hypothesis on an electronic control of selectivity was correct,

TABLE III

Hydrogen transfer reduction of PhCH=CHOMe catalyzed by
$[Ir(cod)(OMe)]_2$ + P–X

P–X	[KOH]/[Ir]	(%) conversion (minutes)	(%) PhCH=CHCH(OH)Me
P–O	2	24(60)	13
P–NH$_2$	–	95(30)	88
P–NMe$_2$	–	95(60)	90

$[Ir] = 4 \times 10^{-4}$; $[P]/[Ir] = 2$; $[sub]/[cat] = 500$; $T = 83°C$; solvent = propan-2-ol.
P–O = $P(o-C_6H_4OCH_3)_3$
P–NH$_2$ = $P(o-C_6H_4NH_2)PPh_2$
P–NMe$_2$ = $P(o-C_6H_4NMe_2)PPh_2$

we could expect an improvement in the selectivity by employing, in the place of P–O ligands, P–N ones, which surely possess a better donating ability.

As can be seen in Table III, a marked increase both in the activity and in the selectivity of the system [Ir(cod)(OMe)]$_2$ + P–X is observed using the ligands P–NH$_2$ and P–NMe$_2$. It is moreover noticeable that when P–N ligands are employed there is no need of any basic cocatalyst, which is an unusual feature in hydrogen transfer reduction from propan-2-ol.

Spectroscopic investigations revealed that the system [Ir(cod)(OMe)]$_2$ + P–NMe$_2$ in the catalytic conditions gives rise to H$_2$Ir(P–NCH$_2$Me) (P–NMe$_2$), which is the same species formed in the hydrogenation reaction [30]. Such complex is formed via intramolecular C-H oxidative addition, and this suggests that in the Ir(I) precursor there is a high electron density on the metal, which promotes oxidative addition of the N-methyl C-H bond. A hydride ligand coordinated to such an electron-rich iridium centre certainly has a highly nucleophilic character, and this would explain its selective attack to the carbonyl group of the substrate.

This metallated compound undergoes partial chlorination by CH$_2$Cl$_2$, to give the compound whose structure is shown in Figure 5 [23]. The dihydride also reacts with CO [30] giving as first step of the reaction an internal reductive elimination to produce HIr(CO)(P–NMe$_2$)$_2$ which has one monodentate and one bidentate P–N ligand.

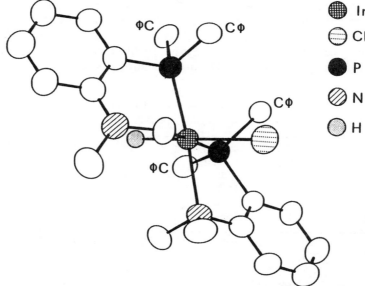

Figure 5. ORTEP drawing of HIrCl (P–NCH$_2$Me)(P-NMe$_2$).

On these basis the reaction pathway shown in Scheme 4 can be proposed. The main feature of this scheme is the «arm off» mechanism due to the presence of uncoordinated amino group which appears to play a rather important role in determining the selectivity of the system. The first step of the reaction is the intramolecular reductive elimination by reaction with the unsaturated substrate to be reduced. At this point an intramolecular nucleophilic attack to the iridium atom by the amino group of the amino-phosphine, promotes the selective migration of the hydride on the carbonyl group to be reduced. The following steps are usual reactions in organometal-lic chemistry and they do not need any further comment.

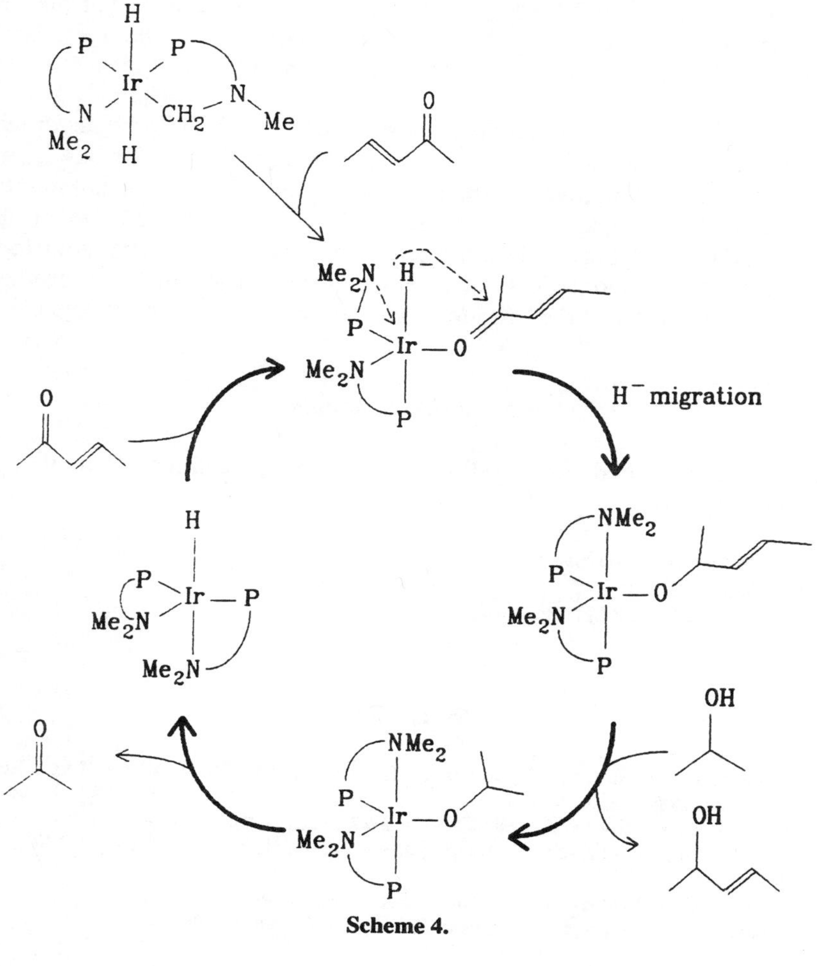

Scheme 4.

3. Conclusions

Some general considerations can be drawn from our results.

The mechanisms operative in the hydrogenation or in hydrogen transfer reduction of α,β unsaturated ketones are very different, and the selectivity appears to be promoted in the two cases by different parameters.

In the hydrogenation reaction the selectivity is sterically controlled and the highly favoured coordination of the C=C bond must be prevented by use of suitable ligands. The basicity of the coordinated ligands, which is problably rather important in determining the catalytic activity of these systems influencing directly the hydrogen addition to the metal, does not play an important role in determining the selectivity: ligands having the same steric requirements, but different basicity give the same selectivity in unsaturated alcohol.

A different situation appears to be operative in the reduction by hydrogen transfer. In this case the steric requirements of the coordinated ligands do not appear to play an important role in determining the selectivity in unsaturated alcohol of the systems we investigated. More important seems to be the basicity of the coordinated ligand. The use of mixed ligands having nitrogen atom as possible donor will create an electron rich metal which will promote the selective transfer of the hydride on the carbonyl group to be reduced.

Acknowledgements

Authors thank C.N.R.-Roma «Progetto Finalizzato Chimica Fine II» for financial support.

Dipartimento di Scienze Chimiche,
Università di Trieste
Via A. Valerio 22, 34127 Trieste (Italy)

References

1. Kaneda, K.; Yasumura, M.; Imanaka, T. and Teranishi, S. *J. Chem. Soc. Chem. Commun.* **1982**, 935.
2. Farnetti, E.; Vinzi, F. and Mestroni, G. *J. Mol. Catal.* **1984**, *24*, 147.
3. Visintin, M.; Spogliarich, R.; Kaspar, J. and Graziani, M. *J. Mol. Catal.* **1985**, *32*, 349.
4. James, B.R. and Morris, R.H. *J. Chem. Soc. Chem. Commun.* **1978**, 929.
5. Bhaduri, S. and Sharma, K. *J. Chem. Soc. Chem. Commun.* **1988**, 173.

6. Nakano, T.; Umano, S.; Kino, Y.; Ishii, Y. and Ogawa, M. *J. Org. Chem.* **1988**, 3752.
7. Farnetti, E.; Pesce, M.; Kaspar, J.; Spogliarich, R. and Graziani, M. *J. Chem. Soc. Chem. Commun.* **1986**, 746.
8. Farnetti, E.; Kaspar, J.; Spogliarich, R. and Graziani, M. *J. Chem. Soc. Dalton Trans.* **1988**, 947.
9. Tolman, C.A. *Chem. Rev.* **1977**, *77* (3), 313.
10. Shaw, B.L. *J. Organometal. Chem.* **1980**, *200*, 307.
11. Fernandez, J.M.; Emerson, K.; Larsen, R.D. and Gladysz, J.A. *J. Chem. Soc. Chem. Commun.* **1988**, 37.
12. Fernandez, J.M.; Emerson, K.; Larsen, R.D. and Gladysz, J.A. *J. Am. Chem. Soc.* **1986**, *108*, 8268.
13. Yang, K.J. and Chin, C.S. *Inorg. Chem.* **1987**, *26*, 2732.
14. Spogliarich, R.; Farnetti, E.; Kaspar, J.; Graziani, M. and Cesarotti, E. *J. Mol. Catal.* **1989**, *50*, 19.
15. Chaloner, P.A. *J. Organometal. Chem.* **1984**, *266*, 191.
16. James, B.R. and Mahajan, D. *J. Organometal. Chem.* **1985**, *279*, 31.
17. Shapley, J.R.; Schrock, R.R. and Osborn, J.A. *J. Am. Chem. Soc.* **1969**, *91*, 2816.
18. Rauchfuss, T.B. and Roundhill, D.M. *J. Am. Chem. Soc.* **1974**, *96*, 3098.
19. Knebel, W.J. and Angelici, R.J. *Inorg. Chem.* **1974**, *13*, 627.
20. Cooper, M.K. and Downes, J.M. *Inorg. Chem.* **1978**, *17*, 880.
21. Fritz, H.P.; Gordon, I.R.; Schwarzhans, K.E. and Venanzi, L.M. *J. Chem. Soc.* **1965**, 5210.
22. Fernandez, M.J.; Esteruelas, M.A.; Covarrubias, M. and Oro, L. A. *J. Organometal. Chem.* **1986**, *316*, 343.
23. Farnetti, E.; Nardin, G. and Graziani, M. *J. Chem. Soc. Chem. Commun.* **1989**, 1264.
24. Farnetti, E.; Nardin, G. and Graziani, M. *J. Organmetal. Chem.*, **1991**, 000.
25. Pez, G.P. and Grey, R.A. *Fundamental Research in Homogeneous Catalysis*, vol.4; M. Graziani and M. Giongo (Eds.), Plenum Press, 1983.
26. Grey, R.A.; Pez, G.P. and Wallo, A. *J. Am. Chem. Soc.* **1981**, *103*, 7536.
27. Wilczynski, R.; Fordyce, W.A. and Halpern, J. *J. Am. Chem. Soc.* **1983**, 2066.
28. Linn, D.E. and Halpern, J. *J. Am. Chem. Soc.* **1987**, *109*, 2969.
29. Crabtree, R.H. and Hamilton, D.G. *J. Am. Chem. Soc.* **1986**, *108*, 3124.
30. Farnetti, E.; Kaspar, J. and Graziani, M. *J. Mol. Catal.* **1990**, *63*, 5.

FRANCO FRANCALANCI[a]*, WALTER CABRI[a],
SILVIA DE BERNARDINIS[a], SERGIO PENCO[a] AND ROBERTO SANTI[b]

PALLADIUM CATALYZED REDUCTION OF ARYL SULFONATES. SELECTIVITY CONTROL AND APPLICATION TO ANTHRACYCLINE CHEMISTRY

The clinically useful antitumor anthracyclines daunorubicin and doxorubicin represent the results of an original research line of Farmitalia-Carlo Erba begun about thirty years ago as screening of antitumor activity of metabolites isolated from the cultures of several strains of *Streptomyces* [1]. As all other anthracyclines these molecules consist of an aglycone moiety and an aminosugar joined through a glycosidic bond at C-7; the two structures differ only at C-14 where daunorubicin bears a methyl group and doxorubicin an hydroxymethyl moiety (Figure 1).

In spite of this structural similarity the two molecules have a quite different behaviour from the biological point of view: doxorubicin displays a much more favourable therapeutic index than daunorubicin in different experimental tumors in laboratory animals and a broader spectrum of activity on human tumors. The drug is however not devoid of toxic side effects, mainly cardiac toxicity, which limit dosages and therefore its effectiveness.

The need for new anthracyclines endowed with more favourable pharmacologic properties has stimulated an intense research activity at industrial and academic laboratories concerning both the modification of the metabolic pathways of productive strains and the search for practical, totally synthetic methods.The result of this effort in Farmitalia-Carlo Erba was the discovery of at least two products very attractive from the therapeutical point of view: epirubicin and idarubicin (Figure 2).

In this case too minor structural changes either in the sugar or in the aglycone resulted in dramatic modification of activity. We will focus on idarubicin because, due to the lack of fermentative processes the preparation of its aglycone represented an important synthetic problem.

Several total syntheses have been described [2], but none of them is particulary straightforward and completely satisfactory because of the numerous steps, low productivity and the need of optical resolution. A representative reaction scheme is reported in Figure 3 [3]. An easier approach was therefore highly desirable.

A. F. Noels et al. (Eds.), Metal promoted selectivity in organic synthesis, 207–222.
© 1991 *Kluwer Academic Publishers. Printed in the Netherlands.*

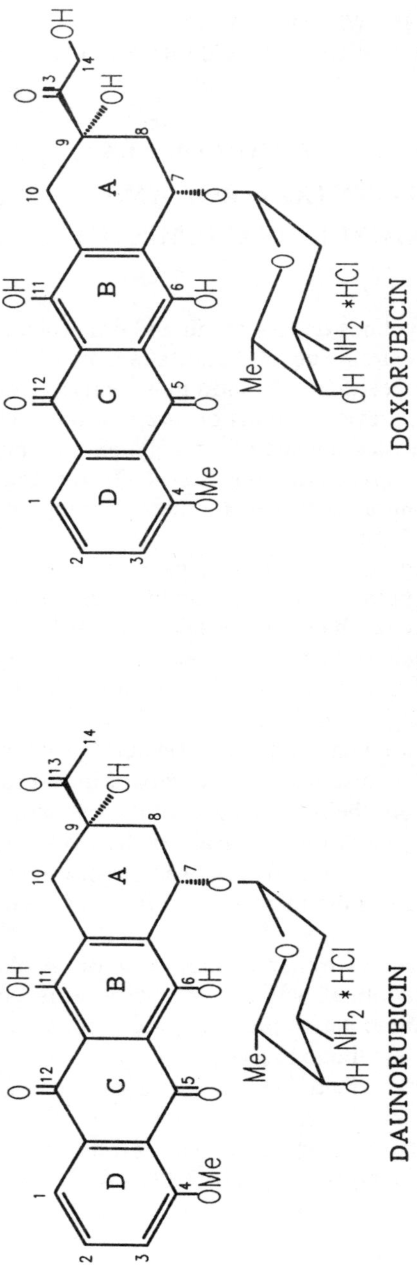

DOXORUBICIN

Figure 1.

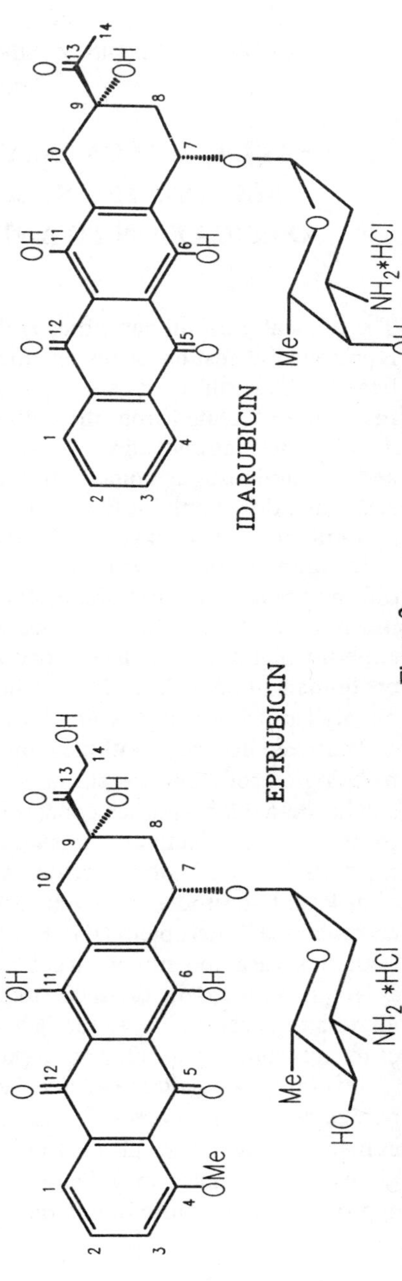

DAUNORUBICIN

IDARUBICIN

Figure 2.

EPIRUBICIN

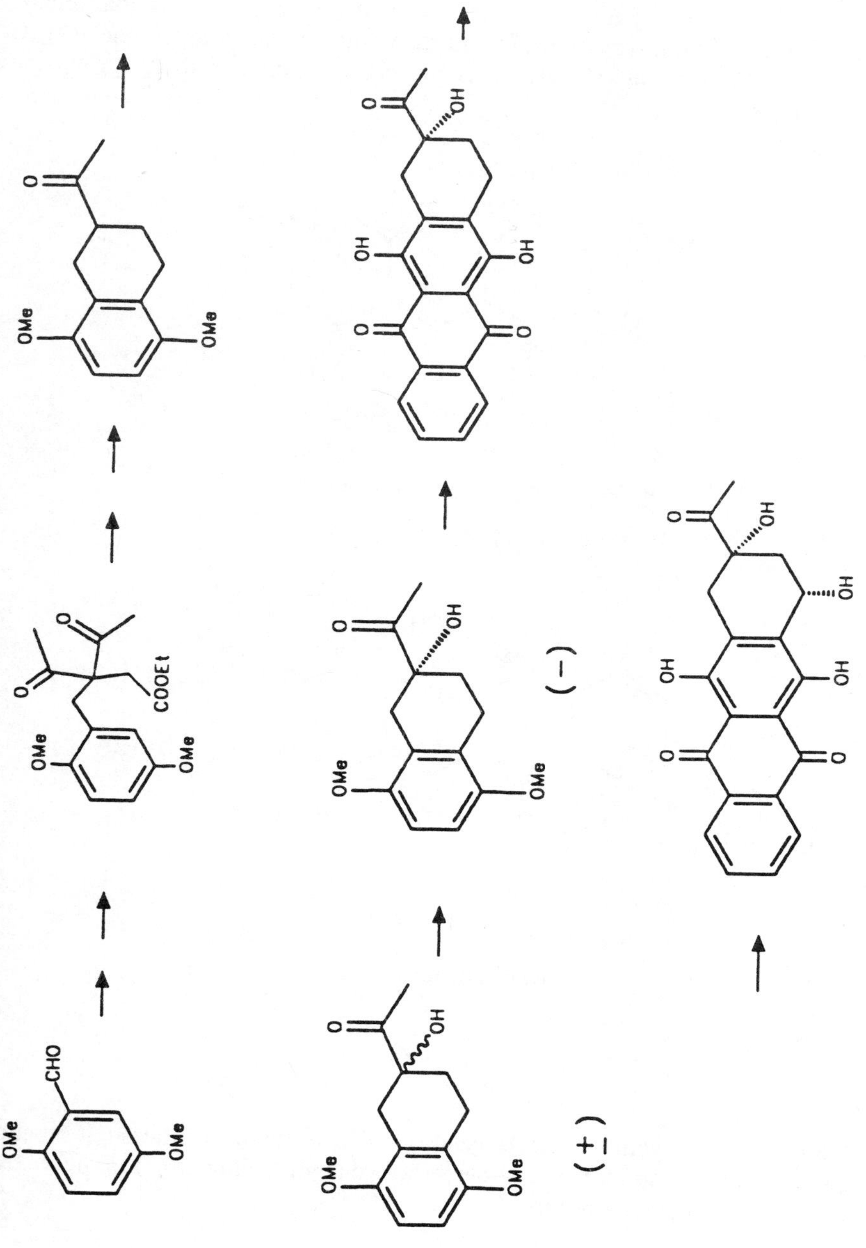

Figure 3.

If we compare (Figure 4) the structures of the aglycones of idarubicin (4-demethoxydaunomycinone) (*1*) and daunomycin (daunomycinone) (*2*) it is evident that the only difference is the presence of a methoxy group in position C-4.

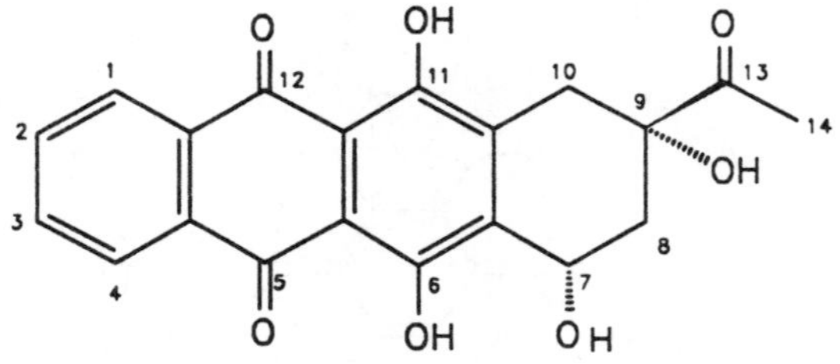

<u>1</u>

4-Demethoxy-daunomycinone

aglycone of idarubicin

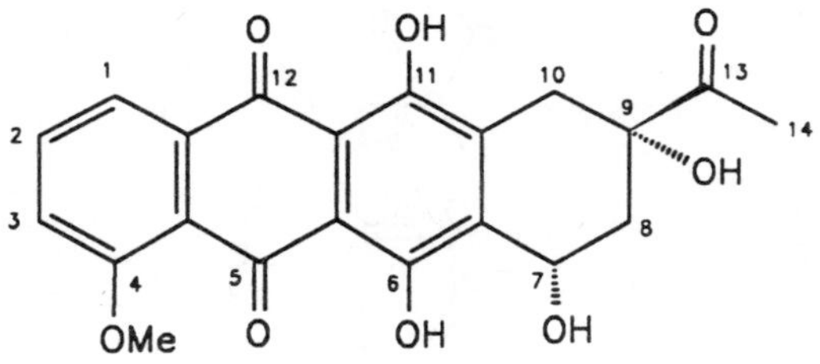

<u>2</u>

Daunomycinone

aglycone of daunomycin

Figure 4.

Since daunomycinone (*2*) is easily available by fermentation it would appear obvious to achieve 4-demethoxydaunomycinone (*1*) simply by removing the methoxy group.

It occurred to us that organometallic chemistry could be very helpful in doing such a reaction. A literature survey revealed that a limited number of papers dealed with phenolic OH selective exchange for hydrogen [4]. Common feature to all of these approaches is the use of phenol derivatives like aryl perfluoroalkanesulfonates in conjunction with either organic (formate salts) or inorganic (e.g. Bu_3SnH, $NaBH_4$) reducing agents in the presence of palladium catalysts.

The use of perfluoroalkanesulfonates could often be a major problem [5] in terms of synthesis, stability and cost, in particular from the industrial point of view. On the other hand much more accessible sulfonates such as mesylates, tosylates etc. are unreactive under mild reaction conditions when formate salts are used in the presence of triphenylphosphine based palladium catalysts [6].

In fact, with these catalysts, 1-naphtyl sulfonates *3(a–d)* and *3f* provided only trace amounts of naphtalene (Figure 5) after 48 hours at 90°C; the concomitant decomposition of the catalyst suggests that the first step in the catalytic cycle, the oxidative addition to palladium, is too slow.

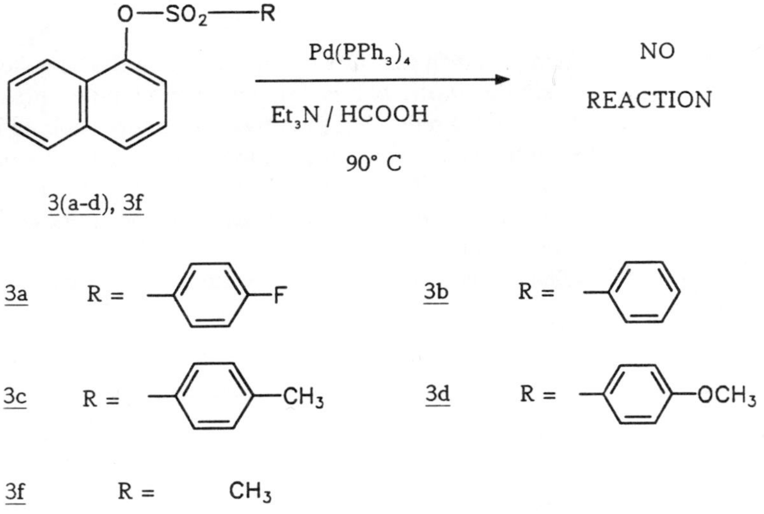

Figure 5.

A possible approach to make such addition easier is to modify the catalyst, for example by using different ligands on palladium. The substitution of triphenylphosphine with the more electron donating methyldiphenyl-phosphine and dimethylphenylphosphine ligands [7] led only to complete decomposition of formate [8].

In searching for a more efficient system we used a 1,3-bis(diphenyl-phosphino) propane (DPPP) containing catalyst that was reported by Dolle [9] to be very active in the alkoxycarbonylation of naphtyl triflate (*3e*) (Figure 6).

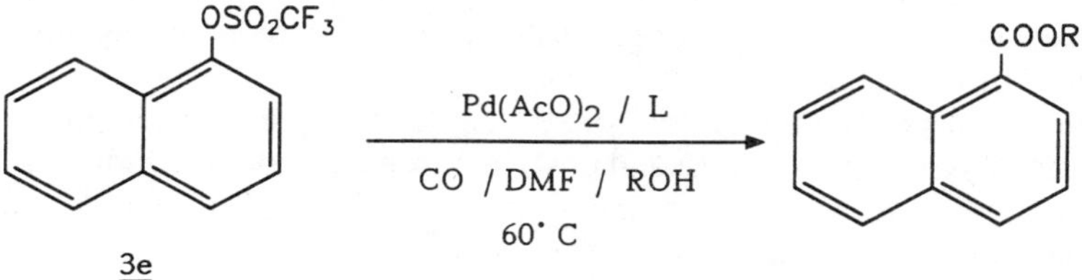

L	Rel.rate
PPh$_3$	1
DPPP	500

Figure 6.

Although the reaction was different it is conceivable that the first step, the oxidative addition, is similar in the two cases. The new complex proved to be a better catalyst for triflates reduction than the triphenylphosphine based one (Table I); its use actually allowed us to extend the reaction to several easily accessible sulfonates, simply by increasing the temperature up to 90°C in DMF [8], (Table II). The nature of substituents on the arenesulfonate moiety affects the reduction rate, the order of reactivity being p–F > H >> p-Me > p–OMe; mesylate turned out to be as reactive as p–OMe benzene sulfonate.

TABLE I

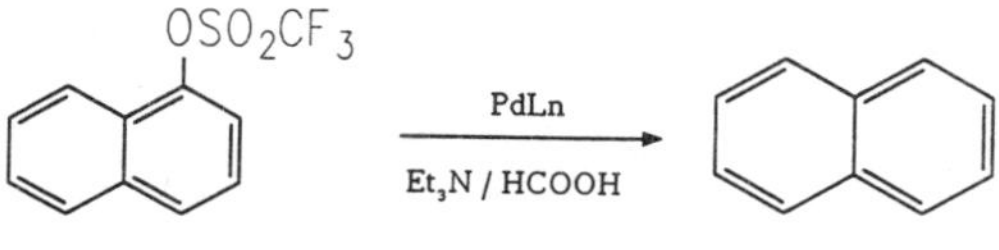

L	(n)	T(°C)	t(h)	Yield (%)
PPh$_3$	4	40	5	90
PPh$_3$	4	18	24	32
DPPP	1	18	0.8	92

TABLE II

Entry	Substrate	t(h)	Yield (%)
1	3a	1.5	95
2	3b	2.5	90
3	3c	24	93
4	3d	45	90
5	3f	48	85
6	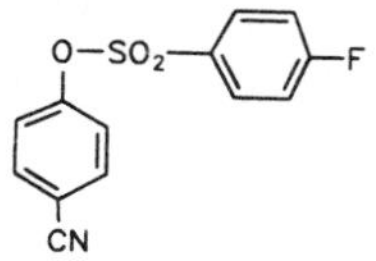	5	88
7	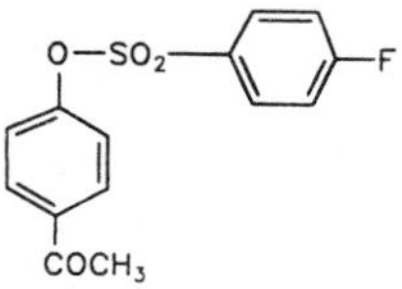	3.5	93

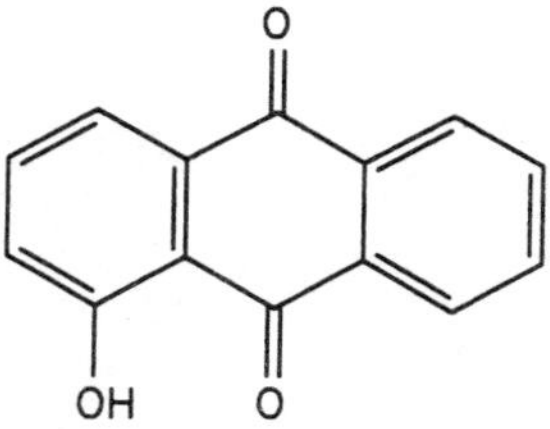

1-Hydroxy-9,10-Anthraquinone

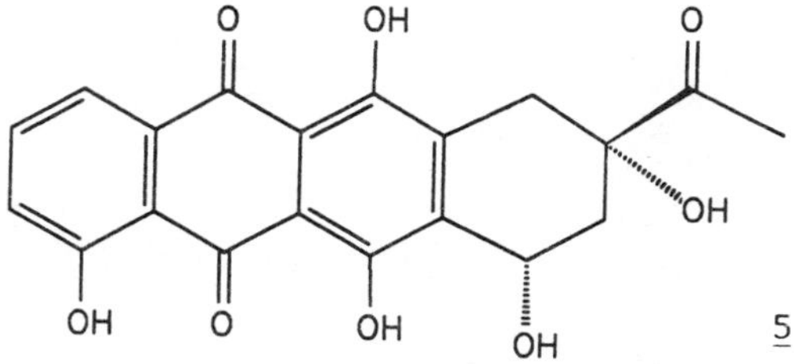

4-Demethyldaunomycinone

(Carminomycinone)

Figure 7.

Substrates bearing potentially reducible groups such as nitrile and acetyl on the phenyl ring, were converted to benzonitrile and acetophenone in almost quantitative yields (Table II, Entries 6–7). These results show the chemoselectivity of the method and the synthetic usefulness of arene sulfonates.

Encouraged by these results we extended the experiments to a more complex substrate like 1-hydroxy-9,10-anthraquinone (*4*); we chose this molecule because it is a simple, but reliable model of 4-demethyl-daunomycinone (carminomycinone) (*5*) (see Figure 7 for comparison of the two structures).

TABLE III

$$\text{Ar-OSO}_2\text{CF}_3 \xrightarrow[\text{HCOOH / DMF}]{\text{Pd(OAc)}_2 - \text{PPh}_3} \text{ArH}$$

Ar	T(°C)	t(h)	Yield (%)
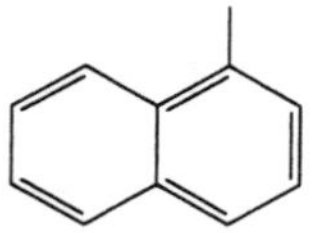 3e	40	5	90
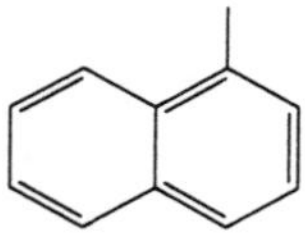 3e	18	24	32
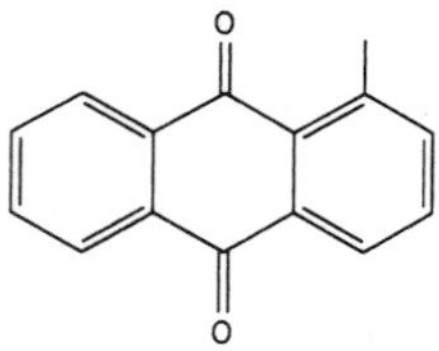 6a	40	1.8	98

The sulfonates of *4* should be, in principle, more activated than simple phenols or naphtols toward oxidative addition because of the presence of the quinone system. In fact triflate *6a* with PPh$_3$ as the ligand was reduced faster than 1-naphtyl triflate (*3e*) (Table III).

On the other hand the reaction failed when 4-fluorobenzensulfonate (*6b*), the most reactive sulfonate in the naphthyl series *3(a–d)*, (Table II, Entry 1) was used as substrate. The reaction carried out at 90°C with the same catalyst afforded only a small amount of 9,10-anthraquinone (*7*); 1-hydroxy-9,10-anthraquinone (*4*), the hydrolysis product, was the major component (Table IV).

TABLE IV

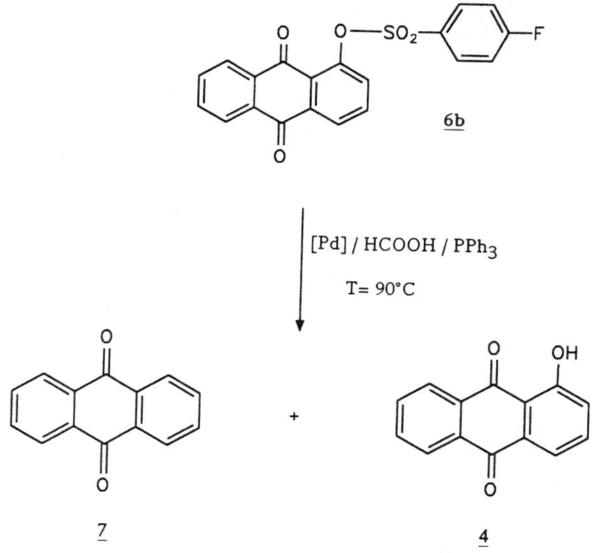

Solvent	t(h)	Conversion (%)	7/4
DMF	20	100	8/92
DIOXANE	52	100	35/65
TOLUENE	48	72	45/55

A solvent effect on the reduction/hydrolysis ratio was observed, with the amount of reduction product increasing in the order DMF > dioxane > toluene. Moreover, in toluene the conversion was incomplete because of decomposition of the catalytic system. The hydrolysis product (4) was not observed following heating of 6b in DMF in the absence of any metal catalyst or with acetate in place of formate, where palladium is present only as Pd^{II} species. These blank experiments suggest that cleavage of the S–O bond is assisted in some way by the presence of the Pd /formate system.

It seems therefore that two competitive Pd-catalyzed reactions are present (Figure 8): the first proceeding through the normal cleavage of the carbon-oxygen bond (path a) leading to the oxidative addition and eventually to 7 and the second one (path b) proceeding through a somehow Pd-catalyzed cleavage of the oxygen-sulfur bond producing 1-hydroxyanthraquinone (4).

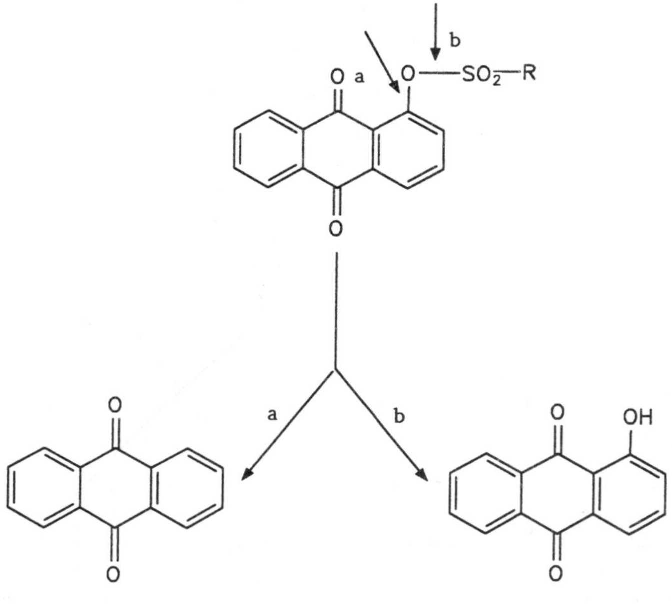

7   Figure 8.   4

To prevent the cleavage of the S–O bond we studied in some detail the influence of palladium ligands on the reaction course, in both dioxane and DMF, using 6b as substrate. The results obtained with monodentate phosphines (Table V) show that the selectivity of the reaction is greatly affected by the ligand.

TABLE V

Reduction of *6b* with monodentate phosphine ligands

Entry	Ligand	(L/Pd)	Cone angle θ, deg	DMF		Dioxane	
				7/4	t(h)	7/4	t(h)
1	P(p–Cl-Ph)$_3$	(2)	145	0/100	1	2/98	24
2	P(p–Cl-Ph)$_3$	(3)	145	0/100	4	8/92	44
3	PPh$_3$	(2)	145	1/99	1.5	8/92	23
4	PPh$_3$	(3)	145	3/97	9	21/79	35
5	PPh$_3$	(4)	145	9/91	20	35/65	52
6	P(p–Tolyl)$_3$	(2)	145	18/82	1.5	80/20	4.5
7	P(p–Tolyl)$_3$	(3)	145	75/25	3.5	97/3	5
8	P(o–Tolyl)$_3$	(2)	194	1/99	1	5/95	4
9	P(o–Tolyl)$_3$	(3)	194	4/96	1.5	10/90	9
10	PMePh$_2$	(2)	136	100/0	0.5	100/0	0.5
11	PMe$_2$Ph	(2)	122	100/0	0.5	100/0	0.5

In particular, the use of phosphines having the same cone angle (Table V, Entries 1–7) allows one to separate the electronic effect from the steric one; the amount of reduction product 7 increases with donor properties of the ligand and the best results were obtained with P(p-tolyl)$_3$ (Entry 7).

Tri-o-tolylphosphine was less effective than the para isomer in spite of a comparable basicity, probably because of the larger cone angle of the former (compare entries 7 and 9). These results suggest that a high basicity of the ligand and a cone angle around 145 or less should be the right combination to prevent the formation of 1-hydroxyanthraquinone (4). This hypothesis is compatible with the results of entries 10 and 11, where PCH$_3$Ph$_2$ and P(CH$_3$)$_2$Ph respectively gave anthraquinone (7) exclusively. In table VI are reported the results obtained with chelating biphosphines.

For the series 1,1-bis(diphenylphosphino)methane (DPPM), 1,2-bis(diphenyl phosphino)ethane (DPPE), 1,3-bis(diphenylphosphino)propane (DPPP) and 1,4-bis(diphenylphosphino)butane (DPPB), which have similar electronic character, the amount of 9,10-anthraquinone (7) increases with the length of the chain that links the phosphorous atoms (entries 1–4). The same selectivity observed with DPPP and DPPB was achieved with the bulkier 1,1′-bis(diphenyl phosphino)ferrocene (DPPF) (entry 5). The presence of electron-releasing groups on phenyl rings, such as in 1,2-bis(di-p-tolylphosphino)ethane (DpTPE) (entry 6) does not improve selectivity; thus, electronic effects seem to play a minor role in comparison to steric ones. On the basis of these results we have extended the reaction to sulfonates 6(c–f)

(Table VII). By working in dioxane and using DPPP as the ligand for palladium, it was possible to obtain the reduction product with complete selectivity in all cases.

This was a satisfactory result because we succedeed to use non perfluorinated sulfonates as substrates for palladium catalysis; however the main target of our research was to use such chemistry for the selective

TABLE VI

Reduction of *6b* with bidentate phosphine ligands

Entry	Ligand	P–Pd–P angle, deg	Solvent			
			DMF		Dioxane	
			7/4	t(h)	7/4	t(h)
1	DPPM	73	0/100	0.5	0/100	5
2	DPPE	85	68/32	0.5	95/5	4
3	DPPP	90	98/2	0.35	100/0	0.5
4	DPPB	90	96/4	0.35	100/0	1
5	DPPF	99	95/5	0.4	100/0	0.4
6	DpTPE	–	55/45	1	85/15	1

TABLE VII

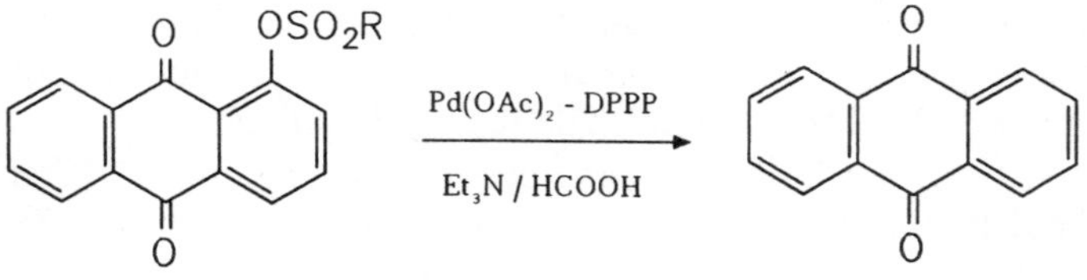

6(c-f)   7

Substrate		t(h)	7 Yield (%)
6c	R=Ph	0.5	92
6d	R=p–Me–Ph	0.5	90
6e	R=p–MeO–Ph	0.5	87
6f	R=Me	0.8	89

removal of the C–4 methoxy group in daunomycinone (2), therefore we tried to exploit what we learnt from our models.

While demethyldaunomycinone (carminomycinone) (5) was easily obtained from daunomycinone (2) by demethylation with AlCl₃, the selective sulfonation in the desired position C–4 was not so straightforward since several OH groups are present in the molecule and all but the tertiary one can compete for the sulfonating agent (Figure 9).

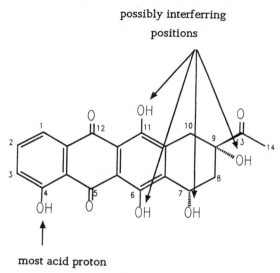

Figure 9.

If the sulfonation is carried out in the conventional way (e.g. methylene chloride, tertiary amine, sulfonating agent) a mixture of several products is always obtained.

Anyway, upon ketalization of the side-chain carbonyl group and working in pyridine in the presence of a strong base as 4-dimethylaminopyridine it was possible to achieve only the desired products 9(*a–e*) in yields ranging from 60% to 90% according to the sulfonating agent (Scheme 1).

The selectivity of this reaction can be explained by the higher acidity of the 4–OH phenol with respect to that of 6–OH and 11–OH. In fact spectrophotometric titration [10] in aqueous solution of carminomycin (4-demethyl daunorubicin) assigns to 4–OH a value of pKa = 8.64; the next ionization step is related to 11–OH with a value of pKa = 10.94.

We tried to extend the most attractive reaction conditions found with the model substrate (dioxane, Pd/DPPP) to the p.fluorophenylsulfonate of carminomycinone (*9b*). Disappointingly at 90°C the deoxygenation took place, but with concomitant hydrogenolysis of the benzylic 7–OH (Figure 10).

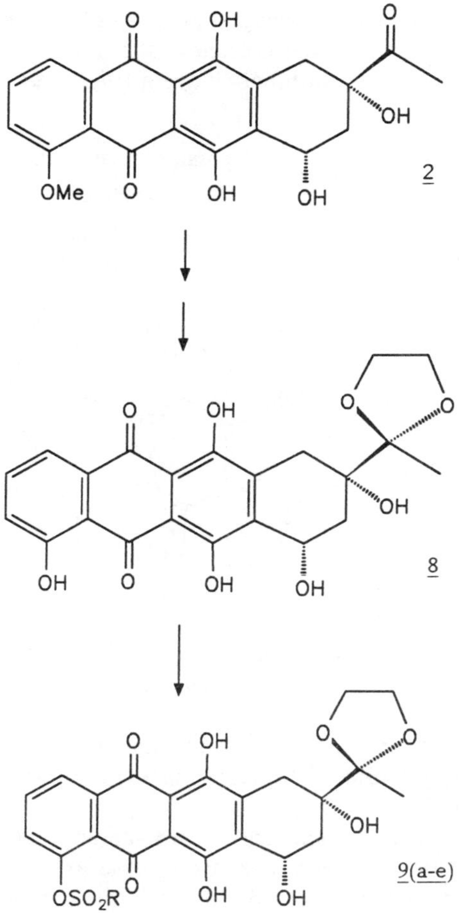

9a   R= CF$_3$;   9b   R= p-F-Ph;   9c   R= Ph;

9d   R= p-Me-Ph;   9e   R= p-MeO-Ph;   9f   R= Me

Scheme 1

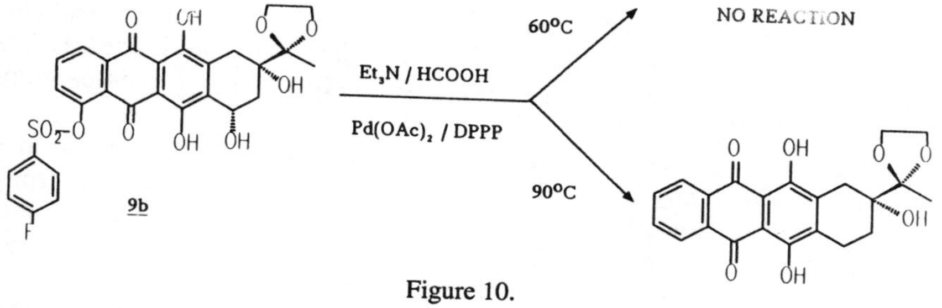

Figure 10.

Although this reaction too is catalyzed by palladium we have not yet succeeded in understanding the mechanism.

This side reaction does not occur at lower temperature, where unfortunately the reduction of sulfonate does not proceed.

On the other hand, this substrate requires very mild reaction conditions to avoid besides the catalyzed hydrogenolysis of 7–OH, also the uncatalyzed aromatization of A ring.

We knew from our previous experience with the model substrate 8 that triflates are much more reactive than other sulfonates; its use therefore seemed to us the only reasonable way to overcome the temperature problem.

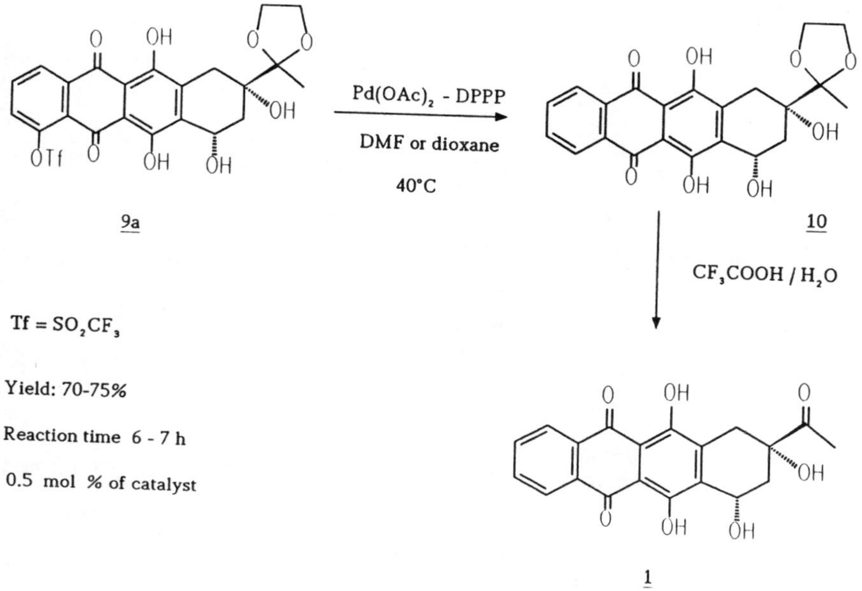

Tf = SO₂CF₃

Yield: 70-75%

Reaction time 6 - 7 h

0.5 mol % of catalyst

Figure 11.

F. Francalanci et al.

The triflation of ketalyzed carminomycinone (*8*) with trifluoromethane-sulfonic anhydride afforded,after washing with methanol the desired triflate *9a* in 67% yield [11]. The latter proved to be highly reactive at a temperature as low as 40°C (Figure 11).

The reduction proceeded in a relatively short time (5-6 hours) at this temperature using only 0.5 mol% of a catalyst generated "in situ" from palladium acetate and 1,1'-bis(diphenylphosphino)ferrocene (DPPF) and triethylammonium formate as hydride source. After deprotection of the carbonyl group and crystallization from methylene chloride we obtained the target compound *1* in 74% yield from *9a* and an overall yield from the starting daunomycinone (*2*) of about 35%.

This semisynthetic approach to 4-demethoxydaunomycinone (*1*) is highly competitive with respect to a total chemical synthesis; besides this the use of palladium catalysis forecasts a widening of the anthracyclinone varieties accessible by means of all the weapons related to homogeneous catalysis.

[a]*Farmitalia-Carlo Erba S.r.l. (Erbamont Group),
R&D, Via dei Gracchi, 35 – 20146 Milano (Italy)*
[b]*Istituto G. Donegani S.p.A.,
Via G. Fauser, 4 – 28100 Novara (Italy)*
*Present address: Istituto G. Donegani
S.p.A. – Via G. Fauser, 4 – 28100
Novara (Italy)*

References

1. Arcamone, F. in *Doxorubicin Anticancer Antibiotics*; Academic Press: New York, *1981*.
2. For a review see: Kelly, T. K. *Annu. Rep. Med. Chem.* **1979**, *14*, 288; and *Tetrahedron Symposia-in-Print* **1984**, *40*, 4539.
3. a) Arcamone, F.; *Cancer Treat. Rep.* **1976**, *60*, 829; b) Penco, S. *Chim. Ind. (It)* **1983**, *65*, 359.
4. Hussey, B. J.; Johnstone, R. A. W.; Entwistle, I. D. *Tetrahedron* **1982**, *38*, 3775.
5. Subramanian, L. R.; Hanack, M. *Synthesis* **1973**, 293.
6. For attempts to hydrogenate aryl tosylates and mesylates, see: Subramamian, L. R.; Martinez, A. G.; Fernandez, A. M.; Alvarez, R. *Synthesis* (**1984**, 481 and references therein.
7. Tolman, C. A. *Chem. Rev.* **1977**, *77*, 313.
8. Cabri, W.; De Bernardinis, S.; Francalanci, F.; Penco, P.; Santi, R. *J. Org. Chem.* **1990**, *55*, 350.
9. Dolle, R. E.; Schmidt, S. J.; Kruse, L. I. *J. Chem. Soc., Chem. Commun.* **1987**, 904
10. Razzano, G.; Rizzo, V.; Vigevani, A. *Farmaco Ed. Sci.* in the press.
11. Cabri, W.; De Bernardinis, S.; Francalanci, F.; Penco, S. *J. Chem. Soc., Perkin Trans. 1* **1990**, 428.

D. KATAKIS, Y. KONSTANTATOS AND E. VRACHNOU

SELECTIVE STOICHIOMETRIC REDUCTIONS OF ORGANIC FUNCTIONAL GROUPS BY AQUEOUS METAL IONS. IMPLICATIONS TO SYNTHESIS AND CATALYSIS

It is well known that the selectivity of a catalyst is a kinetic phenomenon, *i.e.* it is based on differences in rate. The archetype is Maxwell's demon, a creature endowed with the ability of shorting out fast from slow moving molecules.

If the rates are controlled by the matching or dismatching of the size and shape of substrate and catalyst cavity, the critical interactions involved are short range, «contact» interactions, and the selectivity is known as site or key-lock selectivity. In contrast, if the important interactions operate over a considerable part of the pathway – concerted or not – the selectivity is considered to be route controlled.

The great steps in finding rules for route-controlled selectivity were taken by Winger and Witmer [1] and by Woodward and Hoffmann [2].

Another popular differentiation is between the selectivity which is controlled by the distribution in space of the occupied and unoccupied molecular orbitals and that based on the distribution of the nuclei. The first is called electronic, the second steric [3].

It is noted in this connection, that in order to have selectivity, the system must be sufficiently complex to be able to «recognize» alternatives. It is also noted that «recognition» is not only influenced by the intrinsic structure of the interacting species *per se*, but also by the environment and/or by the use of protecting agents.

The two examples of selectivity described in this presentation are not catalytic. They are stoichiometric, but they may help to delineate the factors involved.

The two cases are:

1. Selective reduction of an organic functional group. The practical aspect of these processes is that they provide an alternative to catalytic hydrogenations. The reducing metal ions are of course oxidized, but they can be recycled using cheaper reagents or electrochemically [4]. Low-valent metal ions have been used in synthesis [5] (Table I).

A. F. Noels et al. (Eds.), Metal promoted selectivity in organic synthesis, 223–235.
© 1991 *Kluwer Academic Publishers. Printed in the Netherlands.*

TABLE I
Products and stoichiometry of the reactions between the
low-valent metal ions and the α-keto acids

Metal ion	α-Keto acid	Stoichiometry	Product
Cr^{2+}	Pyruvic acid	2:1	Lactic acid
Eu^{2+}	Pyruvic acid	2:1	Lactic acid
V^{2+}	Pyruvic acid	1:1	Dimethyltartaric acid
Cr^{2+}	Phenylglyoxylic acid	<2:1	Mandelic acid + diphenyltartaric acid
Eu^{2+}	Phenylglyoxylic acid	<2:1	Mandelic acid + diphenyltartaric acid
V^{2+}	Phenylglyoxylic acid	1:1	Diphenyltartaric acid
V^{2+}	Phenylpyruvic acid	1:1	Dibenzyltartaric acid

2. Selective electron transfer either to an organic ligand or to the solvent. The practical aspect of such studies is that they elucidate dihydrogen formation in homogeneous aqua media, and also give information about reductive desulfurization.

One example of the first kind of selectivity is the reduction of the carboxylic group of pyridine carboxylic acids by metal ions of low valence (Eu^{2+} (aq), Cr^{2+*} (aq), V^{2+} (aq)) [6], in acid aqueous solutions:

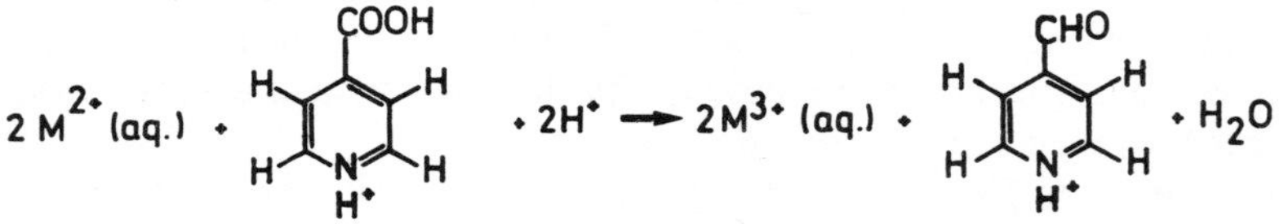

Other reducing agents reduce the pyridine ring rather than the carboxylic group. The interaction of the low-valent metal ion with the carboxylic group is localized. The pyridine nitrogen is effectively protected by protonation and direct interaction with the pyridine ring is non-reactive. In contrast, in the reduction of pyruvic acid by the same metal ions [7], the carboxyl group is unreactive; here it is the carbonyl group that is reduced:

$$2M^{2+}(aq.) + R\overset{\overset{O}{\|}}{C}COOH + 2H^+ \longrightarrow R\overset{\overset{OH}{|}}{C}HCOOH + 2M^{3+}(aq.) \quad (M=Cr, Eu)$$

$$2M^{2+}(aq.) + 2R\overset{\overset{O}{\|}}{C}COOH + 2H^+ \longrightarrow R\underset{\underset{COOH}{|}}{\overset{\overset{OH}{|}}{C}}\underset{\underset{COOH}{|}}{\overset{\overset{OH}{|}}{C}}R + 2M^{3+}(aq.) \quad (M=V, Ti)$$

All four metal ions attack the carbonyl group selectively, but the products differ.

Along the simulation line for CO_2 fixation:

$$O=C=O \cdots X=C=O \qquad \begin{matrix} X \\ | \\ \cdots C=O \cdots \\ | \\ Y \\ \cdots X=C=Y \cdots \end{matrix} \qquad \begin{matrix} X \\ | \\ \cdots C=Z \\ | \\ Y \end{matrix} \qquad X,Y,Z \neq 0$$

ketones lie closer to the end, and the study of the ways they interact with metal ions provide information on how to obtain different reduction products from CO_2.

Reduction of the double bond (carbon-carbon in this case) also occurs in the reactions of substituted olefins:

$$\overset{\diagdown}{\diagup}{=}\overset{\diagup}{\diagdown} + 2M^{n+}(aq.) + 2H^+ \longrightarrow \overset{\diagdown}{\diagup}CH\,CH\overset{\diagup}{\diagdown} + 2M^{n+1}(aq.)$$

These reactions have been investigated for several combinations of substituents [8] and with stable reducing metal ions or transient, generated *in situ* by pulse radiolysis [9,10].

The crucial first attack of the metal ion on the C=O and C=C double bonds (also carbon-carbon triple bond) is directly on carbon. The evidence for that is ample and includes:

— Isolation and characterization of resistant to hydrolysis species containing metal-to-carbon bonds [11].

— Activation of the double bond, as manifested by *cis-trans* isomerization [12], double bond hydrogen exchange [12], labilization of chloride bound to the double bond carbon [8].

— Charge transfer spectra (Figure 1), characteristic of the Cr–C bond [10], with large absorptivities.

— Charge transfer spectra of a complex formed by direct attack of Cr^{2+} on the double bond (Figure 2); pyridine nitrogens were effectively protected by protonation.

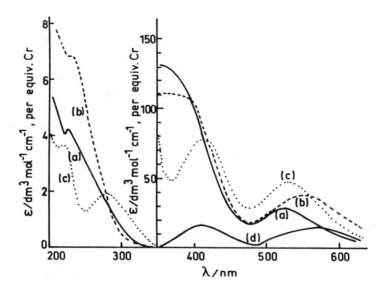

Figure 1. Charge transfer spectra characteristic of the Cr–C bond.

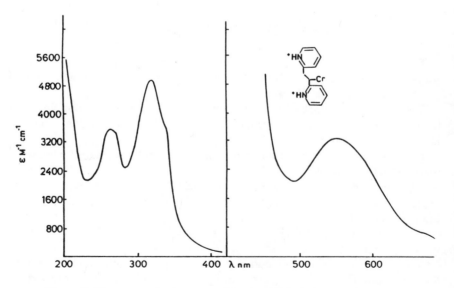

Figure 2. Charge transfer spectrum of the organochromium species shown.

In this connection it is instructive to note that, usually, metal ions behave as Lewis acids, interacting with polar groups. In cases such as those considered here, they behave as donors, and the organic molecule as acceptor. In the examples quoted, the organic ligands are initially free, but even if they are already complexed, they can still act as acceptors. Characteristic kinetic evidence for that is obtained by comparing the rates of the reactions of the pyruvate complexes of Co^{III} with Cr^{2+} and V^{2+}. Chromium (II) reacts one thousand times faster than vanadium (II). With the acetate complexes of Cr^{III} the situation is reversed: it is vanadium(II) now that reacts one thousand times faster. The differences are attributed to a better initial overlap of the Cr^{2+} σ-donor orbital with the empty orbitals of the pyruvate ligand. With acetate, the π-donor orbitals of V^{2+} overlap better [13, 14]. More kinetic evidence [15] for direct interaction of the donor metal ions with multiple bonds is given in Table II, where it is seen that the differences in the values of the activation enthalpies for a diversity of ligands are small, and therefore they can be attributed to a common origin: the interaction of chromium^{2+} with the carbon-nitrogen triple bond.

TABLE II

Summary of kinetic parameters for the reduction of nitrile complexes of pentaamminecobalt(III) by chromium(II)

Nitrile ligand	10^2k (20°C), $M^{-1}s^{-1}$	ΔH^*, kcal.mol^{-1}	$\Delta S^{\neq}$, kcal.mol^{-1} deg^{-1}
$NCCH_3$	0.94 (2.0)	10.3 ± 0.4	−33.1 ± 1.4
$NC(CH_2)_2CN$	2.54	9.5 ± 0.4	−34.1 ± 1.1
$NCCH_2CONH_2$	2.61 (5.3)	9.0 ± 0.6	−35.6 ± 1.9
$NCCH_2CO_2CH_3$	2.34	9.5 ± 0.5	−34.2 ± 1.3
$NCCH_2CO_2$	211	12.5 ± 1.1	−15.2 ± 3.6
Cinnamonitrile	2.2	8.8 ± 0.9	−37 ± 3
Furanacrylonitrile	2.0	9.5 ± 0.8	−34 ± 3
4-Acetoxybenzonitrile	2.1	9.1 ± 1.2	−36 ± 4
Benzonitrile	4.27		

Kupperschmidt and Jordan, *Inorg. Chem.*, **1981**, *20*, 3469.

The few examples given prove beyond any reasonable doubt that there is indeed direct interaction of multiple bonds with aqueous metal ions. More evidence is available, but it is not our objective here to make a complete collection.

The preference for a certain site of attack implied by this evidence should be considered to be controlled by the minimum energy pathway, rather than by the site itself. A classic example of a minimum energy pathway is provided by the nucleophilic attack of hydride on the olefinic double bond (Figure 3), for which it has been shown theoretically [16], and also by using the structure correlation method [17, 18], that the minimum energy route corresponds to an angle of approach of about 110° (Figure 3).

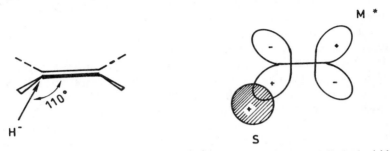

Figure 3. Minimum energy pathway in the nucleophilic attack of an olefin by hydride.

A similar non-symmetric interaction was also postulated [8] to occur during the approach of the first Cr^{2+} to the olefinic plane (Figure 4). In contrast, the interaction of the π-donors V^{2+} and Ti^{3+} is believed to be symmetric.

The carbonyl group (*vide supra*) is also preferentially attacked on carbon, where the π [*] orbital is polarized [19].

Since the reductions of the C=O and C=C groups require two electrons, attack by a second metal ion is necessary.

It has been shown [16] that nucleophilic attack on an olefin coordinated to a metal ion in a symmetric manner requires previous displacement of the ML_n fragment:

$$\underset{ML_n}{\cdots\overset{|}{\cdots}\cdots} \quad \longrightarrow \quad \underset{ML_n}{\cdots\cdots\cdots} $$

With Cr^{2+}, this displacement already exists (Figure 4), with V^{2+} it is necessary to occur, and this contributes to the activation.

In addition to the electronic factors affecting selectivity, there are also other factors that should be taken into account. Thus, in the case of the reaction of pyruvic acid with V^{2+}, there is no attack by a second V^{2+}, presumably because strong repulsions dominate, and it is energetically more favorable to form a free radical:

$$\underset{\text{OOC}}{\overset{H_3C}{\underset{|}{\overset{|}{C}}}}\overset{V^{II}(pyr.)}{\underset{}{=\!=\!=}}O \quad \longrightarrow \quad V^{III}(pyr.) \;+\; \underset{HOOC}{\overset{H_3C}{\diagdown}}\overset{\cdot}{C}{-}O^-$$

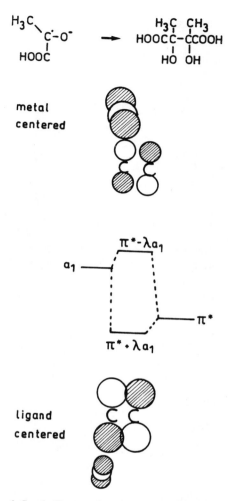

Figure 4. Symbolic presentation of the interaction of a $d_{z^2}Cr^2$ orbital with the olefinic π-antibonding.

In two-electron transfer mechanisms like those just described, the question arises whether after the transfer of the first electron there is a «free» rotation or not. After the transfer of one electron to the antibonding pi orbital, the order of the bond has become 1.5.

A study of the free radicals:

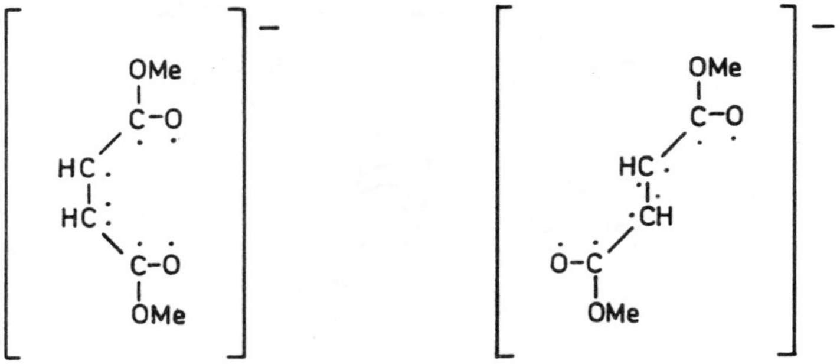

showed [20] that interconversion (rotation about the central carbon atoms) requires considerable activation. There are no data that can lead to unambiguous conclusions about the effect of metal ions on such rotations.

In the reaction between reducing metal ions and unsaturated organic groups, in addition to free radical formation, there are also other possibilities, *e.g.*:

— reductive dimerization:

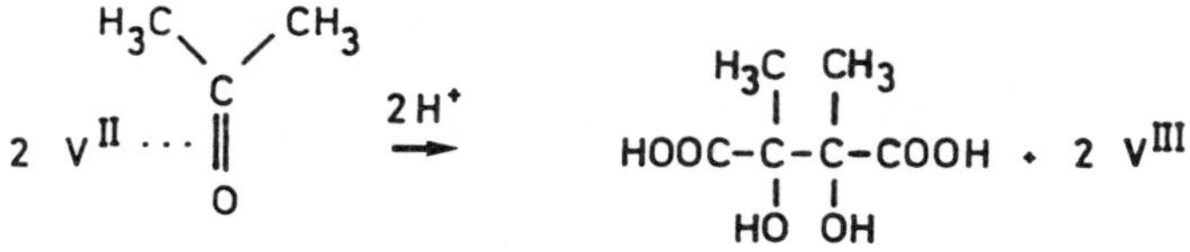

— disproportionation:

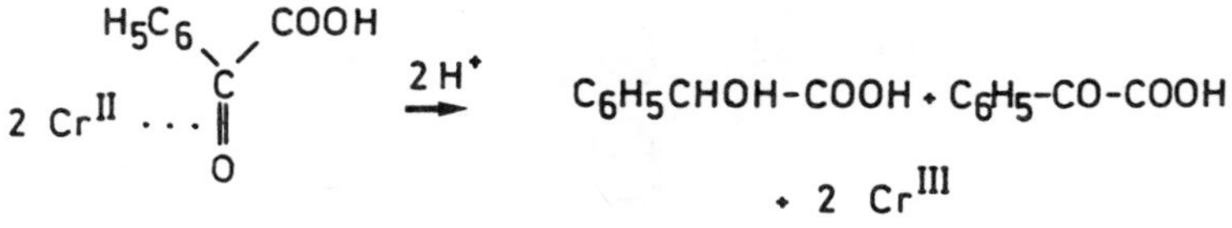

Summary of some of the pathways documented is given in Table III.

A general conclusion to be drawn is that product selectivity for different metal ions and substrates *is largely determined by the crucial first interaction of the donor metal ion orbital with the empty antibonding orbital of the unsaturated system.*

So far in this presentation we addressed the question of what affects localization of the incoming electrons preferably on one site of the molecule only. In the rest of the presentation we will examine briefly some factors that determine whether the electrons stay indeed on the organic substrate or they are transfered elsewhere, *i.e.* to the solvent.

TABLE III
Observed pathways of various ion-radicals

Ion-radical	Coordination	Competing paths	Overall reaction
Cr(III)	η^1	Nucleophilic attack by a second chromiom(II) ion and free radical[a]	Hydrogenation of the carbon-carbon double bond
Cr(III)-mema	η^1	Same as above	Same as above
Cr(III)-pgl	η^1	Dimerization and disproportionation	Hydrogenation of the carbon-oxigen double bond and C–C bond formation
V(III)-ma	η^2	Nucleophilic attack by a second vanadium(II) ion and dimerization	Hydrogenation of the carbon-carbon double bond
V(III)-mema	η^2	Dimerization only	Same as above
V(III)-pyr	η^2	Free radical and dimerization[b]	C–C bond formation
Ti(IV)-pyr	η^2	Dimerization only	C–C bond formation
Eu(III)-pyr	η^1	Free radical only	Hydrogenation of the carbon-oxygen double bond
Eu(III)-pyr	η^1	Free radical only	Same as above

[a] Nucleophilic attack in excess Cr^{2+}. Similarly for Cr(III)-Clma, Cr(III)-Cl$_2$ma.
[b] Assisted by a second pyr.

One of the systems studied [21] is summarized by the following reaction, taking place at pH values around neutral:

$$V^{II}(cys)_3 \xrightarrow{H_2O} V^{III}(cys)_3 + 1/2 \ H_2 \quad (cys = cysteine)$$

In this reaction the electrons are transferred to the solvent, to the protons to be exact, and the mechanism necessarily involves proton oxidative addition *i.e.* formation of an intermediate hydride, which is favored by the accumulation of negative affective charge on the metal ion.

With mercaptoacetic acid [22], the electrons stay on the organic moiety and cause C-S bond scission. The mechanism proposed is:

$$\text{V}^{\text{II}} + \underset{\underset{\text{SH}}{|}}{\text{CH}_2\text{COO}^-} \longrightarrow \text{V}^{\text{III}} + \text{HS}^- \qquad + \overset{\cdot}{\text{C}}\text{H}_2\text{COO}^-$$

$$\downarrow \text{H}^{\cdot} \qquad \qquad \downarrow$$

$$\text{H}_2\text{S} \qquad 1/2 \; \underset{\underset{\text{CH}_2\text{COO}^-}{|}}{\text{CH}_2\text{COO}^-}$$

The practical aspect of this reaction is that it exemplifies a reductive desulfurization process using metal ions as reductants.

Still another case in which the electrons stay on the ligand is the reaction of V^{II} with cystine [23]. Here, the mechanism includes two parallel paths, both involving S-S bond breaking:

$$\text{V(II)} + \text{RSSR} = \text{V(II)(RSSR)} \text{ fast, quantitative equilibrium}$$

$$\text{V(II)(RSSR)} \rightarrow \text{V(III)} + \text{RS} + \text{RS}^- \text{ slow}$$

$$\text{V(II)(RSSR)} + \text{RS} \rightarrow \text{V(III)(RSSR)} + \text{RS}^- \text{ fast}$$

$$\text{V(II)(RSSR)} + \text{(RSSR)} == \text{V(II)(RSSR)}_2 \text{ fast equilibrium}$$

$$\text{V(II)(RSSR)} \rightarrow \text{V(IV)(RSSR)}^- + \text{RS} \text{ slow}$$

This mechanism, in excess cystine, corresponds to the rate law:

$$\text{Rate} = \text{kobs} [\text{V(II)}]_0 = (k + k'[\text{RSSR}]_0)[\text{V(II)}]_0$$

Summary of the results obtained with various organic ligands is given in Table IV.

Two general conclusions emerge from these results:
(a) That the presence of a sulfur atom in the ligand is important in obtaining dihydrogen.
(b) That the formation of a five-membered chelate ring is also important.

Here again, the existence of an effective negative charge on the metal favors proton oxidative addition.

Dihydrogen is also obtained by using dithiolenes as catalysts, in mixed acetone-water solutions [24]. The electron in this case does not come from the complex itself, but from another source, *e.g.*:

$$2\text{MV}^+ + 2\text{H}^+ \longrightarrow 2\text{MV}^{2+} + \text{H}_2$$

TABLE IV

Qualitative comparison[a] of cysteine to related ligands; $[V(II)]_0 = 2 \times 10^{-3}$ M; $[ligand]_0 = 0.1$M; room temperature

Ligand	Formula	pH range	V(II)-V(III)	Gaseous products	Comments
Mercapto-acetic acid	CH_2COOH / SH	3.5-9.5	yes	H_2S	Formation of yellow soluble vanadium(III) complex. Organic products: $(CH_2COOH)_2$
Thiolactic acid	$CH_3CHCOOH$ / SH	ca. 9	yes	H_2S	Formation of yellow soluble vanadium(III) complex. Organic product: $(CH(CH_3)COOH)_2$
3-Mercapto-propionic acid	CH_2COOH / CH_2SH	5.0-9.5	no	–	Formation of insoluble vanadium(I) complex
Ethanethiol	CH_2CH_3 / SH	6.0-8.0	no	–	No complex formation
Cysteamine	CH_2CH_2 / SH NH_2	8.0-10	yes	H_2	Heterogeneous oxidation of vanadium(II)
Cysteine	$CH_2CHCOOH$ / SH NH_2	5.0-9.5	yes	H_2	Formation of soluble yellow vanadium(III) complex. H_2S or NH_3 were not detected
S-methyl-cysteine	$CH_2CHCOOH$ / S NH_2 / CH_3	6.0-9.0	no	–	Formation of soluble pink complex of vanadium(II). Above pH 8.5 colour changes to pale yellow and hydroxo complexes precipitate.
N-acetyl cysteine	$CH_2CHCOOH$ / SH $NHCOCH_3$	5.0-9.0	no	–	Formation of soluble vanadium(II) complex
Cysteine methyl ester	$CH_2CHCOOCH_3$ / SH NH_2	6.0-9.0	yes	H_2	Heterogeneous oxidation of vanadium(II) (see text)
Serine	$CH_2CHCOOH$ / OH NH_2	6.0-9.0	no	–	Formation of pink vanadium(II) complex

[a] Oxidation of V(II) and gaseous product formation within a period approximately two hours. For longer times even solid V(II) hydroxides give H_2.

where MV$^+$ is the free radical obtained from methylviologene

(N,N'-dimethyl 4,4'-dipyridine).

In conclusion, it is noted that the clarification and study of the «dynamic» selectivity factors is perhaps more difficult than the study of the «static» ones. The structures in the former case change with time, involve pathways and molecular trajectories. Yet, it seems worth embarking on such studies, since they may lead to the design of selective routes, an alternative to selective sites.

Department of Chemistry,
University of Athens, Panepistimiopolis,
Kouponia, 157 01 Athens (Greece)
(D.K. and Y.K.)
N.C.R.N.C. Demokritos, Pob 60228,
153 10 Aghia Paraskevi Attiki (Greece)
(E.V.)

References

1. Wigner, E. and Witmer, E.E. *Z. Phys.* **1928**, *51*, 859.
2. Woodward, R.B. and Hoffmann, R. *The Conservation of Orbital Symmetry*; Academic Press, New York, 1970.
3. Tolman, C.A. *Chem. Rev.* **1977**, *77*, 313.
4. Steckhan, E. *Angew. Chem. Int. Ed. Engl.* **1986**, *25*, 683.
5. (a) Hanson, J.P. *Synthesis* **1974**, *1*; (b) Ho, T. L. *Synthesis* **1979**, *1*; (c) Katsaros, N.; Vrachnou-Astra, E.; Konstantatos, J. and Stassinopoulou, S.I. *Tetrahedron Letters* **1979**, *44*, 4319.
6. Vrachnou-Astra, E. and Katakis, D. *J. Am. Chem. Soc.* **1973**, *95*, 3814.
7. Konstantatos, J.; Vrachnou-Astra, E. and.Katakis, D *Inorganica Chimica Acta* **1984**, *85*, 41; more references cited.
8. Katakis, D., Vrachnou-Astra, E. and Konstantatos, J. *J. Chem. Soc., Dalton Trans.* **1986**, 1491; more references cited.
9. Buxton, G.V. and Sellers, R.M. *Coord, Chem. Rev.* **1977**, *22*, 195.
10. Meyerstein, D. *Acc. Chem. Res.* **1978**, *11*, 43.
11. Katakis; D.; Konstantatos, J. and Vrachnou-Astra, E. *J. Organometallic Chem.* **1985**, *279*, 131; more references cited.
12. Vrachnou-Astra, E. and Katakis, D. *J. Am. Chem. Soc.* **1967**, *89*, 6772.
13. Price, H.J. and Taube, H. *Inorg. Chem.* **1968**, *7*, 1.
14. Taube, H. *Electron Transfer Reactions of Complex Ions in Solution*; Academic Press, New York, 1970.
15. Kupferschmidt, W.C. and Jordan, R.B. *Inorg. Chem.* **1981**, *20*, 3469.
16. Eisenstein, O. and Hoffmann, R. *J. Am. Chem. Soc.* **1981**, *103*, 4308.
17. Bürgi, H.B. and Dunitz, J. D. *Acc. Chem. Res.* **1983**, *16*, 153.
18. Katakis, D. and Gordon, G. *Mechanisms of Inorganic Reactions*; Wiley, New York, 1987.

19. Jorgensen, W.L. and Salem, L. *The Organic Chemist's Book of Orbitals*; Academic Press, New York, 1973.
20. Hayon, E. and Simic, M. *J. Am. Chem. Soc.* **1973**, *95*, 2433.
21. Konstantatos, J.; Kalatzis, G.; Vrachnou-Astra, E. and Katakis, D. *J. Chem. Soc., Dalton Trans* **1985**, 2461.
22. Konstantatos, J.; Kalatzis, G.; Vrachnou-Astra, E. and Katakis, D. *Inorg. Chem.* **1983**, *22*, 1924.
23. Kalatzis, G.; Katakis, D.; Vrachnou-Astra, E. and Konstantatos, J. *Inorganica Chimica Acta* **1988**, *151*, 191.
24. Hontzopoulos, E.; Konstantatos, J.; Vrachnou-Astra, E. and Katakis, D. *J. Mol. Catal.* **1985**, *31*, 327.

A. DEMONCEAU, A.J. HUBERT, AND A.F. NOELS

BASIC PRINCIPLES IN CARBENE CHEMISTRY AND APPLICATIONS TO ORGANIC SYNTHESIS

1. Introduction

Carbenes are divalent carbon compounds of considerable chemical importance [1]. Their generation from various precursors is a well established and much used process in organic chemistry. More recently, metal-carbene complexes have been demonstrated in such important catalytic industrial processes as olefin metathesis, Fischer-Tropsch synthesis, CO reductions and suggested as key intermediates in a great deal of other reactions such as the Clemmensen reduction, Ziegler-Natta polymerizations and hydrocarbon cracking.

Most carbenes are very labile species and controlling the reactivity-selectivity pattern of such reactive intermediates is of utmost importance if carbenes are to be used in fine organic synthesis. The concept of selectivity in this context refers essentially to regioselectivity and stereoselectivity. However, chemoselectivity is also of importance for realizing intricate synthesis as carbenes offer a set of straightforward routes to unusual structures of biological interest.

2. Basic principles of carbene reactivity

Free carbenes are divalent species possessing two nonbonding electrons, either paired or unpaired. The multiplicity of their ground state is thus either singlet S_0 (**1**) or triplet T_1 (**2**).

Which of these represents the ground state depends on the relative energies of the σ and p orbitals. If they are comparable in energy, then the triplet will be the ground state. If, however, the σ orbital is sufficiently low so that the increase in electron repulsion associated with going from σp to

A. F. Noels et al. (Eds.), Metal promoted selectivity in organic synthesis, 237–259.

σ^2 is compensated, then a singlet ground state will be obtained. When the electronegativity of the substituents increases, the s character of the orbital increases and thus its energy decreases, favoring the singlet state. Conversely, as the substituents become more electropositive the σ orbital loses s character and the RCR' angle opens as electrons are transferred to the central atom. At some point the σ and p orbitals are close enough in energy that the triplet state is obtained as the ground state.

1   **2**

We therefore anticipate that electronegative substituents will favour a singlet ground state. At the opposite, an electropositive substituent will favour a triplet ground state. Succinctly, as the substituents go from F to Li, the ground state goes from S_0 to T_1.

Experiments have shown that indeed triplet carbenes behave essentially as free radical species (*e.g.* react with poor selectivities in cycloadditions) while singlet carbenes react in a concerted fashion according to the Woodward-Hoffman rules and frontier orbital models.

Frontier molecular orbital theory has been applied successfully to the rationalization and prediction of the behavior of a variety of substituted carbenes. The addition of a singlet carbene to an alkene involves simultaneous interactions of the vacant carbenic p orbital (LUMO) with the filled alkene π orbital (HOMO) and of the filled carbenic σ orbital (HOMO) with the vacant alkene π^* orbital (LUMO). As a singlet carbene can act both as an electrophile and a nucleophile, the decisive features in the transition state for addition is the $LUMO_{carbene}/HOMO_{alkene}$ or $HOMO_{carbene}/LUMO_{alkene}$ interaction: the stronger interaction will determine the electronic distribution in the transition state. Such an interaction is determined both by the differential energies of the «competitive» interactions and by the comparative extents of orbital overlaps.

The concertedness of singlet carbene additions has been scrutinized and proved many times in the past. The experimental probes are based mainly on carbene selectivity and stereospecificity of additions to olefins. However, hints exist that these additions may not necessarily take place via strictly synchronous mechanisms. Moreover a singlet-triplet intercrossing system would permit a portion of the carbene to react in a concerted fashion to give stereospecific additions while the rest reacts with the double bond more as a diradical in a non stereospecific fashion. Addition to conjugated olefins occurs almost exclusively in a 1,2 fashion, although some rare examples of true 1,4 additions have been evidenced [2].

The most synthetically useful reactions of carbenes are:
– cycloaddition reactions to unsaturated systems:
  a) cyclopropanes and cyclopropenes synthesis from alkenes and alkynes;
  b) cycloheptatriene derivatives by cycloaddition to benzene rings;
  c) synthesis of heterocycles, often via ylide formation (*e.g.* by cycloaddition to nitriles).
– insertion reactions into various X–H bonds (Equation 1, X=O, N, S, C, Si): even alkanes can thus be functionalized by carbenes [1]. It is worth noting that while singlet carbenes (*e.g.* S_0CH_2) insert rapidly into C–H bonds, there is no experimental evidence that they can insert into C–C bonds. A recent theoretical calculation suggested that insertion into strained C–C bonds (*e.g.* in cyclopropanes) should occur with a very small energy barrier but this has not been confirmed experimentally [3].

3. Methods for generating carbenes

Many processes are described which yield carbenes from organic precursors. Among these, diazirines **3** and diazo compounds **4** are widely used because of their ready availability from relatively cheap starting materials and because they yield photochemically, thermally or catalytically the carbene free of co-reactants.

$$\tag{1}$$

Other classical sources of carbenes are *inter alia* [1]:
– action of a strong base on an alkyl halide (Reimer-Tilmans reaction) in which dichlorocarbene is produced from chloroform (Equation 2);
– thermolysis of phenyltrihalomethylmercury **5** (Equation 3); this reaction permits the formation of halocarbenes in neutral solution;
– thermal decomposition of trihaloacetate salts, photolysis and pyrolysis of polyhalomethanes, …;

- base-catalysed thermal decomposition of tosylhydrazones **6** via the diazoalkane (Bamford-Stevens reaction) (Equation 4);
- the classical Simmons-Smith reagent, CH_2I_2/Zn generates organometallic species of the type $I-CH_2-ZnI$, which ultimately react as carbenoids.

$$CHCl_3 + B \longrightarrow BH^+ + CCl_3^- \longrightarrow Cl^- + :CCl_2 \qquad (2)$$

$$\underset{5}{PhHgCX_3} \xrightarrow{\Delta} PhHgX + :CX_2 \qquad (3)$$

$$\underset{6}{ArSO_2NHN=CHR} \xrightarrow{B} Ar(R)C=N_2 \xrightarrow{\Delta} :CR-Ar \qquad (4)$$

4. Transition-metal promoted carbene reactions

Actually, the main drawback for using free carbenes in organic synthesis is the problem of controlling the selectivities of such reactive species. In this respect, transition-metal catalysis provided a decisive breakthrough [4–7]. Actually, catalytic methods for the decomposition of diazo compounds have been known for more than 80 years, and their uses for a variety of carbenoid transformations are well established.

The term «carbenoid» was coined to infer that the actual reactive species was neither a free carbene nor an activated diazo compound but rather a carbene ligated to a transition-metal complex (as originally suggested by Yates in 1952). Support for the intermediacy of carbenoid comes mostly from the observation of asymmetric induction in the cyclopropanation of olefins when using chiral catalysts. The discovery of stable metal-carbene complexes and the demonstration of their compatibility for stereospecific olefin cyclopropanation lent further support to transient metal carbene intermediates in catalytic processes. Moreover, transient methylene-metal complexes (Cu, Fe,...) have been unambiguously characterized (IR) in argon matrices [8].

It was soon recognized that the reactivity patterns exhibited by metal-carbene complexes were divided into two classes: electrophilic carbenes which add nucleophiles at the carbene carbon and nucleophilic carbenes which add electrophiles at the carbene carbon, hence the distinction between the «carbenoids» and «alkylidenes» when designing carbene ligands corresponding respectively to the first or second class. A carbon complex can in fact be considered as a singlet carbene-metal complex, whereas an alkylidene would be a triplet carbene high spin metal complex.

These two greatly different modes of reactivity reflect a dramatic difference in the metal-carbon bonding. Conventional design prescriptions call for «low valent» metal fragments for carbenes and «high-valent» metal moieties (high spin) for alkylidenes, in order to maximize stability of the resultant complex. The presence of a heteroatom on the divalent carbon stabilizes carbenoids, while alkyl or hydrogen groups are thought to stabilize alkylidene ligands. The combination of a «low-valent» metal site with a C(OR)X carbene ligand corresponds to the now classical concept of a σ-donor bond from the carbene and donor π back-bond from the «low-valent» metal. These contrasting bonding structures (donor/acceptor for carbene and covalent for alkylidene) are given physical justification via the valence bond view of metal carbene or alkylidene bonds [9].

This view is however an oversimplification. The complexes on which the comparisons have been based often differ greatly in either the molecular charge and/or the substituents on the carbene carbon. Actually, variation of the metal and the ancilliary ligands in a complex can alter the reactivity of the carbene carbon center from electrophilic to nucleophilic. This can be substantiated by the observation that, for example, the same ligand (CF_2) bonded to the same metal (Ru) exhibits reactivity dependent on the oxidation state of the metal [10] and by the exceptional basicity of a methylene ligated to an iridium complex [11].

The most typical reaction of carbenoids is the transfer of the electrophilic carbene ligand to a nucleophilic substrate (*e.g.* to an olefin to yield a cyclopropane) while alkylidenes undergo Wittig-type reactions or catalyse the metathesis of olefins, especially when the metal is highly oxidized.

Only terminal carbene complexes are considered here. Informations on metal cluster complexes containing substituted carbene ligands can be found in reference 12.

4.1. REACTIONS OF CARBENOIDS

4.1.1. The catalytic cycle

It is now generally accepted that the mechanism involves a sequence of elementary steps, one of the many variations of which is sketched in Figure 1. Globally, it includes:

i) formation of a metal-diazo compound complex with the possible interaction of the nucleophilic carbon atom of the diazofunctional group with the metal center;

ii) elimination of dinitrogen with formation of a metal carbene complex yielding a so-called «carbenoid»;

iii) attack of the substrate by the carbenoid and elimination of the product with regeneration of the original catalyst.

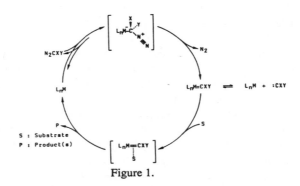

Figure 1.

Moreover, the hitherto neglected possibility of an equilibrium carbenoid ⇌ free carbene might also be operative in certain cases, especially when a stabilized electrophilic carbene species is ligated to an electron-deficient metal complex [13].

4.1.2. Factors governing the catalytic activity

The efficiency and the selectivity of transition-metal-promoted carbene reactions are governed by the classical principles involved in homogeneous transition-metal catalysis. Besides the intrinsic peculiarities of the metal, of particular importance are the ancilliary ligands, responsible, *inter alia*, for the control of the stereoselectivity (particularly the enantioselectivity of these reactions), steric effects being decisive for such control.

However, the details of the stereoelectronic effects in carbene reactions is far from being completely understood. For instance, recent investigations in the field of Rh-catalysed alkane functionalisation have revealed shape selective interactions [14] as well as unexpected influence of the catalyst and diazoester substitution on the outcome of the reactions [15].

4.1.3. Typical catalysts

The oldest known catalysts for promoting carbene reactions from diazoesters are copper-based systems (the metal, copper bronze, salts such as CuCl,

CuBr, CuSO$_4$, Cu triflates, and organometallic complexes such as Cu(acac)$_2$, iminocomplexes,...).

Generally, the simple copper promoters require high temperature; they are often handicaped by a lack of selectivity. However, enantioselective reactions leading to high ee have been successfully realized with some particular chiral complexes (see further). In fact, the VIII group metals afforded, more recently, a set of improved and unusually efficient catalysts, for example palladium(II) and particularly rhodium(II) carboxylates [4]. These systems led to high yields and unusual regioselectivities. Some example of interesting *cis-trans* selectivities have been reported but up to now, a total control of the stereoselectivities is still out of reach in most cases.

4.1.4. Typical ligands

The ligands constitute a key factor as far as the control of the selectivities is concerned. With one same metal, very different reaction pathways can be promoted by modifying the ligands associated to the metal center. Particularly, the problem of enantioselectivity is completely dependent on the ligand control: the use of optically active ligands is a basic requirement to solve the problem of chiral induction as no chiral metal centers are up to now available for catalytic systems. Regioselectivities are modified according to the ligand situations around the metal: classically, strong ligands (acetylacetonate) in opposition to weak ligands (triflate ion) promote a mechanistically different pathway in copper-catalysed cyclopropanation of differently substituted olefins [16]: strong ligands (acac) promote carbenoid attack of free olefins whereas the weak triflate ligand leads to an inner-sphere attack of the coordinated olefinic substrate (coordinative mechanism).

The particular case of palladium catalysis is related to some extent to the above aspect, electronic factors being however important in this case [17].

In the particular case of rhodium catalysed C–H insertion in alkanes and of ring enlargement reactions of aromates, the presence of strongly electron-withdrawing ligands determines a high catalytic activity: it appears, at least in the diazomalonate case, that destabilisation of the carbene-metal complexes would occur in such cases leading to practically free carbenes [13].

On the other hand, restricted access to the metal in a family of ortho-substituted tetrakis-benzoato complexes of rhodium(II) permitted either to deactivate the catalyst (no access to the metal center) or to observe a high *cis-trans* ratio in cyclopropanation reactions, according to the size and number of ligands around the metal [18].

4.2. SELECTIVITIES IN CATALYSED CARBENE REACTIONS

4.2.1. Stereoselectivity

4.2.1.1. Enantioselectivity

The first report on (weak) optical activity induction in a copper-catalysed cyclopropanation reaction when using chiral phosphite ligands was of theoretical importance in proving for the first time the occurrence of copper carbenoids rather than free carbenes in copper-catalysed decomposition of diazoesters. Since these pioneer works, some remarkable achievements have been realized in the field of enantioselective cyclopropanation of olefins with diazoester precursors [19, 20, 21].

The role of the ligand is essential: it controls the approach of the olefin to the carbene center. Therefore, the use of a chiral ligand is required and the steric requirements around the active center are determining as shown in Equation 5 [22] when comparing the enantioselectivities of ligands 9 and 10.

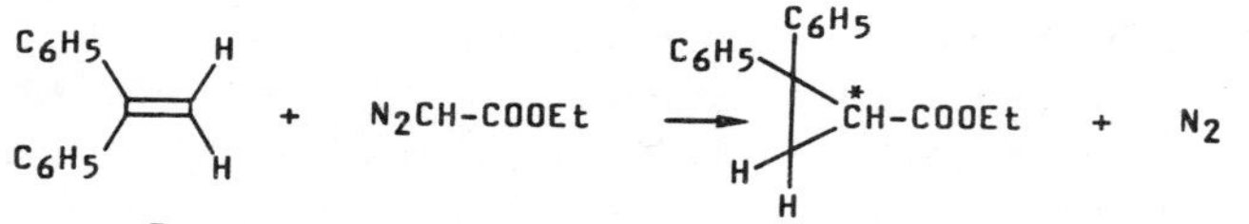

R :	a	b	c	d
	CH_3	$CH(CH_3)_2$	$CH_2C_6H_5$	CH_2OH
% e.e.	52.3(S)	45.0(S)	65.6(S)	37.9(S)

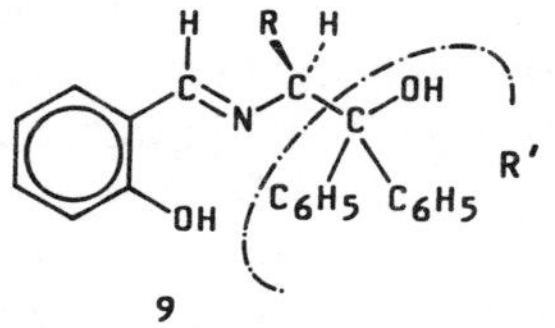

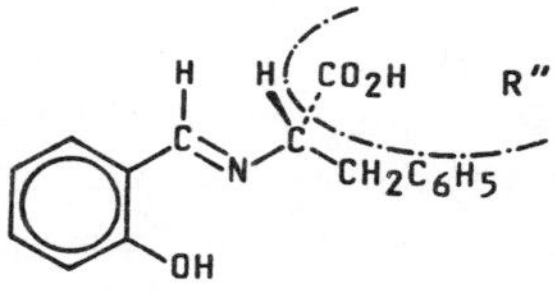

However, if the steric effect of group R' relatively to R'' is predominant (compare R' to R'' in 9 and 10), the impact of R' in 9 does not appear as purely correlated to the size of this substituent (compare complexes 9-a, 9-b and 9-d: b and d are more crowded than a); the kinetic control responsible for the enantioselection appears therefore as extremely sensitive to the precise geometry of the active center and to subtle (e.g., lipophilic, polar,...) interactions.

Thus, beside purely steric effects, the importance of lipophilic interactions has to be considered. For example, we have reported that the cyclopropane yield from *trans*-4-octene increases from 7% with methyl diazoacetate and rhodium acetate up to 80% with *n*-butyl diazoacetate and rhodium pivalate: the presence of large lipophilic groups both in the diazoester and the catalyst led to a dramatic increase of the yields [23]. Subtle discriminations in carbene insertion in alkanes offer additional striking examples [24, 25]. In fact, the impact of the ligand appears as being the result of a particularly delicate balance as the investigation of a large set of chiral ligands (thirty seven imino or amino ligands were tested by H. Brunner [22]) showed that only two of them (**9-a** and **9-c**) led to enantiomeric excess higher than 50%; this demonstrates clearly that the choice of the proper ligand is often the result of a tedious screening.

Interesting results have been recently reported when using a semi-corrin copper complex **11** as catalyst. High enantioselectivities have been observed in this case: up to 97% ee for the *cis* cyclopropanes in the styrene and butadiene cases when using **11** with R = *d*-menthyl [26].

Co(α-cqd)₂

11   **12**

The importance of the metal is not fundamentally determining as far as enantioselectivity is concerned. The best results reported up to now have however been obtained by Aratani [20] in the preparation of pyrethroid precursors (see Applications). High enantioselectivity was also obtained by Nakamura when using the camphorquinone-β-dioximate cobalt complex **12** but these processes remained mostly limited to the enantioselective cyclopropanation of styrene and conjugated diolefins.

The most efficient cyclopropanation catalysts, rhodium(II) carboxylates (so far, early 1990), do not lead to appreciable induction of activity when using chiral carboxylate moieties: in fact, the structure of the complex (**13**) does not allow appreciable interactions between the chiral group and the active metal-carbene center. For example, even with the relatively bulky R′

ligands **13** (the corresponding optically active carboxylic acids are readily available respectively from L-(+)alanine and L-(–)-phenylalanine) the rhodium(II) complexes displayed only modest e.e. (e.e. < 10%).

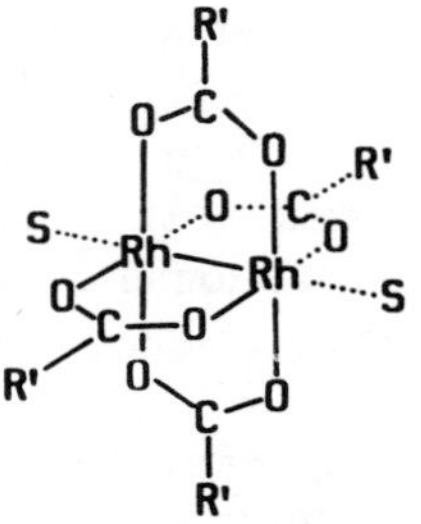

13

chiral R' groups :

e.g., R' = $-^*CH-N$

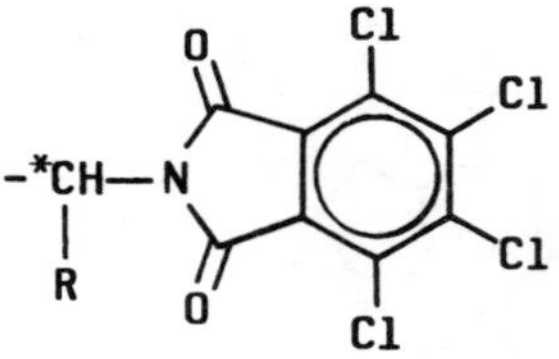

13-a   R = CH$_3$

13-b   R = CH$_2$C$_6$H$_5$

4.2.1.2. *Cis* to *trans* selectivity

The problem of *cis-trans* selectivity in cyclopropane synthesis from olefins and diazoesters is of practical importance in some particular circumstances in terms of biological activity. It is the case, *inter alia*, for the important chrysanthemic and permethric acids related insecticides; the *E* isomer displays the biological activity in the chrysanthemic acid case whereas the *Z* isomer is the active form with permethric acid.

However, controlling the *cis-trans* selectivity in cyclopropanation reactions of olefins cannot be easily achieved. In general, no clear correlations appear between such selectivities and one particular factor. Moreover the preference for one particular geometrical isomer is generally low. The observed effects result obviously from a balance between steric and electronic interactions around the coordination sphere of the metal. The nature of the metal and of the ligands are particularly determining. In general, the trans isomer is preferred, stressing the importance of the steric factors in the transition state.

For example, the palladium acetate-catalysed reaction of styrene (**14**, $R=C_6H_5$) with ethyl diazoacetate (Equation 6) displays a notable selectivity in favour of the *trans* isomer (*trans/cis* = 2.0), but the ratio decreases (down to *trans/cis* = 1.0) when triphenylphosphite (3 moles/mole of metal) is added to the system [27].

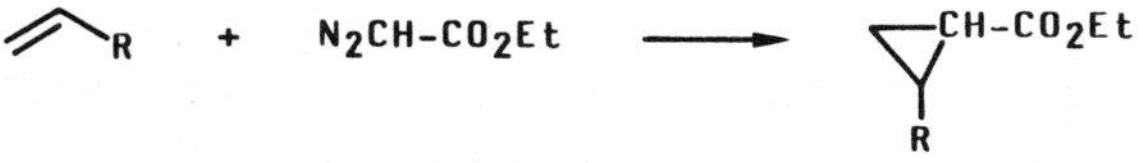

$$\text{R} + N_2CH-CO_2Et \longrightarrow \overset{CH-CO_2Et}{\underset{R}{\triangle}} \qquad (6)$$

14   **15** (cis + trans)

On the other hand, with rhodium carboxylates as catalysts, the *cis* to *trans* ratio displays a rather wide gap of distribution according to the nature of the carboxylate group and the olefinic substrate. The cyclopropanation reaction of various olefins affords mixtures of isomers (*Z:E* or *endo:exo*) ranging from 1:50 up to about 5:1 according to the nature of the reactants and the catalyst. The crowded rhodium porphyrin catalyst **16** displays a slightly larger preference in favour of the *Z* isomer (Table I).

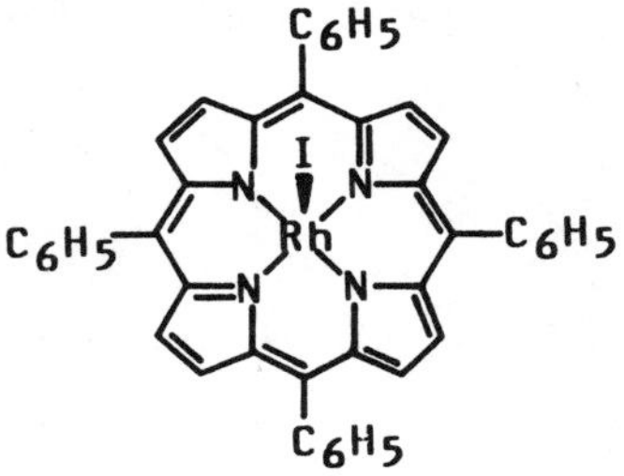

16 (RhTPPI)

TABLE I
Cis/trans or *endo/exo* ratio for cyclopropane carboxylates

Catalysts	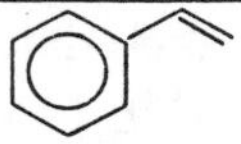 ⌬⁀	 ⬡	 ⟁	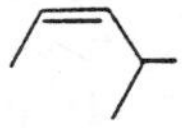 ⧸⧹
RhTPPI(**16**)	1,13	0,84	1,85	4,9
Rh(II)	1,22	0,94	0,86	4,2
CuCl	–	0,12	0,02	0,56

Similarly, the copper-semicorrin complex **11** displays a significant preference for the formation of the *E*-cyclopropane from various olefins (styrene, butadiene, 1-heptene); a *E/Z* ratio as high as 85/15 being obtained from styrene. Here again, large steric requirement in the ligand appears beneficial as far as both enantioselectivity and *E* to *Z* selectivity are concerned [26].

4.2.2. Regioselectivity and discrimination between different substrates and functions

a. *Regioselectivity* control of carbene reactions has been particularly successfully handled by application of transition-metal catalysis. In some cases, interesting applications resulted from these studies.

For example, polyolefins containing differently substituted unsaturations display different patterns of distribution of the products according to the catalysts (Scheme 1) [28]. Similarly, competitive cyclopropanation reactions between pairs of differently substituted olefins show that the discrimination between the double bonds appears as deeply dependent on various factors such as the catalyst (nature of the metal, its oxidation state and the ligands). Moreover, the olefinic discrimination seems to be essentially related to steric factors and to coordinating ability in some cases, whereas the electron density together with the ability to stabilize charge development appear determining in other cases.

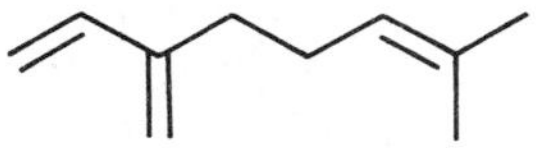

$Rh_2(OAc)_4$	39	48	0
$Pd(OAc)_2$	25	8	0

Scheme 1

b. *Discrimination* in olefin pairs has been classically applied in the carbene field since the pioneering approach of the mechanistic problems: the recognition of the electrophilic nature of classical carbenes and carbenoids (particularly copper carbenoids) has thus been established.

More recently, a duality of mechanisms (coordination (Equation 7) versus carbenoid (Equation 8)) has been recognized by comparing copper triflate to copper acetylacetonate [16]: the former contains the weak triflate ligand which allows therefore the participation of the olefin as a ligand inside the coordination sphere of the metal, in opposition to the strong acetylacetonate ligand which cannot be exchanged with the olefin substrates.

COORDINATION MECHANISM :   (7)

$$XCuL \; + \; S \; \rightleftharpoons \; XCu(S) \; + \; L$$
$$17$$

$$XCu(S) \; + \; N_2CHCO_2R \; \xrightarrow{-N_2} \; (S)XCu=CHCO_2R$$
$$18$$

$$\longrightarrow \; cyclopropanes \; + \; CuX$$

CARBENOID MECHANISM :   (8)

$$XCuL \; + \; N_2CHCO_2R \; \longrightarrow \; XCu=CHCO_2R \; + \; L$$
$$19$$

$$\xrightarrow{+ \; S} \; cyclopropanes \; + \; CuX$$

S = olefin

It appeared that rhodium and palladium carboxylates displayed very different reaction pathways and were acting according to two essentially different mechanisms. The general pattern in the palladium case is reminiscent of the coordination mechanism of Kochi (Equation 7) [16] whereas rhodium presents a pattern indicative of a carbenoid out-of-sphere process (Equation 8) [1c, 5]. Some recent observations indicate however that other parameters such as the development and stabilization of charges are other important factors [17].

c. *Competitions* in C–H insertion reactions of alkanes (Equation 9) allowed to stress the importance of «solvation effects» of the catalytic center in determining the discrimination between different positions inside an alkane molecule (primary, secondary and tertiary CH bonds (Scheme 2)) as well as between alkane pairs.

$$AlkH \; + \; N_2CHCO_2R \; \xrightarrow{[Rh(II)]} \; Alk-CH_2CO_2R \; + \; N_2 \quad (9)$$
$$20$$

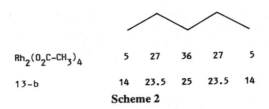

Rh$_2$(O$_2$C–CH$_3$)$_4$	5	27	36	27	5
13–b	14	23.5	25	23.5	14

Scheme 2

Subtle effects were thus discovered [14, 15, 24, 25], some of them being reminiscent of the well known lipophilic interactions in enzymes. For example, notable discrimination between very similar molecules (such as *n*-hexane and *n*-decane or cyclopentane and cyclooctane) was thus established (Table II).

TABLE II
Selectivities of C-H insertions in intermolecular competitions
between pairs of alkanes [15, 24]

Alkane pair	Catalyst	Ratio of C–H insertion
c–C$_5$H$_{10}$/c–C$_8$H$_{16}$	Rh$_2$(O$_2$CCF$_3$)$_4$	0.60
	Rh$_2$(O$_2$C-n–C$_7$F$_{15}$)$_4$	1.80
n–C$_6$H$_{14}$/n–C$_{10}$H$_{22}$	Rh$_2$(O$_2$CCF$_3$)$_4$	0.5
	Rh$_2$(O$_2$CCH$_3$)$_4$	0.6
	Rh$_2$(O$_2$C-9-triptycenyl)$_4$	1.0

Ethyl diazoacetate; equivalent weight of alkanes.

Among other considerations, theoretical calculations lead to the prediction that the activation energy for the insertion process will depend on the angle of approach of the intermediate Rh-carbenoid to the target C–H bond. Moreover, electron-withdrawing substituents decrease the reactivity of the adjacent C–H bonds (α- and β-methylene in the case of ester **21**) (Equation 10) [34].

$$
\text{21} \xrightarrow{\ \text{Rh}_2(\text{O}_2\text{C–CH}_3)_4\ } \text{22} \qquad 83\,\% \qquad (10)
$$

21   22   (10)

With α-diazoketones, the order of reactivity of the C–H sites was found to follow the decreasing sequence: methine, methylene, methyle. Allylic and benzylic C–H were found to be less reactive than aliphatic C–H bonds. It is thus apparent that even if the mechanism of Rh-mediated C–H insertions is not fully established, the selectivity is governed by a delicate balance of both steric and electronic factors. In some cases, it has been possible to tune this balance by modifying the electronic demand and steric bulk of the ligands on the rhodium.

d. *Regioselectivity and discrimination in the Büchner's reaction*: ring enlargement reaction of the benzene ring (Equation 11) offers a straightforward access to the cycloheptatriene system [29].

$$\text{(11)}$$

23

The original reaction catalysed by copper-based systems (poor yield, poor regioselectivity in favour of the isomer 23) was dramatically improved by the use of rhodium carboxylate catalysts: quantitative yields together with 95% selectivity in favour of the kinetic isomer 23 was thus achieved with benzene (R^1=H). Rhodium trifluoroacetate was particularly active, even at room temperature, but was irreversibly deactivated at 0°C; rhodium acetate was also active but at temperature above 50°C. Definite regioselectivities were observed in the case of substituted benzenes (Table III); for instance, one single positional isomer was formed from *p*-xylene.

The mechanism seems to involve essentially very active electrophilic rhodium carbenoids. It is noteworthy that the formation of free carbenes seems probable in some cases, particularly with rhodium trifluoroacetate and diazomalonate, owing to the destabilizing effect of the electroattracting trifluoroacetate ligand towards the electrophilic carbene moiety. In opposition to ethyl diazoacetate (EtDA) (Table IV), ethyl diazomalonate (EtDM) gives practically the same distribution of products with the three modes of generation of the carbene [13, 24].

e. *Discrimination between different functionalities: chemoselectivity*. Competition reactions between olefins and alcohols have been investigated with allylic alcohols [30] whereas the acetylene-alcohol case was considered with propargylic alcohols.

A. Demonceau et al.

TABLE III
Regioselectivities in ring enlargement of aromates [29]

Substrate	Isomers	Distribution of isomers (%)	
		$R^1=CH_3$	$R^1=Cl$
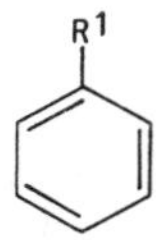	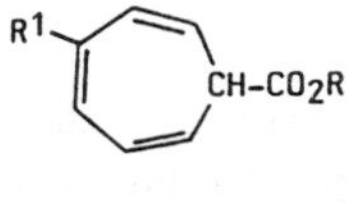 CH–CO$_2$R	56	80
	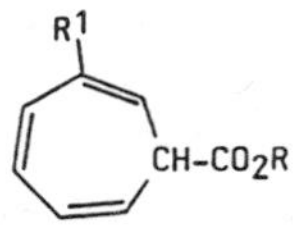 CH–CO$_2$R	23	15
	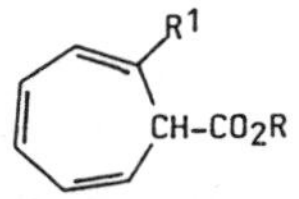 CH–CO$_2$R	17	5
	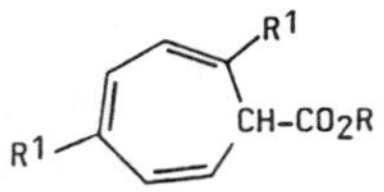 CH–CO$_2$R	95	–

TABLE IV
Comparison of catalytic, thermal and photolytic decompositions
of diazoesters in benzene-cyclohexane (1:1) mixtures

Generation mode of carbene	EtDA	EtDM
$Rh_2(O_2CCF_3)_4$	5.7	0.28
Pyrolysis	0.86	0.33
Photolysis	0.82	0.26

The insertion/cycloaddition ratio in both cases depends essentially on the substitution of the allyl/propargyl CH_2 group. Interestingly, the cyclopropane lactone **26** resulting from the intramolecular transesterification of the *cis*-cyclopropane isomer, was obtained (Equation 12).

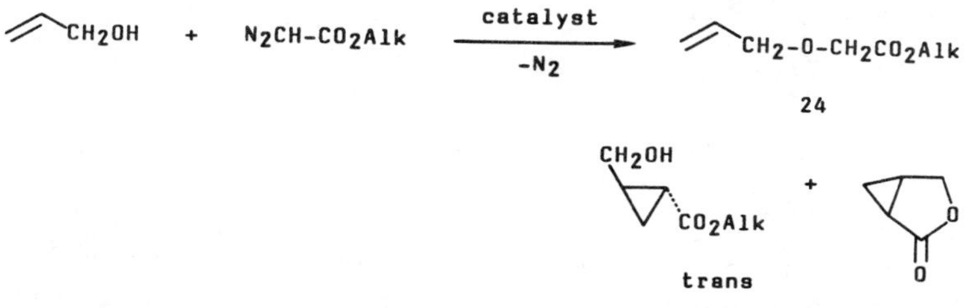

$$\text{(12)}$$

24

trans

25   26

Discrimination between alcohol and thiol groups occurs in favour of the thiols [31] (despite the fact that the thiols give lower rates than the corresponding alcohols when the reaction is run separately rather than in competition experiments with both compounds mixed together in a 1:1 molar ratio).

5. Applications

Some typical useful applications of stereoselective reactions of carbenes are presented in this section.

a. Chrysanthemate and permethrate synthesis.Pyrethroids derived from chrysanthemic acid **28-a** (R′=H) and permethric acid **28-b** (R′=H) (Equation 13) exhibit exceptionally potent insecticidal activity together with very low mammalian toxicity and fast biodegradability. Therefore, their synthesis from readily available precursors provides an attractive commercial target.

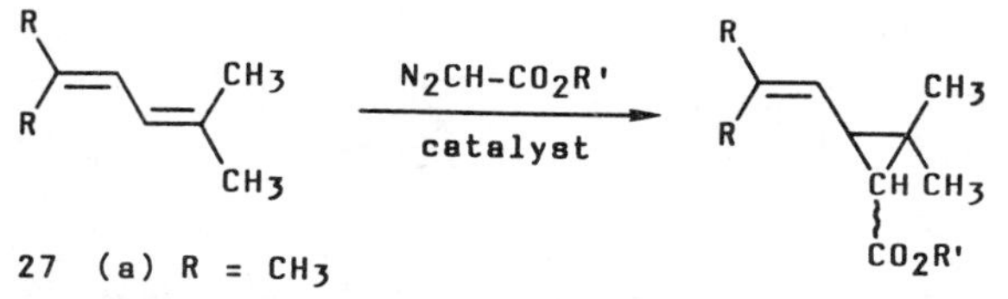

$$\text{(13)}$$

27 (a) R = CH3

(b) R = Cl

28

The best catalysts for induction of chirality in these molecules were discovered by Aratani [20]: they are based on complexes bearing optically imino ligands (29 and 30).

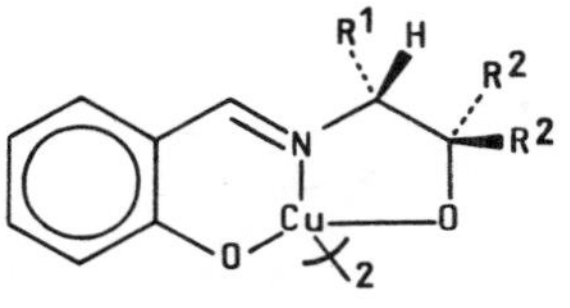

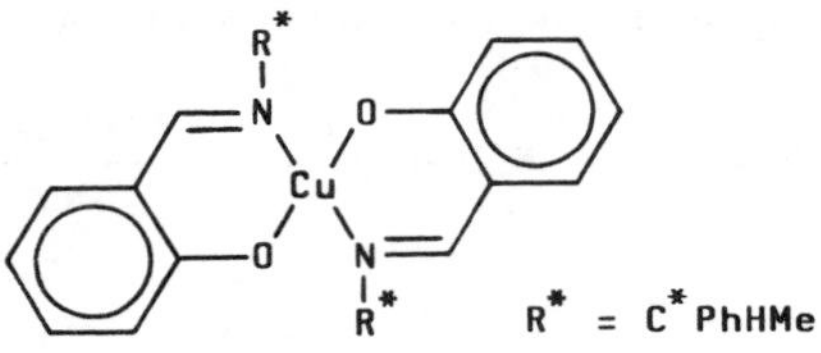

29   30

$R^* = C^* PhHMe$

For instance, a mixture of *cis* and *trans* isomers (28) (R=CH$_3$) was obtained with an enantiomeric excess of 62 and 68%, respectively. Moreover, it has been observed that the optical activity of the products increases with the bulkiness of both the substituent R^2 of 29 and the alkyl group R' of the diazoester. As a consequence, e.e. higher than 90% have been attained with a proper choice of the diazoacetate-catalyst pair.

Rhodium carboxylates are also very efficient and regioselective catalysts for this reaction but they are not suitable for induction of optical activity [32]. However, some control of *cis* to *trans* ratio can be obtained by using suitable bulky carboxylate groups: indeed, the Z cyclopropane isomer 28-b presenting the desired biological activity is formed with a notable selectivity.

b. Stereoselective synthesis of cyclopentane systems (32) (potentially useful for the synthesis of steroids, prostaglandins and various natural and biologically active compounds) are obtained by intramolecular insertion of carbenes in C–H bonds (Equation 14, and Scheme 3) [33, 35].

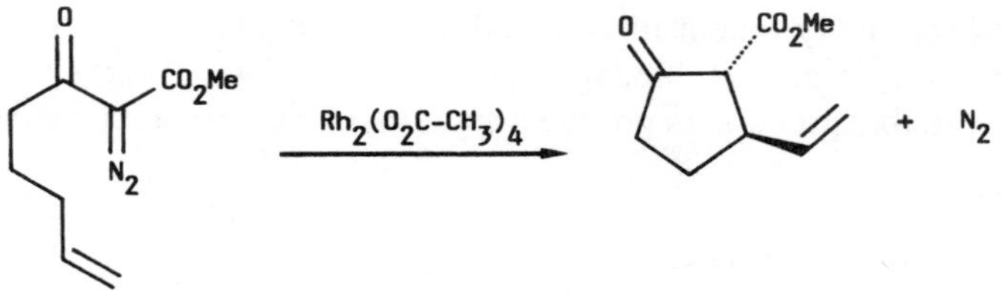

$$Rh_2(O_2C-CH_3)_4 \quad (14)$$

31   32

c. The reaction of ethyl diazoacetate with 2-phenyloxetane 35 in the presence of chiral copper complex results in ring enlargement yielding a *cis-trans* mixture of ethyl 3-phenyl-tetrahydrofuran-2-carboxylate 26 (Equation 15) [36].

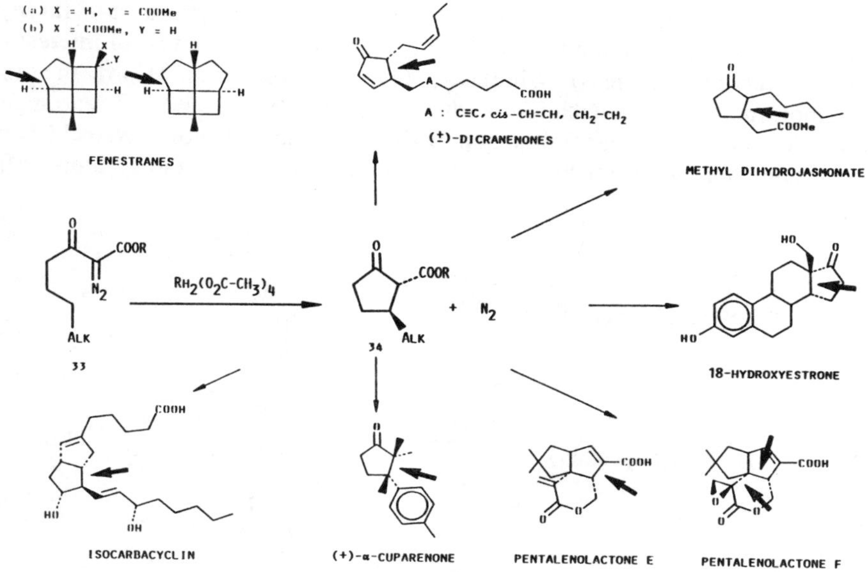

(a) X = H, Y = COOMe
(b) X = COOMe, Y = H

FENESTRANES

A : C≡C, *cis*-CH=CH, CH$_2$-CH$_2$

(±)-DICRANENONES

METHYL DIHYDROJASMONATE

RH$_2$(O$_2$C-CH$_3$)$_4$

33

34

+ N$_2$

18-HYDROXYESTRONE

ISOCARBACYCLIN

(+)-α-CUPARENONE

PENTALENOLACTONE E

PENTALENOLACTONE F

Scheme 3

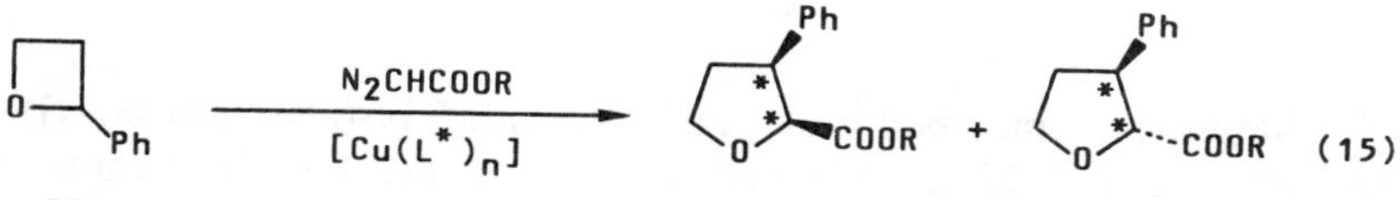

35

N$_2$CHCOOR

[Cu(L*)$_n$]

36-a

+

36-b

Ph

Ph

COOR

COOR

(15)

Recently, rhodium carboxylates have received additional applications in therapeutic compound synthesis.

d. For example, the highly potent antitumor antibiotic CC-1065 has received much attention as it covalently alkylates DNA in a site-selective way by cyclopropane ring opening. Rhodium(II) pivalate affords the best yield (73% at 40°C) in the key cyclopropanation step of its synthesis, whereas a maximum yield of 62% is reached by copper catalysis (Equation 16) [37].

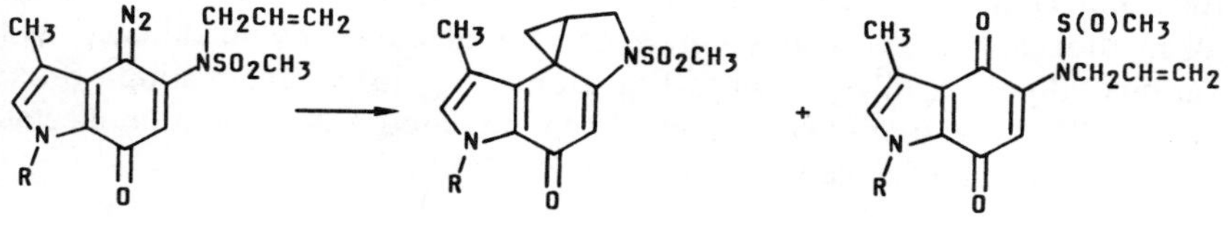

CH$_3$ N$_2$ CH$_2$CH=CH$_2$

NSO$_2$CH$_3$

37

CH$_3$

NSO$_2$CH$_3$

38

CH$_3$ O   S(O)CH$_3$

NCH$_2$CH=CH$_2$

39

+

(16)

e. On the other hand, intramolecular cyclopropanation reaction of furanyl diazoketones leads to the formation of cyclopentenone systems (of interest in prostaglandin synthesis). Similarly, thienyldiazoketones afford the intramolecular cyclopropanation product which can be isolated and rearranges in acidic medium to benzothiophen-5(4*H*)-one. The carbenoid cycloaddition-rearrangement sequence has been extended to diazobenzofuran systems [39].

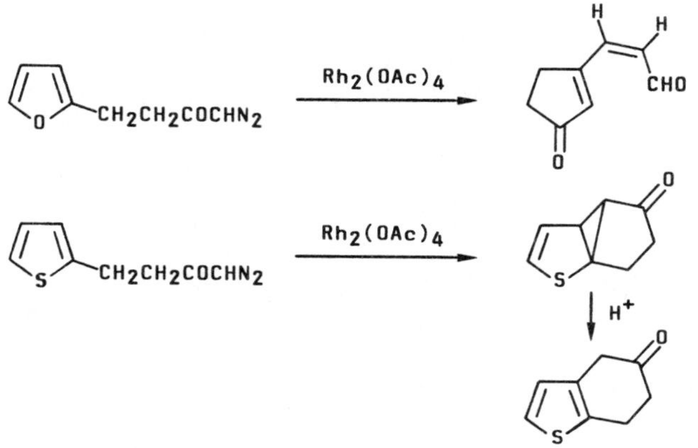

$$(17)$$

f. Moreover, members of the 4-aryl-3,5-bis(alkoxycarbonyl)-1,4-dihydropyridines (**42**) are clinically useful agents for the treatment of cardiovascular diseases such as angina pectoris and hypertension. One key step in analogue synthesis is based on rhodium(II) acetate-catalysed double insertion of carbalkoxycarbenes into substituted aniline N–H bonds (Equation 18) [38].

g. Rhodium carboxylates have also proved to be exceptionally effective catalysts for promoting the addition of carbenes onto the oxygen atom of carbonyl groups to give 1,3-dipoles via an intermediate carbonyl ylide. Further intramolecular and intermolecular 1,3-dipolar additions to dipolarophiles provide a straightforward access to various complex heterocyclic systems from cheap starting materials. In a typical example, rhodium(II) acetate catalyses the formation of ylides **44** from 1-diazo-alkanediones; these ylides undergo bimolecular cycloaddition with various dipolarophiles X=Y (Equation 19) [41]. Interesting 8-oxa-bicyclo(3.2.1)oct-6-en-2-ones have been prepared by a rhodium acetate-catalysed insertion sequence (Equations 20, 21) [40].

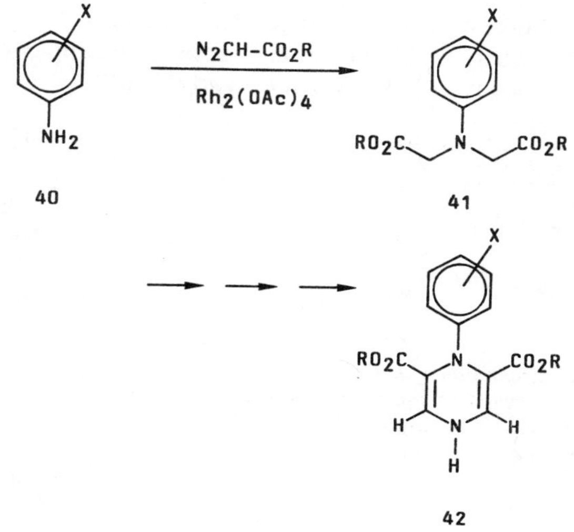
X
NH2
N2CH-CO2R
Rh2(OAc)4
X
RO2C
N
CO2R
40
41
X
RO2C
N
CO2R
H
N
H
H
42

(18)

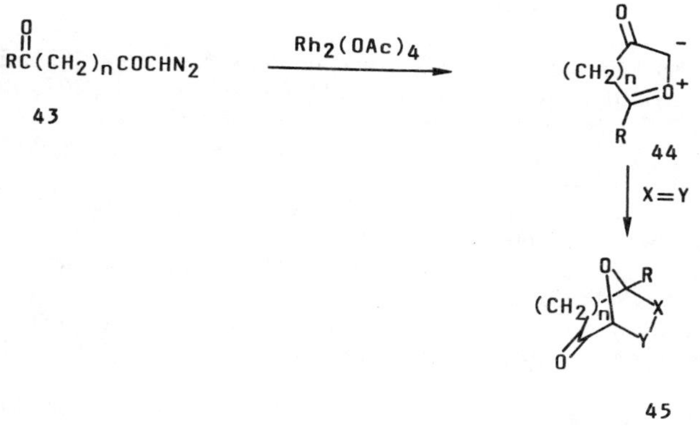
O
RC(CH2)nCOCHN2
43
Rh2(OAc)4
O
(CH2)n
O+
R
44
X=Y
O
R
(CH2)n
X
Y
O
45

(19)

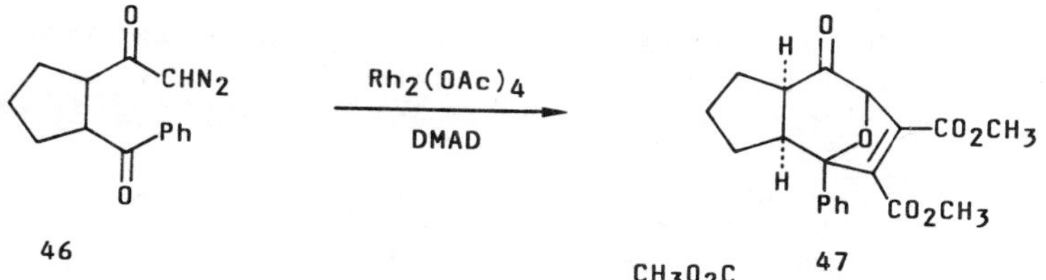
O
CHN2
Ph
O
46
Rh2(OAc)4
DMAD
H
O
O
CO2CH3
H
Ph
CO2CH3
47

(20)

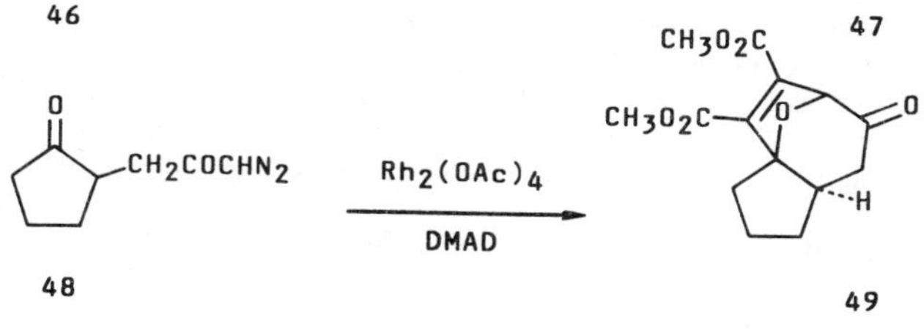
O
CH2COCHN2
48
Rh2(OAc)4
DMAD
CH3O2C
CH3O2C
O
O
H
49

(21)

Furthermore, the synthesis and rearrangement reactions of cyclic sulphonium ylides (**51**, Equation 22) constitutes another example of useful application of rhodium catalysts in heterocycle synthesis [42]

Laboratoire de Synthèse Organique et de Catalyse
Laboratoire de Chimie Macromoléculaire et de Catalyse Organique
University of Liège, Institute of Chemistry (B.6),
SART TILMAN, B-4000 Liège (Belgium)

References

1. (a) Kirmse, W. *Carbene Chemistry*, 2nd ed.; Academic Press, New York, 1971; (b) *Carbenes*; Vol. I and II; Moss, R.A. and Jones, M. Jr, Wiley, New York, 1973 and 1975; (c) Hubert, A.J. *Catalysis in C-1 Chemistry*; W. Keim, Reidel, 1983.
2. Le, N.A.; Jones, M. Jr; Bickelhaupt, F. and de Wolf, W.H. *J. Am. Chem. Soc.* **1989**, *111*, 8491.
3. Gordon, M.S.; Boatz, J.A.; Gano, D.R. and Friederichs, M.G. J. Am. Chem. Soc. 1987, 109, 1323. Becerra, R. and Frey, H.F. *Chem. Phys. Lett.* **1987**, *138*, 330.
4. Demonceau, A.; Noels, A.F. and Hubert, A.J. *Aspects of Homogeneous Catalysis*, Vol. 6; R. Ugo, Reidel, 1988; 199.
5. Maas, G. *Top. Curr. Chem.* **1987**, *137*, 75. Hubert, A.J.; Demonceau, A. and Noels, A.F. *Industrial Applications of Homogeneous Catalysis*; A. Mortreux and F. Petit, Reidel, 1988; 93.
6. Chang, S.C.; Kafafi, Z.H.; Hauge, R.H.; Margrave, J.L.; Billups, W.E. *Tetrahedron Lett.* **1987**, *28*, 1733.
7. Dötz, K.H.; Fischer, H.; Hofmann, P.; Kreissl, F.R.; Schubert, V. and Weiss, K. *Transition Metal Carbene Complexes, Verlag Chemie*, **1983**.
8. Chang, S.C.; Kafafi, Z.H.; Hauge, R.H.; Billups, W.E. and Margrave, J.L. *J. Am. Chem. Soc.* **1987**, *109*, 4508 and *Ibid* **1988**, *110*, 7975.
9. Carter, E.A. and Goddard, W.A. *J. Am. Chem. Soc.* **1986**, *108*, 2180.
10. Brothers, P.J. and Roper, W.C. *Chem. Rev.* **1988**, *88*, 1293.
11. Klein, D.P. and Bergman, R.G. *J. Am. Chem. Soc.* **1989**, *111*, 3079.
12. Adams, R.D. *Chem. Rev.* **1989**, *89*, 1703.
13. Demonceau, A.; Noels, A.F.; Costa, J.-L. and Hubert, A.J. *J. Mol. Catal.* **1990**, *58*, 21.
14. Demonceau, A.; Noels, A.F.; Teyssié, P. and Hubert, A.J. *J. Mol. Catal.* **1988**, *49*, L13.
15. Demonceau, A.; Noels, A.F. and Hubert, A.J. *J. Mol. Catal.* **1989**, *57*, 149.

16. Salomon, R.G. and Kochi, J.K. *J. Am. Chem. Soc.* **1973**, *95*, 3300.
17. (a) Doyle, M.P.; Tamblyn, W.H. and Bagheri, V. *J. Org. Chem.* **1981**, *46*, 5094; (b) Doyle, M.P.; Dorow, R.L.; Buhro, W.E.; Griffin, J.H.; Tamblyn, W.H. and Trudell, M.L. *Organometallics* **1984**, *3*, 44; (c) Doyle, M.P.; Dorrow, R.L.; Tamblyn, W.H. and Buhro, W.E. *Tetrahedron Lett.* **1982**, *23*, 2261.
18. Callot, H.J.; Albrecht-Gary, A.M.; Al Joubbeh, M. and Metz, B. *Inorg. Chem.* **1989**, *28*, 3633.
19. (a) Nakamura, A.; Konishi, A.; Tatsuno, Y. and Otsuka, S. *J. Am. Chem. Soc.* **1978**, *100*, 3443, 6544; (b) Nakamura, A.; Konishi, A.; Tsujitani, R.; Kudo, M.A. and Otsuka, S. *Ibid* **1978**, *100*, 3449.
20. Aratani, T.; Yoneyoshi, Y. and Nagase, T. *Tetrahedron Lett.* **1975**, *16*, 1707; *Ibid* **1977**, *18*, 2599; *Ibid* **1982**, *23*, 685.
21. (a) Nozaki, H.; Moriuti, S.; Takaya, H. and Noyori, R. *Tetrahedron Lett.* **1966**, *17*, 5239; (b) Nozaki, H.; Takaya, H.; Moriuti, S. and Noyori, R. *Tetrahedron* **1968**, *24*, 3655; (c) Noyori, R.; Takaya, H.; Nakanisi, Y. and Nozaki, H. *Tetrahedron* **1969**, *47*, 1242.
22. Brunner, H. and Miehling, W. *Monatsh. Chem.* **1984**, *115*, 1237.
23. Hubert, A.J.; Noels, A.F.; Anciaux, A.J. and Teyssié, Ph. *Synthesis* **1976**, *9*, 600.
24. Demonceau, A., Ph. D. Thesis, Liège, 1987.
25. Demonceau, A.; Noels, A.F.; Teyssié, Ph. and Hubert, A.J. *Bull. Soc. Chim. Belg.* **1984**, *93*, 945.
26. Fritschi, H.; Leutenegger, V. and Pfaltz, A. *Angew. Chem. Int. Ed. Engl.* **1986**, *25*, 1005.
27. Paulissen, R.; Hubert, A.J. and Teyssié, Ph. *Tetrahedron Lett.* **1972**, *15*, 1465.
28. Petiniot, N.; Noels, A.F.; Anciaux, A.J.; Hubert, A.J. and Teyssié, Ph. *Fundamental Research in Homogeneous Catalysis*, Vol. 3; M. Tsutsui, Plenum Publishing Corporation, 1979; 421.
29. Anciaux, A.J.; Demonceau, A.; Hubert, A.J.; Noels, A.F.; Petiniot, N. and Teyssié, Ph. *J. Chem. Soc., Chem. Commun.* **1980**, 765.
30. Noels, A.F.; Demonceau, A.; Petiniot, N.; Hubert, A.J. and Teyssié, Ph. *Tetrahedron* **1982**, *38*, 2733.
31. Hayez, E., Ph. D. Thesis, Liège, 1975.
32. Petiniot, N., Ph. D. Thesis, Liège, 1981.
33. Tsuji, J.; Okumoto, H.; Kobayashi, K. and Takahashi, T. *Tetrahedron Lett.* **1980**, *21*, 1475.
34. (a) Taber, D.F.; Salem, S.A. and Korsmeyer, A.W. *J. Org. Chem.* **1980**, *45*, 4699; (b) Taber, D.F.; Ruckle, R.E. Jr and Henessy, M.J. *J. Org. Chem.* **1986**, *51*, 4077.
35. Mori, K. *Tetrahedron* **1988**, *44*, 2835.
36. Nozaki, H.; Takaya, H.; Moriuti, S. and Noyori, R. *Tetrahedron* **1968**, *24*, 3655.
37. Sundberg, R.J.; Baxter, E.W.; Pitts, W.J.; Ahmed-Schoffield, R. and Nishiguchi, T. *J. Org. Chem.* **1988**, *53*, 5097.
38. Chorvat, R.J. and Rorig, K.J. *J. Org. Chem.* **1988**, *53*, 5779.
39. Padwa, A.; Wisnieff, T.J. and Walsh, E.J. *J. Org. Chem.* **1989**, *54*, 299.
40. Padwa, A.; Fryxell, G.E.; Zhi, L. and Hornbuckle, S.F. *Tetrahedron Lett.* **1988**, *29*, 6889.
41. Padwa, A.; Chinn, R.C.; Hornbruckle, S.F. and Shi, L. *Tetrahedron Lett.* **1987**, *30*, 301.
42. Moody, C.J. and Taylor, R.J. *Tetrahedron Lett.* **1988**, *29*, 6005.

S. MAIORANA, C. BALDOLI AND E. LICANDRO

CARBONYLCHROMIUM(O) COMPLEXES
IN ORGANIC SYNTHESIS

The title topic is very extensive and many papers are published in this field every year. The aim of this presentation is to pick out from such literature information necessary to give an overall picture of the chemical behaviour of carbonylchromium(O)complexes and of their utilization in synthesis. Two main classes of organometallic compounds can be referred to as carbonylchromium complexes:

1 – Arenetricarbonylchromium complexes **1**

$$\text{arene-}Cr(CO)_3 \quad R$$

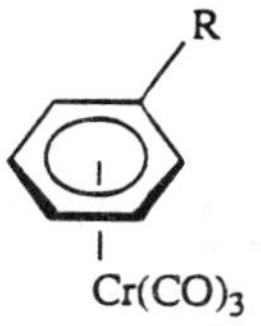

1

2 – Pentacarbonylcarbenechromium complexes **2**

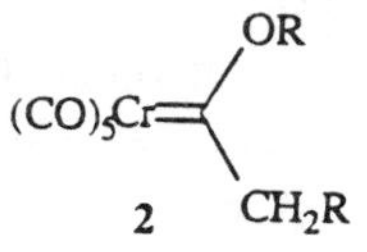

$$(CO)_5Cr= \overset{OR}{\underset{CH_2R}{\big\langle}}$$

2

1. Arenetricarbonylchromium complexes

Some of the reasons why these compounds have become subject of extensive research are: *i)* the preparation is easy and reagents are readily available; *ii)* complexes are easily handled and are quite stable although reactive; *iii)* decomplexation of the arene ligands is easy.

261

A. F. Noels et al. (Eds.), Metal promoted selectivity in organic synthesis, 261–286.
© 1991 *Kluwer Academic Publishers. Printed in the Netherlands.*

1.1. METHODS OF PREPARATION AND DECOMPLEXATION [1]

The main preparation methods include the direct complexation of the arene ring with $Cr(CO)_6$ in donor solvents, *eq.a*, and ligand exchange from $L_3Cr(CO)_3$ complexes, *eq.b*.

$$\text{Arene} + Cr(CO)_6 \rightarrow \text{Arene } Cr(CO)_3 + 3CO \qquad eq.\ a$$
$$L_3Cr(CO)_3 + \text{Arene} \rightarrow \text{Arene } Cr(CO)_3) + 3L \qquad eq.\ b$$

When $L=CH_3CN$, NH_3 the exchange is thermally promoted, while if L=Pyridine the presence of BF_3 as Lewis acid is required.

Ligand decomplexation proceeds generally through an oxidation process achieved, for example, by hv/O_2, I_2, Ce(IV). This destroys the $Cr(CO)_3$ unit to give Cr(III) and CO or CO_2.

The arene ligand can also be displaced by heating with pyridine [2]. This decomplexation method is particularly useful in the presence of groups sensitive to oxidation.

$$\text{Arene } Cr(CO)_3 \; ^{Py} \rightarrow \text{Arene} + \text{Py } Cr(CO)_3$$

Moreover, by this method the $Cr(CO)_3$ unit is recovered as a complex with pyridine which can be recycled to make fresh arene complex.

1.2. EFFECTS OF COORDINATION OF ARENES TO $Cr(CO)_3$

The presence of the $Cr(CO)_3$ group on the arene ring poses two main question: i) how does the $Cr(CO)_3$ unit change the properties of the arene ring? ii) how can the $Cr(CO)_3$ effects be exploited in organic synthesis?

The main effects on the arene ring are summarized in Figure 1.

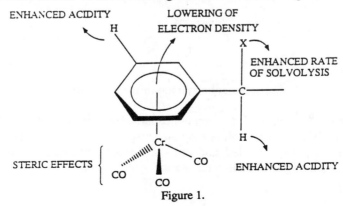

Figure 1.

The chromiumtricarbonyl group acts as a powerful electron withdrawing group. It lowers the electron density of the ring and enhances the acidity of ring and benzylic protons. Moreover, the rate of solvolysis of appropriate substituent at carbon α of the ring is enhanced. Finally the chromiumtricarbonyl group exerts steric effects which are very important in determining the selectivity of substitution reactions.

The electronwithdrawing effect is a consequence of both inductive and resonance mechanisms. The electron donation from the arene σ and π orbitals to the metal is greater than the retro donation from the metal to the arene. As shown (benzoic acid)$Cr(CO)_3$ and (phenylacetic acid)$Cr(CO)_3$ are stronger acids than the corresponding uncoordinated compounds [3]. The electron withdrawing effect of the $Cr(CO)_3$ group is very similar to that of a nitro substituent (Table I)

TABLE I

	pK$_a$		pK$_a$
$C_6H_5CO_2H$	5.68	$C_6H_5CH_2CO_2H$	5.64
$(C_6H_5CO_2H)Cr(CO)_3$	4.77	$(C_6H_5CH_2CO_2H)Cr(CO)_3$	5.02
$p{-}NO_2C_6H_4CO_2H$	4.48	$p{-}NO_2C_6H_4CH_2CO_2H$	5.01

For $Cr(CO)_3$ complexes the chromium atom is relatively electron poor because of the three CO ligands which are good π-acceptors and remove electron density from the metal. The donation from the arene to the metal is then of greater importance than the back donation. However, if one of the CO ligands is replaced by a donor ligand (*e.g.* phosphine or phosphite) then the chromium atom has a relatively higher electron density, back donation becomes more important and related carboxylic acids are less acidic [4], as shown in Table II.

TABLE II

	pK$_a$
$(PhCO_2H)Cr(CO)_3$	4.77
$(PhCO_2H)Cr(CO)_2P(OMe)_3$	5.52
$(PhCO_2H)Cr(CO)_2P(OEt)_3$	5.62
$PhCO_2H$	5.68
$(PhCO_2H)Cr(CO)_2PPh_3$	6.15

1.3. STABILIZATION OF BENZYLIC CARBANION

As shown before, the coordination of arenes to Cr(CO)$_3$ leads to an increase in the acidity of benzylic protons. In other words, the Cr(CO)$_3$ group stabilizes the benzylic carbanions.

As a consequence, deuteration of 1,4-diphenylbutane, where only one of the phenyl rings is complexed, occurs only on the benzylic protons tied to the complexed ring. It is interesting to notice that, in the rigid structure of the indane complex deuteration occurs, for steric reasons, only on protons lying on the side opposite the Cr(CO)$_3$ group [5].

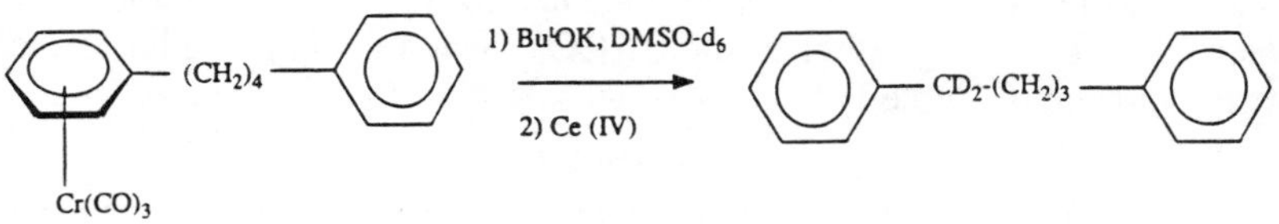

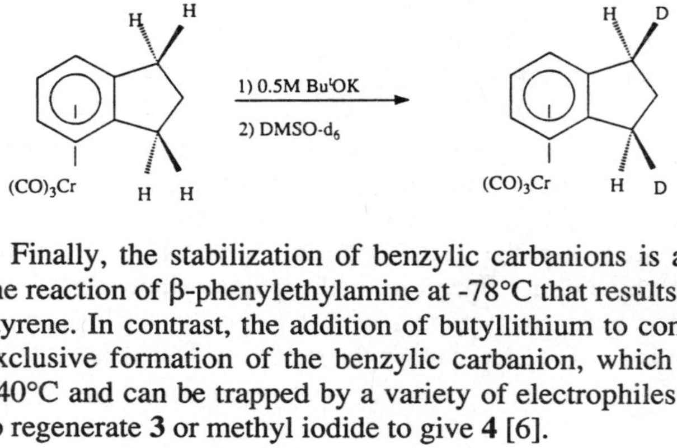

Finally, the stabilization of benzylic carbanions is also demonstrated by the reaction of β-phenylethylamine at -78°C that results in rapid formation of styrene. In contrast, the addition of butyllithium to complex **3** results in the exclusive formation of the benzylic carbanion, which is now stable below −40°C and can be trapped by a variety of electrophiles including H$^+$/MeOH to regenerate **3** or methyl iodide to give **4** [6].

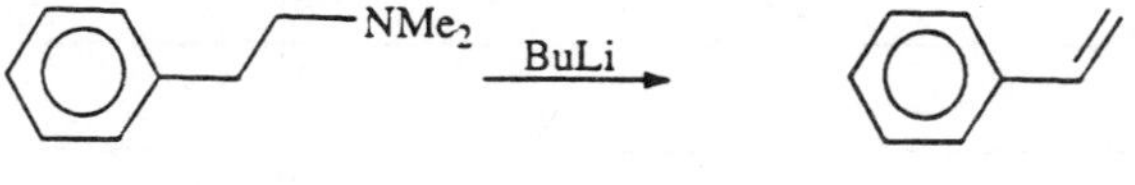

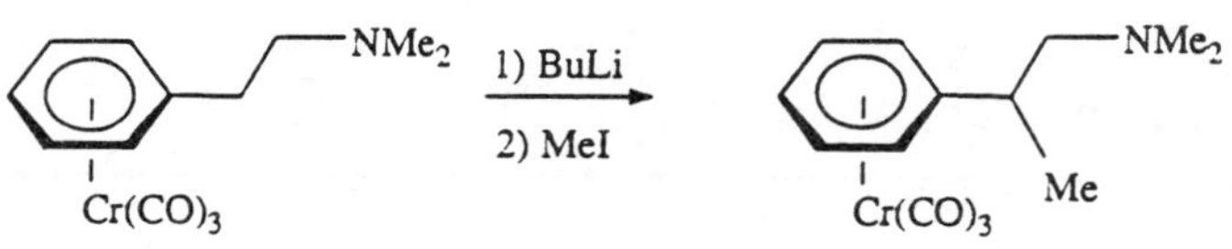

1.4. METALLATION OF THE COMPLEXED ARENES

Metallation of the complexed ring at -78°C is both easier and faster than the metallation of the uncomplexed arene. In the case of anisole as expected, lithiation occurs ortho to the electronegative oxygen function.

Similarly, the fluorobenzene complex is readily lithiated under the same conditions to give a 2-lithio-species that is stable up to –20°C and does not transform into a benzyne complex [8]. In the uncomplexed arenes the reported order of directing abilities is that methoxy is greater than fluorine.

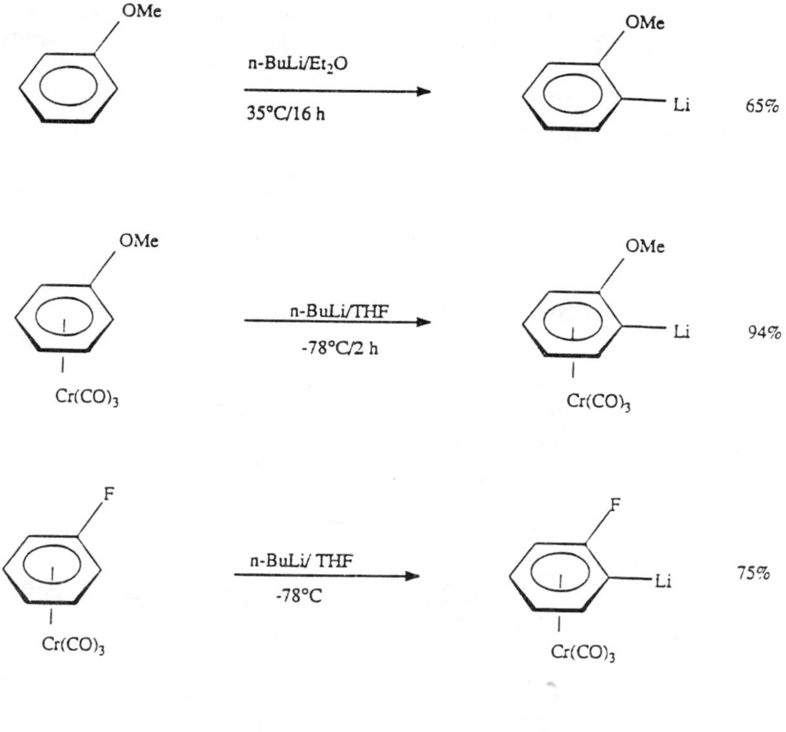

Therefore it was surprising to find out that lithiation of the complexed para-fluoroanisole occurred on a position adjacent to the fluorine atom to an extent greater than 98%. The reason for this reversal of regiospecificity is that the reduced electron density on the oxygen atom does not allow the coordination of the incoming BuLi. This means that the electronegativity of the substituent becomes the main effect directing the lithium ortho to the fluorine atom [9, 10].

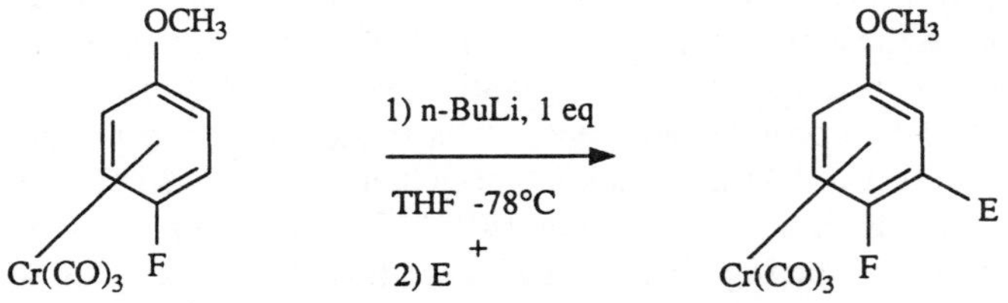

It has been possible to determine experimentally the sequence of substituents of decreasing activating ability for lithiation. Comparing it with the reported sequence for uncomplexed arenes, it is evident that the results are not conventional.

Sequence for complexed arenes

$$F > NH\text{-}CO\text{-}Bu^t > OMe \approx CH_2NMe_2 > CH_2OMe(> NR_2, SR)$$

Sequence for uncomplexed arenes (Wakefield, 1974)

$$CONHR > CONR_2 > NHCOBu^t > CH_2NMe_2 > OMe > F > NR_2 \approx SR$$

Since lithiation of the aromatic ring can be followed by reaction with electrophiles the sequence gives access to polyfunctional arenes with expected and varied regiocontrol.

1.5. NUCLEOPHILIC ATTACK ON η^6-ARENE $Cr(CO)_3$ COMPLEXES

We have seen that the attachment of a $Cr(CO)_3$ unit to an aromatic ring results in a definite lowering of electron density on the ring and enhances its tendency to undergo nucleophilic attack.

When a leaving group is present on the aromatic ring, the reaction results in an addition-elimination process as in nucleophilic aromatic substitution S_NAr. The product is a substituted arene complex that in turn can be oxidized. Carbanions, sulphur and oxygen anions behave in such a way. When the formal leaving group is a hydride ion, nucleophiles are always carbanions and the intermediate cyclohexadiene complex can be differently elaborated to give substituted arenes or, on treatment with acids, monosubstituted cyclohexadienes derivatives [11, 12].

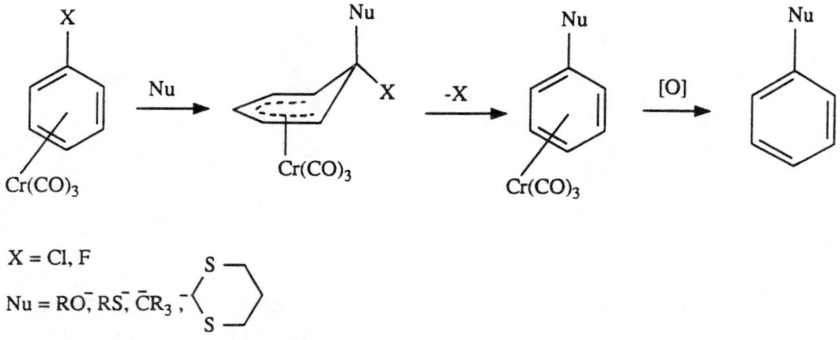

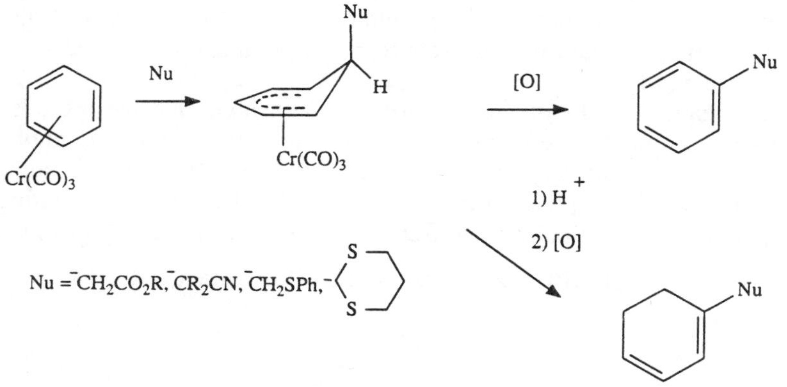

X = Cl, F

Nu = RO⁻, RS⁻, C̄R₃ , ⁻⟨S...S⟩

Nu = ⁻CH₂CO₂R, C̄R₂CN, ⁻CH₂SPh, ⁻⟨S...S⟩

When a substituent, which is not a leaving group, is present on the benzene ring, new problems of regioselectivity arise. The electronic effects of substituents are not varied by $Cr(CO)_3$ group however substituents can influence the adoption of a preferred conformation of the $Cr(CO)_3$ group [13].

For the benzene $Cr(CO)_3$ complex, the staggered form is 0.3 Kcal/mole more stable than the eclipsed one and is the conformation found by X-ray analysis.

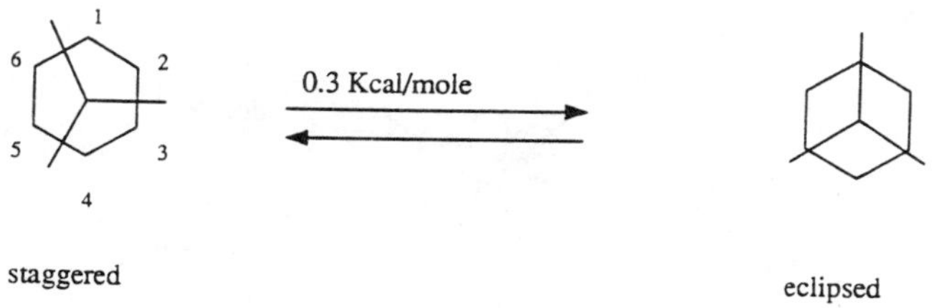

0.3 Kcal/mole

staggered   eclipsed

anti-eclipsed   sin-eclipsed

   Substituted benzene-Cr(CO)$_3$ complexes fall neatly into two classes: in fact X-ray analysis shows that when an electron donating group is present the syn-eclipsed structure is normally found. Conversely the anti-eclipsed structure is observed when an electron withdrawing group is present. Despite the low rotational barriers, NMR studies made by Solladiè-Cavallo seem to show that in solution the main conformeric population is the same as in the solid state [14].

   A simple model explaining these facts considers that the orbitals of the Cr(CO)$_3$ group can be represented as a system of bonding and antibonding hybridized orbitals. Superimposing the system onto the benzene ring, empty antibonding orbitals should point to the region of the molecule with greater negative charge which is o,p to ED substituents, meta to EW substituent.

HYBRIDIZED ORBITALS RELATED TO Cr-CO BONDS

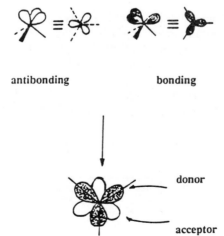

DISPOSITION OF THE Cr(CO)$_3$ GROUP ON THE ARENE RING IN RELATION TO THE NATURE OF THE SUBSTITUENT

In agreement with this model, calculations on eclipsed benzene-Cr(CO)$_3$ show that carbons eclipsed by Cr-CO bonds are electrondeficient compared to those not eclipsed. On the contrary, the conformation adopted by the Cr(CO)$_3$ group seems to have little effect on the arene complex Frontier Molecular Orbitals; in fact the pattern of coefficients in the relevant arene centered LUMO, parallels the patterns of the uncomplexed arene LUMO and is relatively insensitive to the conformation of the Cr(CO)$_3$ unit, as shown in Table III [15].

TABLE III

EHT coefficients for the lowest unoccupied arene-centered molecular orbital in mono-substituted (arene)Cr(CO)$_3$ complexes

Complex conformant	LUMO coefficients						total energy, eV
	1	2	3	4	5	6	
	0.00	-0.55	0.41	0.00	-0.41	0.55	-1311.885
	0.00	-0.41	0.55	0.00	-0.55	0.41	-1311.865

-0.03	-0.54	0.52	0.08	-0.49	0.46	-1311.899
0.00	-0.55	0.41	0.00	-0.41	0.56	-1628.845
0.00	-0.41	0.55	0.00	-0.55	0.41	-1628.347
0.01	-0.55	0.51	0.10	-0.50	0.45	-1628.198

Semmelhack [18].
Estimated LUMO coefficients for substituted arenes.

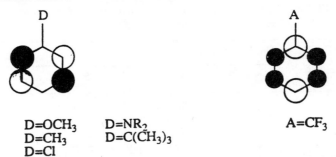

D		A
D=OCH$_3$	D=NR$_2$	A=CF$_3$
D=CH$_3$	D=C(CH$_3$)$_3$	
D=Cl		

Considering all the parameters mentioned before, the experimental data on the regiochemistry of nucleophilic attack on Cr(CO)$_3$ complexed arenes can, in many cases, be rationalized by considering a balance between charge and frontier orbital control [16].

The above interpretation of the observed regiochemistry of the addition of carbanions to arene Cr(CO)$_3$ complexes is limited to reactions that are under kinetic control. The study of the question of kinetic or thermodynamic control of the regioselectivity of addition of carbon nucleophiles to arene $Cr(CO)_3$ complexes has been the subject of several and interesting papers that cannot be reviewed here [17, 18].

Nucleophilic substitution on arene chromiumtricarbonyls has been extensively used for synthetic purposes. The one pot synthesis of 18-dibenzo crown-6 starting from the ortho-dichlorobenzene complex and the sodium salt of diethylene glycol in phase transfer conditions has been reported with an overall yield of 40% [19].

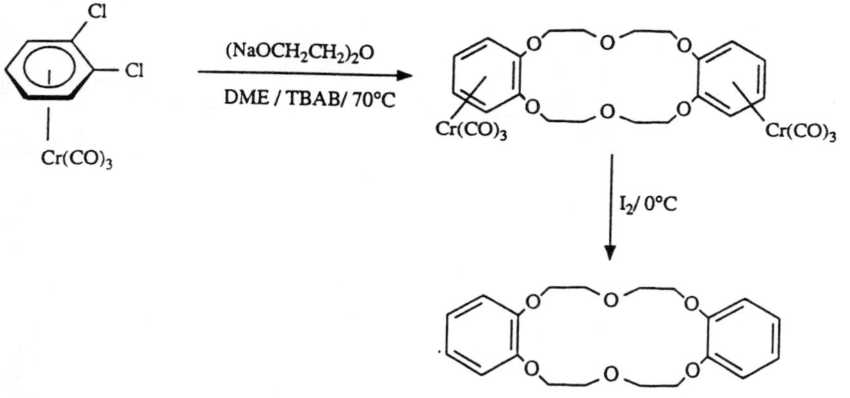

Another example is the synthesis of O-aryloximes which are interesting compounds because they can be easily transformed into benzofuran derivatives through a rearrangement similar to that of the Fischer indole synthesis. The reaction has been carried out in phase transfer conditions at room temperature with very good yields [20]. This method leads to O-arylated oximes 5, substituted also with electron donor groups which are difficult to prepare by different methods.

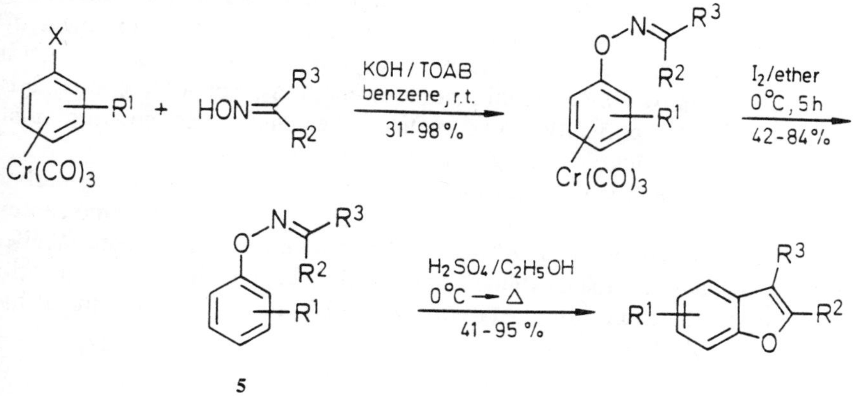

In a similar way, starting from a temporary protected hydroxylamine, it has been possible to synthesize O-aryl-t-butoxycarbonylhydroxylamines [21]. Oxidation with iodine followed by hydrolysis leads to the final O-arylhydroxylamines which are difficult compounds to make, but are useful in the amination reaction of carbanions or azanions. Complexed O-arylhydroxylamines have not been obtained because the hydrolysis of the t-butoxy carbonyl group causes decomplexation of the organic ligand.

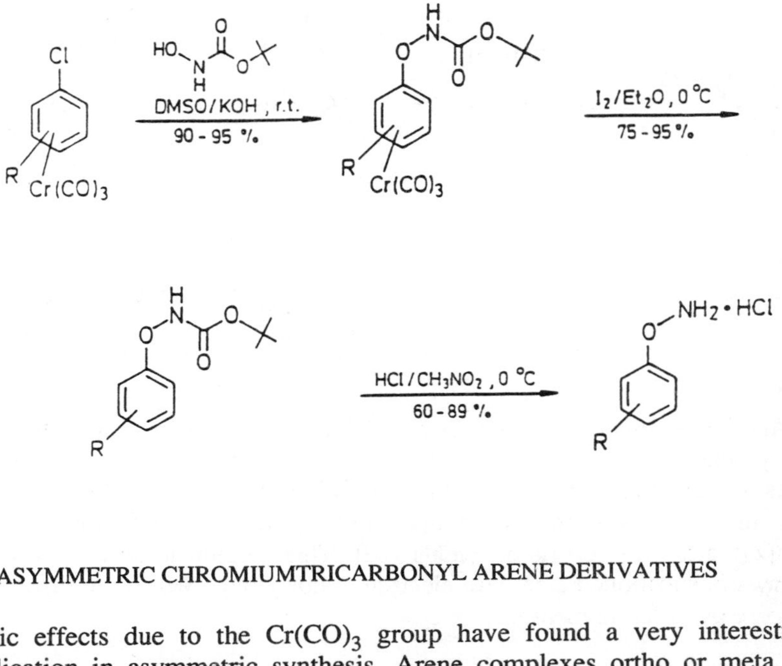

1.6. ASYMMETRIC CHROMIUMTRICARBONYL ARENE DERIVATIVES

Steric effects due to the $Cr(CO)_3$ group have found a very interesting application in asymmetric synthesis. Arene complexes ortho or meta di-substituted with different substituents are asymmetric and can be resolved in the enantiomers by conventional resolution methods. For these compounds, and generally for optically active metallocenes, the stereo notation is an extension of Cahn-Ingold-Prelog sequence rules.

The hapto attachment to the metal is arbitrarily split into metal-carbon bonds and all the ring carbon atoms can be considered as stereogenic centers and be identified by the symbol R or S. The whole molecule is then labelled as R or S based on the absolute configuration having the highest priority according to the same rules. Applying the C.I.P. rules at C_1, the metal has priority; C_2 is second and C_5 is third. Looking at the molecule along the indicated axis, the stereonotation is 1S [22].

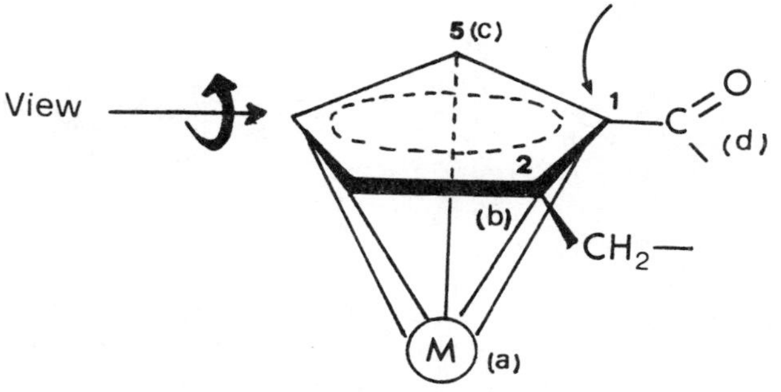

$$(a) > (b) > (c) > (d) : (1S)$$

Optically active complexed aldehydes have been the subject of more recent research, utilizing them as chiral synthons. Resolution of complexed aldehydes can be accomplished by transforming them into optically active oxamazones, separating the two diastereoisomers by column chromatography and hydrolyzing them with sulphuric acid (Scheme 1) [23].

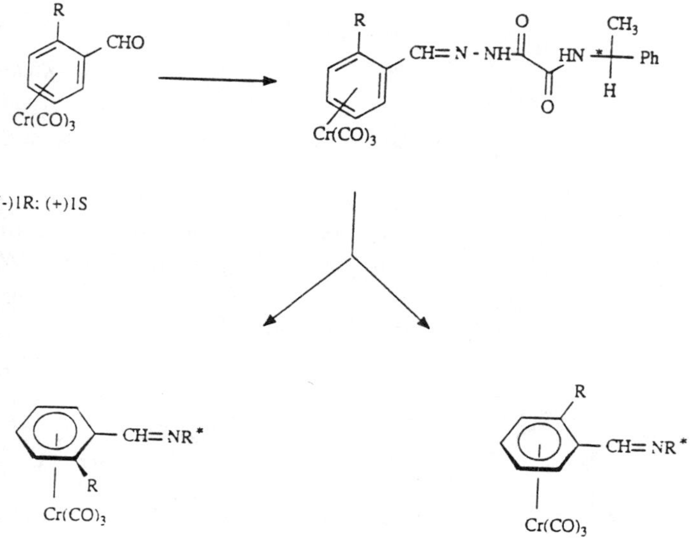

IS

1R

Recently the first example of enantioselective microbial kinetic resolution of planar chiral metallocenic aldehydes has been reported. [24]. One of the two enantiomers was preferentially reduced by baker's yeast. Best results were obtained with 2-methoxy, 2-trimethylsilyl and 2-chloro benzaldehydes, (e.e. were in the range of 60–81%) while meta substituted derivatives gave lower e.e. Further crystallization notably improved (about 15%) the e.e.

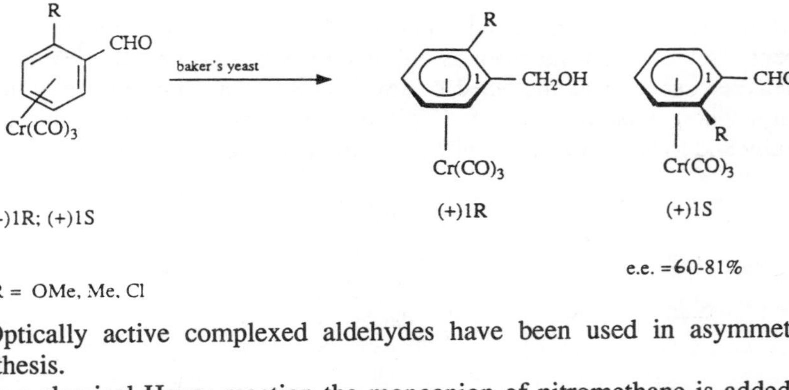

(-)1R; (+)1S

(+)1R

(+)1S

e.e. =60-81%

R = OMe, Me, Cl

Optically active complexed aldehydes have been used in asymmetric synthesis.

In a classical Henry reaction the monoanion of nitromethane is added to the aldehyde at –40°C and in these conditions only one diastereoisomer is obtained. Starting from the optically active aldehyde the e.e. was 92%. After decomplexation the hydroxyl group is protected by TBDMSCl and the nitro group is reduced to give the final aminoalcohol 6 [25].

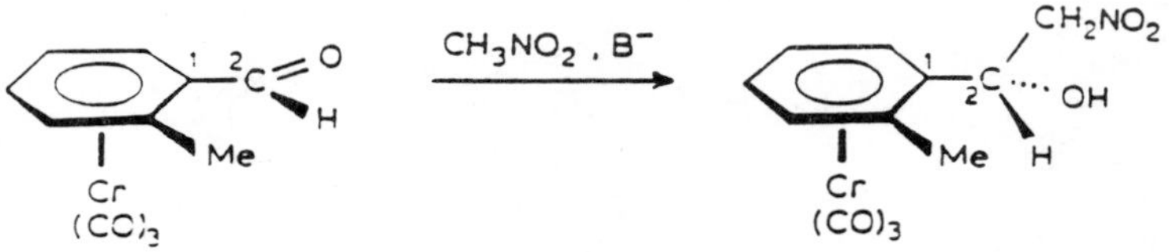

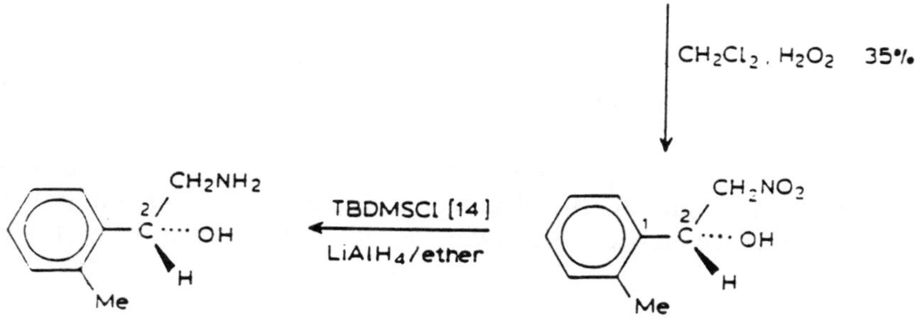

CH₂Cl₂ . H₂O₂   35%.

6

The absolute configuration assigned to **6** is consistent with the enantioselection model generally accepted for this kind of substrates: the aldehyde group undergoes attack by the anion from the side opposite the bulk Cr(CO)₃ group, in that conformation stabilized by the ortho substituent.

The Darzens condensation of ortho-substituted optically active aldehydes gives after decomplexation the optically active epoxides with the highest e.e. ever reported for the Darzens reaction [26]. The e.e. reaches the maximum in the case in which R is chlorine or methoxy group.

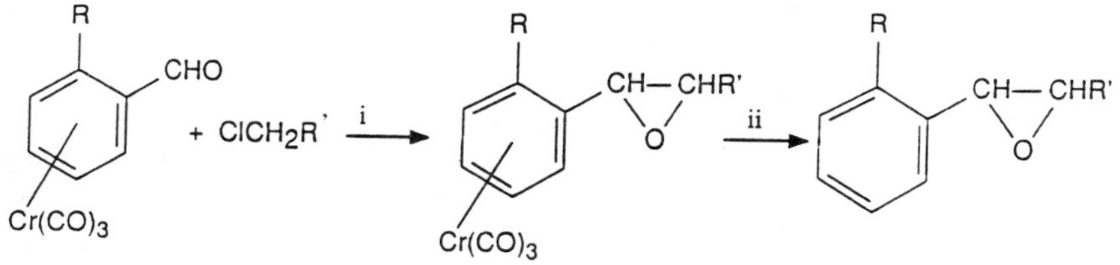

R= CH₃
R= OCH₃
R= Cl

R'= CN
R'= COOC(CH₃)₃
R'= COPh

i= Bu^{t}OH/Bu^{t}OK; r.t.
ii= 30% H₂O₂, TOAB, CH₂Cl₂; r.t.

Also in these cases the absolute stereochemistry at carbon epoxide is in agreement with the same model of enantioselection described before. The substitution by the chlorine atom is interesting since this group can be easily removed if necessary.

2. Pentacarbonylcarbene chromium complexes

Carbonyl-metal carbene complexes of the Fischer type are well represented by the three structures **A, B** and **C** where X is an heteroatom and provides stabilization to the carbene structure by donating its lone pair, as in structure **C**.

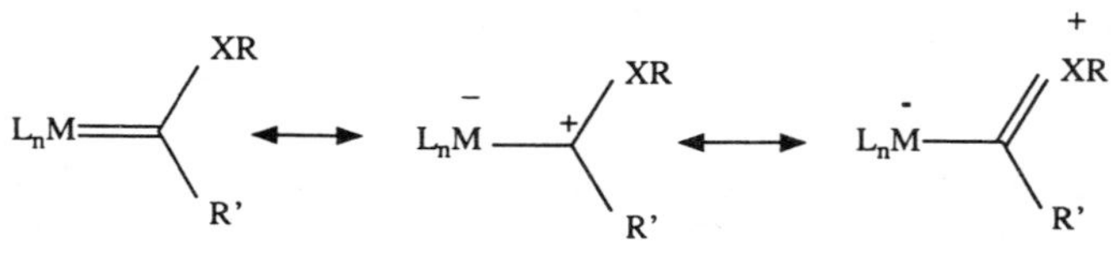

The main chemical features of chromium carbene complexes are reported in the Figure 2:

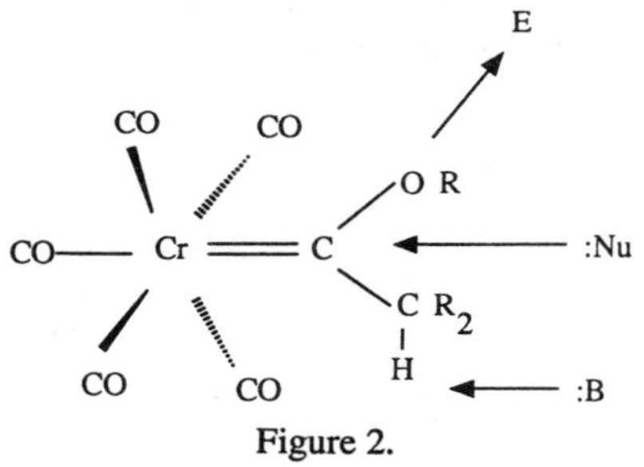

Figure 2.

i) protons α to the carbene carbon atom are acidic; *ii)* the carbene carbon atom is electrophilic; *iii)* electrophiles attack the heteroatom of the alkoxy group; *iv)* CO ligands can be substituted by other ligands.

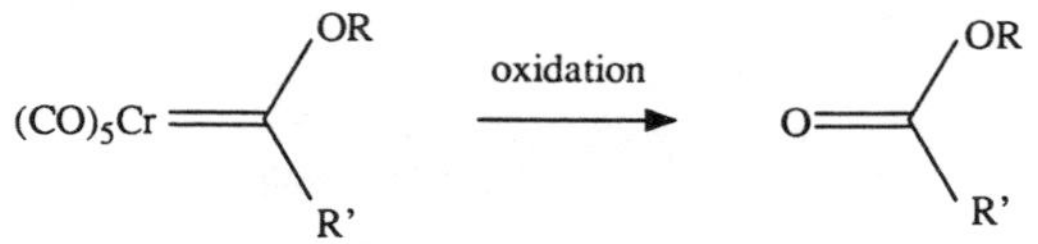

oxidants = CAN, Py-N→O, PhIO, DMSO

From a synthetic point of view it is also important that the bond between the metal and the carbene carbon atom can be transformed into a carbonyl group by oxidation with various agents. Therefore, carbene complexes are formally analogous to carboxylic acid esters but much more reactive.

2.1. SYNTHESIS OF CARBENE COMPLEXES

The synthesis of pentacarbonylcarbene chromium complexes can be performed as represented in Scheme 2.

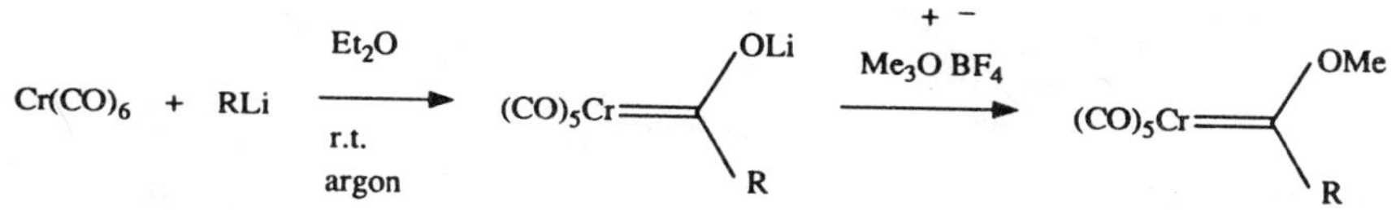

Scheme 2

The reaction of $Cr(CO)_6$ with organolithium compounds gives the corresponding acylmetallates that can be transformed into the stable pentacarbonyl(methoxymethylcarbene)chromium by alkylation with trimethyloxonium tetrafluoborate [27, 28].

2.2. REACTIVITY OF CARBENE COMPLEXES

A very interesting and synthetically useful reaction of carbene complexes was discovered by Doetz in 1975 and has been exploited since then by many other authors. Doetz reacted chromium carbene complexes with alkynes obtaining $Cr(CO)_3$ complexed hydroquinone derivatives. A typical set of examples is reported in Scheme 3.

The alkyne is incorporated at the 2,3-position of the hydroquinone ring. The C_4 ring position comes from the carbene ligand, whereas C_1 is contributed by the carbonyl ligand. The cyclization takes place under mild conditions (40–60°C) in donor solvents.

This reaction has been exploited in the synthesis of vitamins A and E and antracyclinone derivatives [29, 30].

The generally accepted rationalization of the mechanism of this interesting reaction is outlined in Scheme 4.

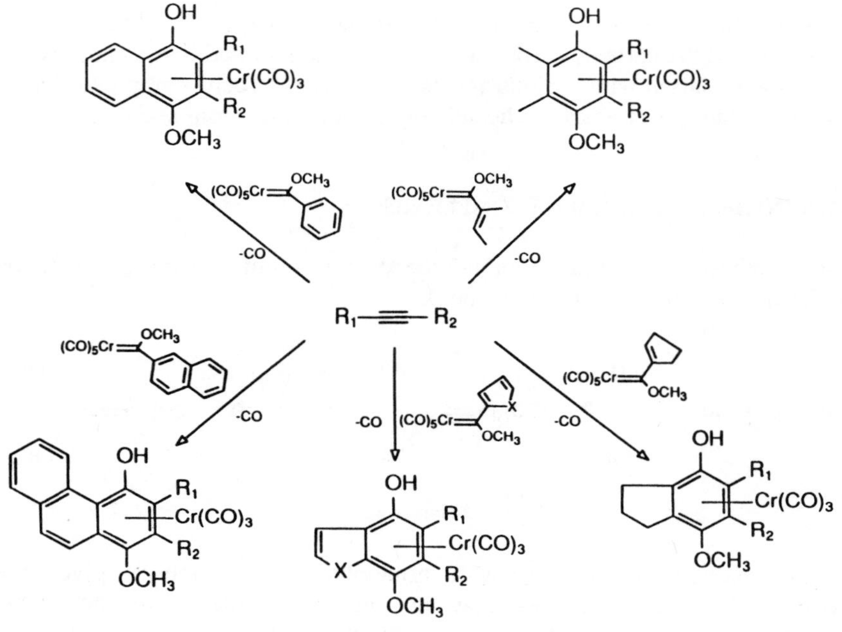

Scheme 3

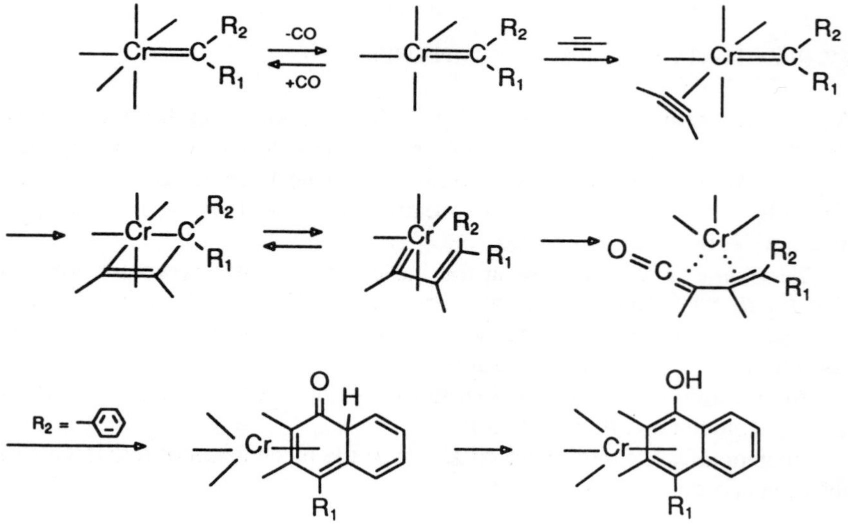

Scheme 4

The thermally promoted cleavage of a CO group in the first step allows the coordination of the alkyne to the metal, followed by the formation of a four membered metallacycle which isomerizes to a new alkenyl carbene complex. The carbonylation of the alkenyl carbene carbon atom by a CO ligand leads to a vinylketene system which is kept in s-cis conformation by a diene-like coordination to the metal. Cyclization of the vinylketene species, gives a metal-coordinated bicyclic cyclohexadienone which readily isomerizes to the final hydroquinone derivative.

Regioselectivity in alkyne incorporation has several aspects [31]. In the case of naphthyl substituted carbene complexes like **7**, annulation can occur in the position 1 or 3 but only the product from ring closure in 1 is isolated.

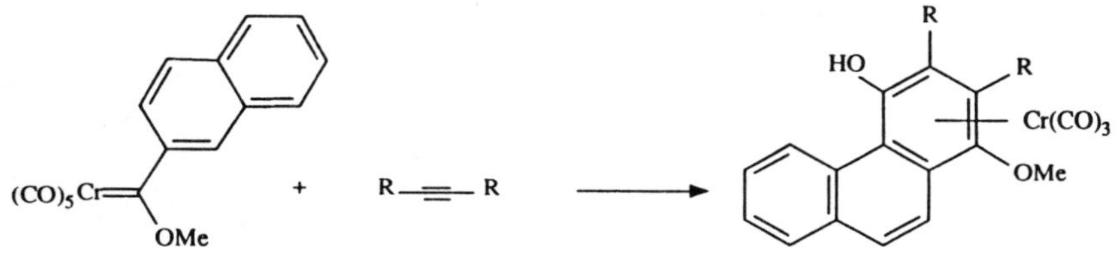

When a phenyl ring is present together with a furane or a naphthalene ring, the alkyne incorporation always occurs on the phenyl ring.

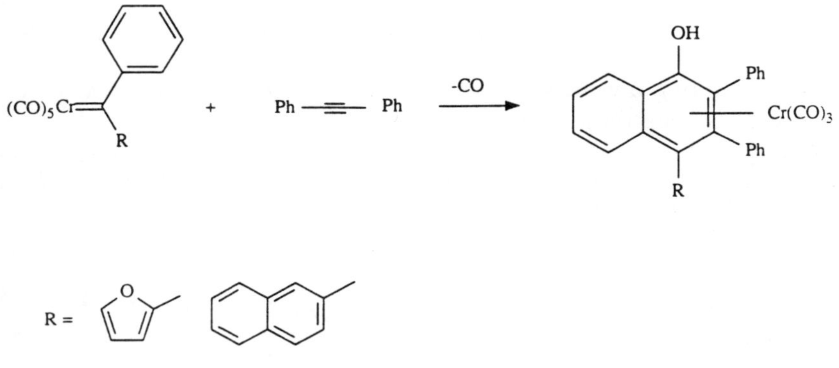

When two differently substituted phenyl rings are present as in the case of carbene **8**, the phenyl ring bearing the electronwithdrawing substituent is preferentially annulated.

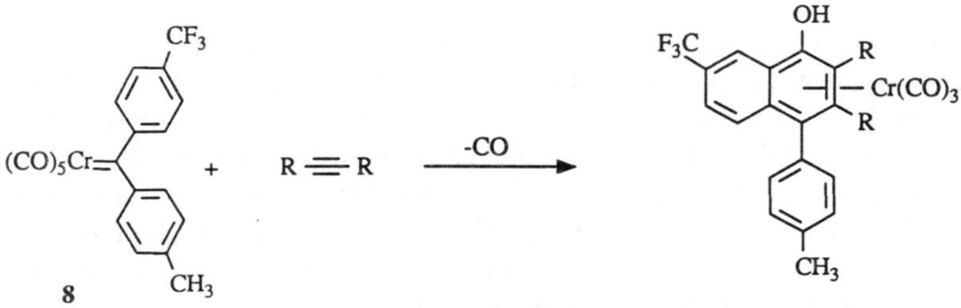

Finally, with unsymmetrically substituted alkynes, the alkyne incorpora-tion always brings the bulky R^1 group at position 2.

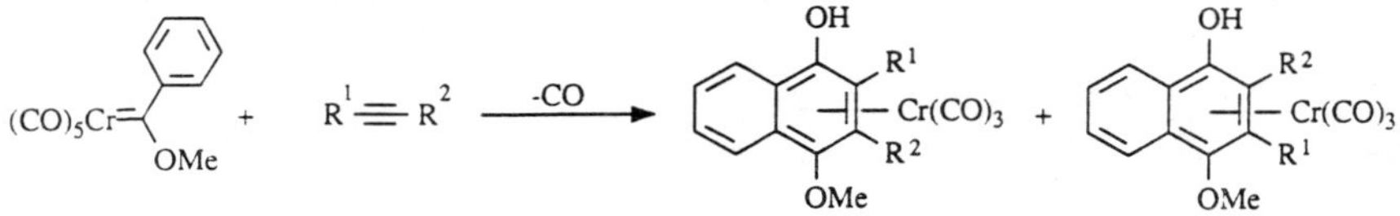

R^1 =	Et	^{t}Bu	Pr	Bu	Ph	Pr	Oc
R^2 =	Me	Me	Et	Pr	Me	H	H
A /B	69/31	92/8	100/0	100/0	>97/<3	>97/<3	>97/<3

An interesting and new aspect of carbene chemistry has been recently developed by Hegedus.

Sunlight photolysis of a solution of methoxy carbene complexes and imines directly produces β-lactam derivatives [32]. The reaction is remarkably stereoselective and leads to a single diastereoisomer.

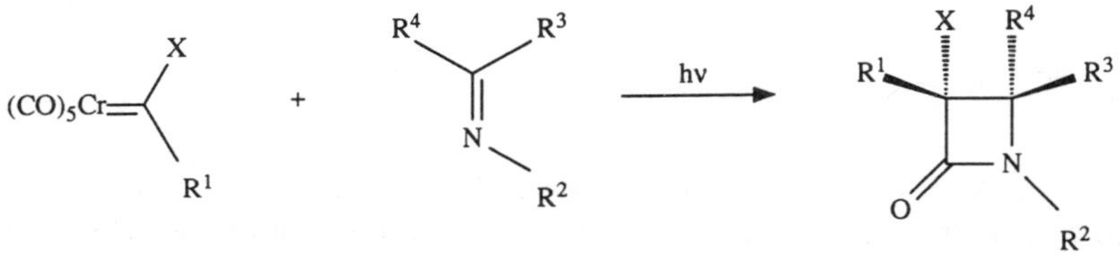

(R^1 = Me, Ph ; X = MeO)

Since biologically active β-lactams have hydrogen and acylamino groups in position 3, it was necessary to make appropriate carbene starters having a carbene carbon atom substituted with a protected amino group and a hydrogen atom. This was achieved [33] by the reaction sequence outlined in Scheme 5.

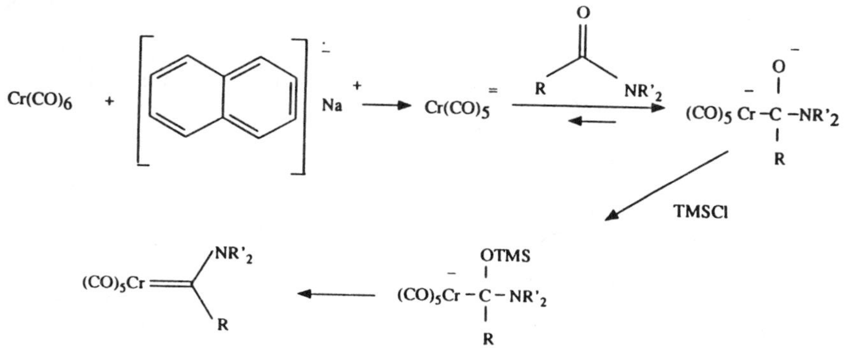

Scheme 5

Reacting the new N,N-dibenzylsubstituted aminocarbene complexes with the appropriate imines, cephalosporine and penicillin derivatives were obtained stereoselectively and in high yield [34]. Starting from the optically pure thiazoline the e.e. was greater than 98%.

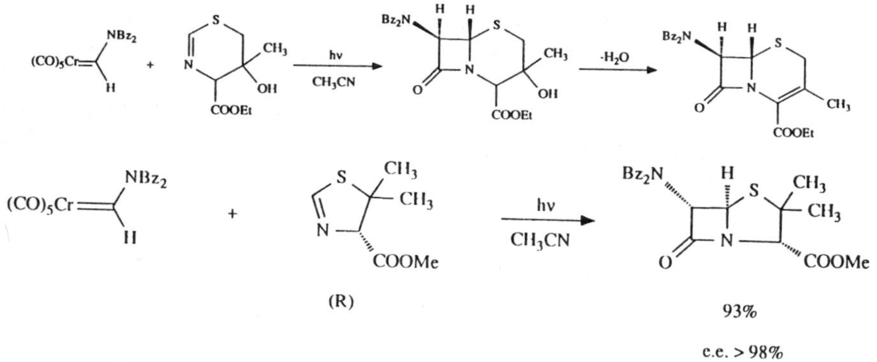

Catalytic hydrogenation of the intermediate gives access to the corresponding aminoazetidinone derivatives.

The mechanism proposed for the formation of the β-lactam ring implies the light promoted CO insertion in the chromium-carbon double bond to give the intermediate metallacyclopropenone **9** which can be considered the equivalent of a complexed ketene **9a**.

This intermediate would react with the imine to give the zwitterionic derivative **10**, precursor of the final β-lactam derivative. The formation of intermediate **9a** is supported by the nature of its derivatives obtained by nucleophilic trapping.

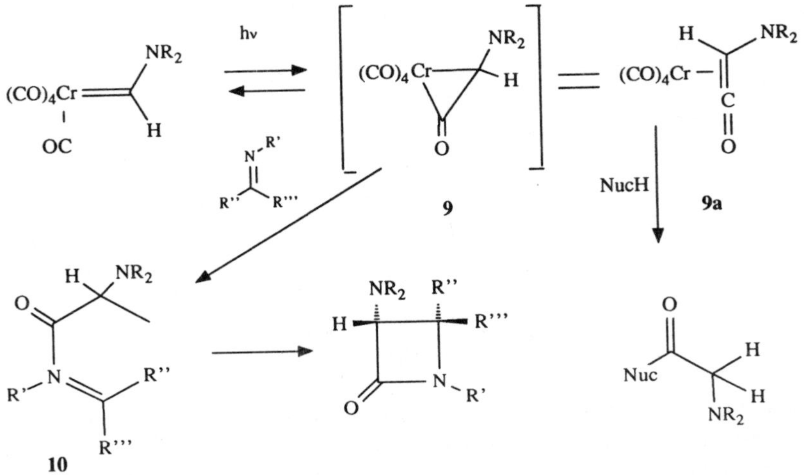

Fischer type carbene complexes can also be utilized for carbon-carbon bond formation by exploiting the reactivity connected with the acidity of the hydrogens on the carbon α to the carbene carbon atom. In fact it is easy to make the anion in this position and react the conjugate base of the carbene with electrophiles.

The reaction of carbene complex **11** with dimethylformamide dialkylacetals takes place in very mild conditions affording the corresponding enamine derivative **12** [35].

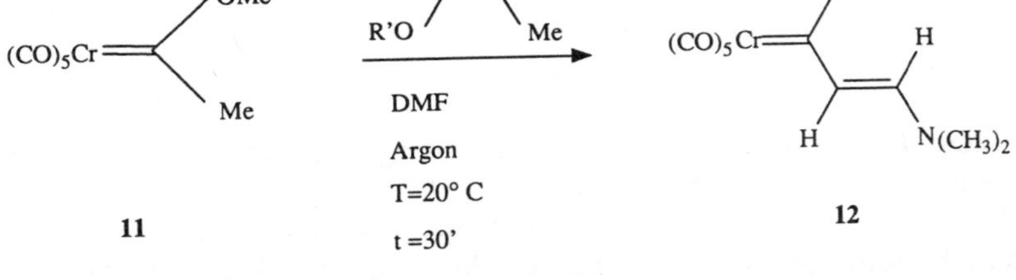

However this compound does not behave as enamine since the nitrogen lone pair is completely delocalized. This delocalization has been shown by the ^{13}C NMR chemical shift value of the carbene carbon atom compared with that found in the original carbene and was confirmed by the X-ray analysis of a single crystal of compound **12**.

On studying the scope of this reaction, it has been found that passing from the methyl group to the ethyl, propyl and butyl groups a different reaction takes place. In fact, in all these cases the corresponding dialkoxy carbene derivatives **13** were obtained.

$$(CO)_5Cr = \underset{R}{\overset{OMe}{<}} \quad \underset{R'O}{\overset{R'O}{>}} N \underset{Me}{\overset{Me}{<}} \quad \xrightarrow[\substack{DMF \\ Argon \\ T=20°\,C}]{} \quad (CO)_5Cr = \underset{OR'}{\overset{OMe}{<}} \quad \textbf{13}$$

R= Et

R= n-Pr

R= n-Bu

R'= Me

R'= Et

R'= -CH₂ Ph

R'= -C₆H₁₁

R'= -CH₂-C(CH₃)₃

The intermediate of this reaction is probably compound **14** from which, when R is hydrogen, the enamino derivative **12** is obtained by alkoxide elimination. When R is an alkyl group a different reaction takes place. In this case molecular models of the intermediate **14** show that the α-hydrogen is hindered and inaccessible. Attack of R'-O⁻ to the carbene carbon atom gives products **13** and **15** through a carbon-carbon bond break.

$$(CO)_5Cr = \underset{CH_2R}{\overset{OCH_3}{<}} \xrightarrow{DMF\text{-}DAA} (CO)_5Cr = \underset{\underset{N(CH_3)_2}{\overset{|}{R'O}}}{\overset{OCH_3}{<}}_{CH\text{-}R}$$

14

R = H → **12**

R = alkyl → **13** + R - CH == CH –N(CH₃)₂

15

Another reaction leading to the formation of a carbon-carbon bond is the opening of the ring of epoxides by the conjugate base of carbene complexes.

Casey [36] was the first in 1974 to produce such a reaction with ethylene and propylene oxides and obtained the corresponding oxacyclopentylidene derivatives.

The carbanion opens the epoxide ring and the oxyanion thus formed displaces the methoxy group leading to the formation of the 5-membered ring.

The extension of this reaction to a wider range of epoxides gave an unsatisfactory yield of products.

Since BF_3Et_2O is known to activate the ring opening of epoxides in reaction with carbanions, it has been employed in this reaction [37].

This resulted in improved yields of chromium pentacarbonyl-2-oxacyclopentylidenes thus producing quite a large number of new products.

The reaction is stereospecific; in fact, starting from cis epoxides, only the trans 3,4-disubstituted compounds **16** are obtained.

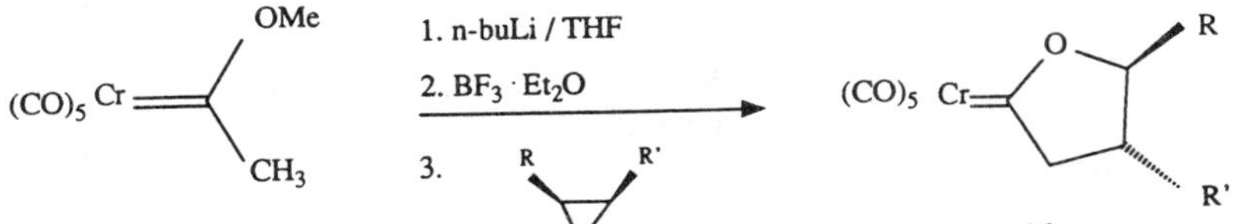

R= R'= Me	R = CH$_2$Cl	R' = H
R= R'= Ph	R = CH$_2$Br	R' = H
R - R'=-(CH$_2$)$_5$-	R = CH$_2$OCOC$_6$H$_4$-p-NO$_2$	R' = H
R - R'= -(CH$_2$)$_4$		

The exo-methylene group is an important substituent which is, in fact, present in several lactone derivatives with biological activity.

The transformation of chromium pentacarbonyl-2-oxacyclopentylidenes into the corresponding exo-methylene derivatives, precursors of exo-methylene-γ-butyrolactones, was achieved by reaction of the conjugate base of the complexes **16** with Eschenmoser salt.

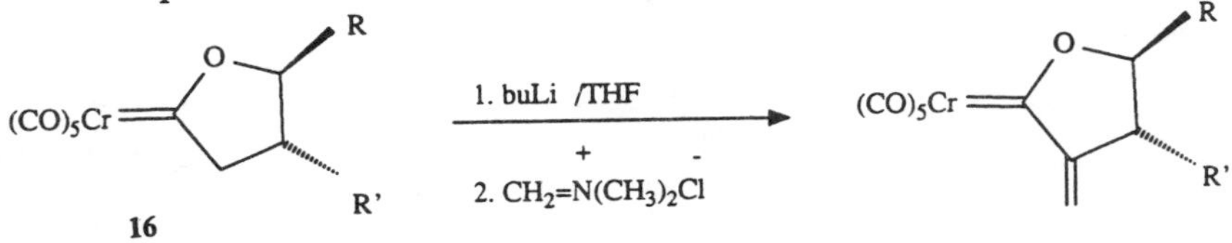

R = R' = Me

R-R'= -(CH$_2$)$_5$-

R=CH$_2$Cl R'=H

R=CH$_2$Br R'=H

R-R'= -(CH$_2$)$_4$-

Dipartimento di Chimica Organica e Industriale
Università di Milano
Via Golgi, 19 I-20133 Milano (Italy)

References

1. S.G. Davies, *Organotransition Metal Chemistry: Application to Organic Synthesis*, **1984**, Pergamon Press.
2. G. Carganico, P. Del Buttero, S. Maiorana, and G. Riccardi, *J. Chem. Soc. Chem. Commun.* **1978**, 989.
3. B. Nicholls and M.C. Whiting, *J. Chem. Soc. Chem. Commun.*, **1959**, 551.
4. G. Jaouen and R. Dabard, *J. Organomet. Chem.*, **1973**, *61*, C36.
5. W.S. Trahanovsky and R.J. Card, *J. Am. Chem. Soc.*, **1972**, *94*, 2987.
6. S.G. Davies and J. Blagg, *J. Chem. Soc. Chem. Commun.*, **1985**, 653.
7. S.G. Davies and J. Blagg, *Tetrahedron*, **1987**, *43*, 4463.
8. M.F. Semmelhack, J. Bisaha and M. Czarny, *J. Am. Chem. Soc.*, **1979**, *101*, 768.
9. D.A. Widdowson and J.P. Gilday, *J. Chem. Soc. Chem. Commun.*, **1986**, 1235.
10. D.A. Widdowson and J.P. Gilday, *Tetrahedron Lett.*, **1986**, *27*, 5525.
11. M.F. Semmelhack, H.T. Hall, Jr., R. Farina, M. Yoshifuji, G. Clark, T. Bargar, K. Hirotsu and J. Clardy, *J. Am. Chem. Soc.*, **1979**, *101*, 3535.
12. M.F. Semmelhack, G.R. Clark, J.L. Garcia, J.J. Harrison, Y. Thebtaranonth, W. Wulff and A. Yamashita, *Tetrahedron*, **1981**, *37*, 3957.
13. T.A. Albright and B.K. Carpenter, *Inorg. Chem.*, **1980**, *19*, 3092.
14. A. Solladiè-Cavallo and J. Suffert, *Org. Magn. Res.*, **1980**, *14*, 426.
15. M.F. Semmelhack, J.L. Garcia, D. Cortes, R. Farina, R. Hong and B.K. Carpenter, *Organometallics*, **1983**, *2*, 467.
16. M.F. Semmelhack, *Selectivity – a Goal for Synthetic Efficiency*, 14th Workshop Conference Hoechst, Schloss Reisenburg, Verlag Chemie GmbH, Weinheim, **1984**; p. 227.
17. E.P. Kunding, V. Desobry, D.P. Simmons and E. Wenger, *J. Am. Chem. Soc.*, **1989**, *111*, 1804.
18. F. Rose-Munch, E. Rose and A. Semra, *J. Organomet. Chem.*, **1989**, *363*, 103.
19. S. Maiorana, C. Baldoli, P. Del Buttero and A. Papagni, *J. Chem. Soc. Chem. Commun.*, **1985**, 1181.
20. S. Maiorana, A. Alemagna, C. Baldoli, P. Del Buttero and E. Licandro, *J. Chem. Soc. Chem. Commun.*, **1985**, 417.
21. S. Maiorana, C. Baldoli, P. Del Buttero and E. Licandro, *Synthesis*, **1988**, *4*, 344.
22. K. Schlögl, *Topics in Stereochemistry*, **1967**, *1*, 39.
23. A. Solladiè-Cavallo, G. Solladiè and E. Tsamo, *J. Org. Chem.*, **1979**, 44, 4189.
24. G. Jaouen, S. Maiorana, C. Baldoli, J. Gillois and S. Top, *J. Chem. Soc. Chem. Commun.*, **1988**, 1284.
25. A. Solladiè-Cavallo, S. Colonna, G. Lapitajs, P. Buchert, A. Klein and A. Manfredi, *J. Organomet. Chem.*, **1987**, *330*, 357.
26. S. Maiorana, C. Baldoli, P. Del Buttero, E. Licandro and A. Papagni, *J. Chem. Soc. Chem. Commun.*, **1987**, 762.
27. E.O. Fischer, *Angew. Chem.*, **1974**, *86*, 651.
28. D.F. Shriver, *Inorg. Chem.*, **1979**, *19*, 164.

29. K.H. Doetz, *Angew. Chem. Int. Ed. Eng.*, **1984**, *23*, 587.
30. K.H. Doetz and M. Popall, *Tetrahedron*, **1985**, *41*, 5797.
31. K.H. Doetz, *Pure and Applied Chemistry*, **1983**, *55*, 1689.
32. L.S. Hegedus, M.A. McGuire, L.M. Schultze, C. Yijun and O.P. Anderson, *J. Am. Chem. Soc.*, **1984**, *106*, 2680.
33. L.S. Hegedus and R. ImWinkelried, *Organometallics*, **1988**, *7*, 702.
34. L.S. Hegedus, G. de Weck and S.D.'Andrea, *J. Am. Chem. Soc.*, **1988**, *110*, 2122.
35. S. Maiorana, L. Lattuada, E. Licandro, A. Papagni, A. Chiesi Villa and C. Guastini, *J. Chem. Soc. Chem. Commun.*, **1988**, 1092.
36. C.P. Casey and R.L. Anderson, *J. Organomet. Chem.*, **1974**, *73*, C28.
37. S. Maiorana, L. Lattuada, E. Licandro and A. Papagni, *Advances in Metal Carbene Chemistry*, Nato ASI Series, Kluwer Academic Publishers **1989**; p. 149.

G. BRACA AND A.M. RASPOLLI GALLETTI

ROLE AND IMPLICATIONS OF H⁺ AND H⁻ ANIONIC HYDRIDO CARBONYL CATALYSTS ON ACTIVITY AND SELECTIVITY OF CARBONYLATION REACTIONS OF UNSATURATED AND OXYGENATED SUBSTRATES

The most famous and industrially important homogeneous catalytic carbonylation reactions show two common features (Scheme 1):

— the involvement of an anionic transition metal species
— the involvement of a hydrido metal species.

In the related literature great emphasis has been generally laid on the ligands nature and properties, on the metal coordination sphere and on the nature of the intermediate species involved in the catalytic path, whereas minor attention has been paid to the H^+ and H^- nature of the hydridic species [14] and to the effect of the counter ion and ion pairs involved.

In this contribution just the importance of the H^+ and H^- nature of some anionic hydridocarbonyl catalysts and the effects of their counterions on the kinetics and selectivity of the reaction will be discussed with some appropriate examples taken from the literature and from our work.

1. Hydroformylation of olefins

The importance of the nature of the hydridic species for the kinetics and the chemo- and regio-selectivity is well documented in the hydroformylation of olefins.

In this process, together with linear and branched aldehydes, significant amounts of byproducts, *i.e.* hydrocarbons (formed by direct olefin hydrogenation), alcohols (formed by hydrogenation of aldehydes), acetals, formates and aldol condensation products can be formed.

It has been shown that the selectivity as well as the kinetic of the reaction are strongly affected by the type of metal catalyst and certainly by the acidity of the hydrides involved (Scheme 2). In the homologous series of the VIII group metal hydridocarbonyls, a decrease in the acidity dramatically decreases the activity and also the chemo-selectivity to aldehydes, simul-

A. F. Noels et al. (Eds.), Metal promoted selectivity in organic synthesis, 287–310.

taneously increasing the hydrogenation to hydrocarbons and alcohols (Scheme 2).

Process	Hydridic species involved	Ref.
Hydroformylation of olefins	$H^+[Co(CO)_4]^-$, $HCo(CO)_3PR_3$ $H^+[Rh(CO)_4]^-$, $HRh(CO)(PPh_3)_3$	1, 2
Hydroesterification of olefins	$H^+[Co(CO)_4]^-$, $H^+[Rh(CO)_4]^-$, $H^+[HRu(CO)_4]^-$	3
Carbonylation of methanol to acetic acid	$H^+[Rh(CO)_2I_2]^-$	4
Carbonylation of methyl acetate to acetic anhydride	$H^+[Rh(CO)_2I_2]^-$	5
Homologation of alcohols	$H^+[Co(CO)_4]^-$, $H^+[Ru(CO)_3I_3]^-$ $H^+[Rh(CO)_2I_2]^-$	6, 7
Homologation of methyl to ethyl acetate	$H^+[Ru(CO)_3I_3]^-$	8
Hydrogenation of CO to methanol and ethylene glycol	$H^+[HRu_3(CO)_{11}]^-$, $H^+[Ru(CO)_3I_3]^-$ $H^+[Rh(CO)_4]^-$, $HRh(CO)_3(PR_3)$	9, 10
Hydrogenation of CO to ethanol	$H^+[HRu_3(CO)_{11}]^-$, $H^+[Ru(CO)_3I_3]^-$	11, 12
Hydrocarbonylation of formaldehyde to glycol aldehyde	$H^+[Rh(CO)_2X_2]^-$, $[Rh_5(CO)_{15}]^-$	13
Hydrocarbonylation of acetylene to acrylic derivatives	$H^+[Ni(CO)_nX]^-$	1

Scheme 1. Anionic H^+ and H^- hydridic species of transition group metals involved in homogeneous catalytic reactions.

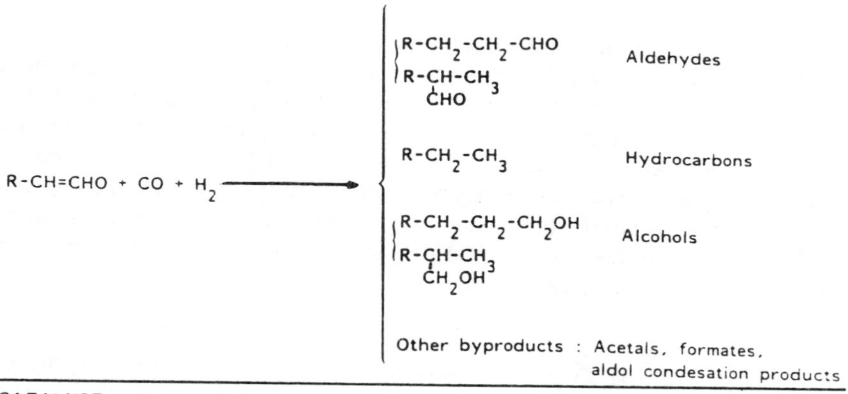

$$R-CH=CHO + CO + H_2 \longrightarrow$$

$\begin{cases} R-CH_2-CH_2-CHO \\ R-CH-CH_3 \\ \quad \overset{|}{C}HO \end{cases}$ Aldehydes

$R-CH_2-CH_3$ Hydrocarbons

$\begin{cases} R-CH_2-CH_2-CH_2OH \\ R-\overset{|}{C}H-CH_3 \\ \quad \overset{|}{C}H_2OH \end{cases}$ Alcohols

Other byproducts : Acetals, formates, aldol condesation products

CATALYST	Nature of the hydrido group [a]	pK$_a$	Relative activity	Products Selectivity %	
HRh(CO)$_4$	strong acid — Decreasing in acidity	-	680	Aldehydes Alcohols	70-85 5-10
HCo(CO)$_4$		8.3	1	Hydrocarbons Byproducts	1-5 5-15
HIr(CO)$_4$		-	0.1	Aldehydes Hydrocarbons	20 80
H$_2$Fe(CO)$_4$	weak acid — Decreasing in acidity	11.4	0.0003		
H$_2$Ru(CO)$_4$		18.7	0.2	Aldehydes Hydrocarbons Alcohols	20-50 30-50 5-30
H$_2$Os(CO)$_4$		20.8	0.01		

[a] See Ref. 15 and 16

Scheme 2. Hydroformylation of olefins with VIII group metals hydrocarbonyl catalysts.

When the carbonyl catalysts are modified by replacing the carbonyl ligands with phosphines, the acidity of the hydrides strongly decreases, until they become H- hydrides, and correspondingly the activity decreases and the chemo- and regio-selectivity of the aldehydes and alcohols produced is varied.

In order to rationalize this behaviour a lot of work has been done sometimes with contradictory conclusions which from time to time invoke:

– the formation under the reaction conditions of different metal hydrido carbonyl species [1]
– the isomerization of the olefin [17]
– the reversibility and isomerization of the metal-alkyl intermediates [18]

– the isomerization of the metal-acyl intermediates [19]

Actually, even though the reaction paths are very complicated and differ from metal to metal, the effect of the nature of the hydrides can be roughly traced for instance in the regioselectivity obtained in the α-olefins hydroformylation (normal to branched ratio of the aldehydes and/or alcohols produced). The H$^-$ hydrides in fact by a Markovnikov addition to the double bond give preferentially the linear alkyl intermediate and then the linear aldehydes whereas the more acid H$^+$ hydrides give the branched ones and favour the isomerization reactions:

$$H_2C=CH-R +
\begin{array}{l}
H^-[M] \xrightarrow{} [M]-CH_2-CH_2-R \xrightarrow{CO/H_2} \text{linear products} \\[2em]
H^+[M] \xrightarrow{} \underset{\underset{R}{|}}{[M]-CH-CH_3} \xrightarrow{CO/H_2} \text{branched products}
\end{array}$$

Accordingly, the Co-phosphine (1) and Rh-phosphine (20) catalytic systems, suppressing or minimizing the olefin isomerization, afford the highest regioselectivities toward linear products (l/b = 12 and 25 respectively). In contrast unmodified proton acid hydrido catalysts like HCo((CO)$_4$ and HRh(CO)$_4$ promote an extensive isomerization and give l/b ratios of 1–4 (Table I).

TABLE I

Propylene hydroformylation processes with different catalytic systems [2, 16]

Metal	Cobalt		Rhodium	
Catalyst	HCo(CO)$_4$	HCo(CO)$_3$(PBu$_3$)	HRh(CO)$_4$	HRh(CO)(PBu$_3$)$_3$
Acidic dissociation constant	1	1.1×10^{-7}	> 1	n.d.
Temperature (°C)	110–180	160–200	100–140	60–120
Pressure (MPa)	20–30	5–10	20–30	0.1–5
Metal concentration (%)	0.1–1	0.6	10^{-4}–10^{-2}	0.01–0.1
Hydrocarbons formation	low	high	low	high
Main reaction products	aldehydes	alcohols	aldehydes	aldehydes
Linear/branched	4	7.3	1	11.5

By varying the reaction conditions (temperature, catalyst concentration, H$_2$/CO/olefin ratios), some significant unexpected modifications on the regioselectivity have been sometimes observed contradicting the mechanistic hypotheses. Some of these not well clarified behaviours may be related in our opinion to the involvement of hydrides of different nuclearity and acidity, sometimes not detectable under the reaction conditions due to their low concentration and to the overlapping of the spectral absorptions of the species present (see for instance the case of $H^+[Rh(CO)_4]^-$ and $H^+[Rh_6(CO)_{15}]^-$ [21].

One illustrative example of this event is given by the hydroformylation of α-olefins with ruthenium hydridocarbonyl catalysts (Table II). The mononuclear and trinuclear hydrido carbonyl ruthenium species in equilibrium under the hydroformylation conditions (Scheme 3) have the two hydrido ligands of different nature: one is of acid type, protonates the phosphine oxide [12] and can be replaced by alkaline or bis(triphenylphosphine)iminium ions with formation of tight or solvent separated ion pairs [25], the second is a H^- as indicated by its proton NMR spectrum [26] and by its transfer to K^+ ions [27].

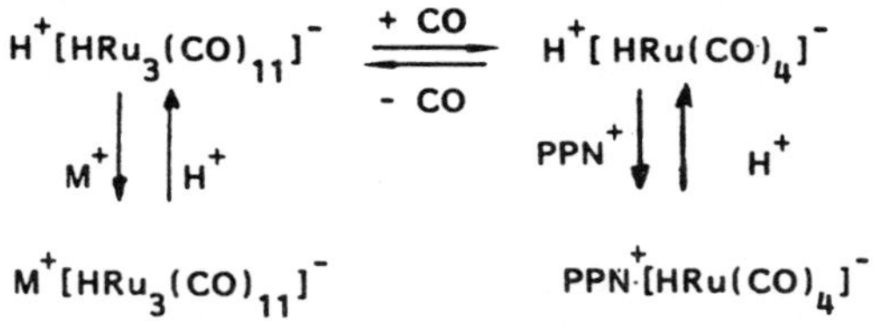

Scheme 3. Nature of the hydridocarbonyl ruthenium species involved in the catalysis.

Under the conditions where H^+ and H^- hydrido groups are present, the regioselectivity is not particularly high (aldehydes + alcohols linearity 72–82%) and the isomerization of olefins is significant. In contrast when the acid hydride protonates a basic solvent, such as dimethylformamide, or it is replaced by the bis(triphenylphosphine) iminium cation, only the H^- hydride is actually active and by a Markovnikov addition to the olefin produces the linear alkyl intermediate and increases the regioselectivity toward the linear products up to a linearity of 90–95%.

Another peculiarity of the hydroformylation processes related to the ionic nature of the catalytically active species is their sensitivity to the presence of ionic soluble or unsoluble additives, nucleophilic promoters and solvents with different solvating power.

Some examples cited by Falbe [2] and Pino [1] for the cobalt catalyzed olefin hydroformylation are reported in Table III. The most part of this information comes from patents and it is difficult to correlate the addition of

TABLE II

Hydroformylation of α-olefins in the presence of different ruthenium carbonyl systems at 150°C

Precursor	α-olefin	Solvent	Ruthenium species under reaction conditions	Conv. %	Aldehydes (Linearity %)[b]	Selectivity %[a] Alcohols (Linearity %)[b]	Hydrocarbons	Ref.
$Ru_3(CO)_{12}$	1-pentene	Toluene	$H^+[HRu(CO)_4]^-$ $H^+[HRu_3(CO)_{11}]^-$	32.2.	74.3 (72.5)	traces	25.7	22
$Ru_3(CO)_{12}$[c]	1-hexene	Benzene	$H^+[HRu(CO)_4]^-$ $H^+[HRu_3(CO)_{11}]^-$	24.2	99.2 (82.5)	traces	0.8	23
$Ru_3(CO)_{12}$	1-pentene	DMF[d]	$[DMFH]^+[HRu(CO)_4]^-$ $[DMFH]^+[HRu_3(CO)_{11}]^-$	94.3	42.7 (84.9)	0.8 (94.9)	5.9	24
$PPN[HRu(CO)_4]$[e]	1-pentene	DMF	$PPN[HRu(CO)_4]$	90.5	61.7 (90.1)	2.9 (93.4)	3.6	24
$PPN[HRu_3(CO)_{11}]$	1-pentene	DMF	$PPN[HRu_3(CO)_{11}]$	90.8	60.6 (94.8)	0.1 (100)	2.2	24

[a] The % difference is constituted by isomerization products.
[b] Linearity = n-aldehyde/all isomeric aldehydes.
[c] Reaction temperature: 120°C.
[d] DMF = Dimethylformamide.
[e] PPN = bis(triphenylphosphine)iminium.

the promoters with the nature of the active species present in solution and with the reaction mechanism because of the possible overlapping of different chemical and physical effects (presence of different metal carbonyl species, solubility of reagents and catalyst, phase separation, etc.).

TABLE III

Additives responsible for improving activity
and/or selectivity of cobalt oxo-catalysts [1][2]

Additives	Improvement of		Patents by (year)
	Reaction rate	Selectivity	
Alkali metals carboxylates	X		BASF (1971)
Alkali metal hydroxides	X		BASF (1972), Celanese (1974)
Alkyl-aryl sulphonates	X	X	Rhone-Poulenc (1976)
Protonic acids	X		Degussa (1964), BASF (1975)
Ziegler type additives	X		Inst. Franc. Petr. (1969)
Zeolites		X	UOP Inc. (1975)
Water	X	X	Shell Oil Co. (1975); Standard Oil Co. (1953); Montedison (1971)
Pyridine and nitrogen bases	X	X	Wender (1957)
Ethers, ketones, alcohols	X	X	BAFS (1972)

Anyway the presence of more or less solvated counterions of the anionic metal carbonyls affects their reactivity and the products selectivity; accordingly the anionic species free of ion pair interactions provide the more active catalyst.

On the other hand, the presence of strong electrophiles like Lewis acids or unsolvated cations, which are known to accelerate the carbonyl insertion into the metal-alkyl bond, may result in the improvement of the activity and selectivity of the catalytic system.

2. Hydrocarbonylation of oxygenated substrates with ruthenium catalysts

The hydrocarbonylation of oxygenated substrates in the presence of carbonyl and iodocarbonyl ruthenium catalysts is another interesting group of reactions where the H^+ or H^- nature of the anionic hydrides involved in the catalysis appears determinant for the selectivity of the process.

The main reactions studied in the literature are reported in Scheme 4 together with the yields to the different products. The processes are not based on a single reaction path but on a number of parallel and successive steps which are represented in Scheme 5.

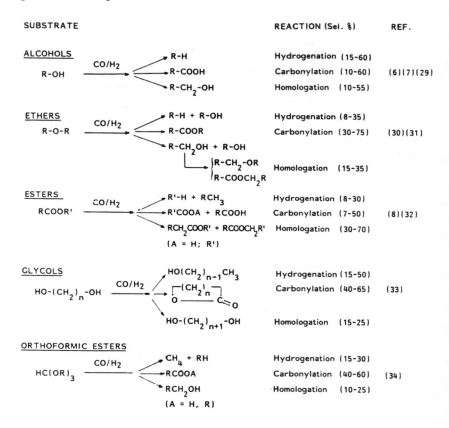

SUBSTRATE   REACTION (Sel. %)   REF.

ALCOHOLS
                        CO/H₂   R-H   Hydrogenation (15-60)
    R-OH   ─────────────   R-COOH   Carbonylation (10-60)   (6)(7)(29)
                                     R-CH₂-OH   Homologation (10-55)

ETHERS
                        CO/H₂   R-H + R-OH   Hydrogenation (8-35)
    R-O-R   ─────────────   R-COOR   Carbonylation (30-75)   (30)(31)
                                     R-CH₂OH + R-OH

                                         R-CH₂-OR
                                         R-COOCH₂R   Homologation (15-35)

ESTERS
                        CO/H₂   R'-H + RCH₃   Hydrogenation (8-30)
    RCOOR'   ─────────────   R'COOA + RCOOH   Carbonylation (7-50)   (8)(32)
                                     RCH₂COOR' + RCOOCH₂R'   Homologation (30-70)

                        (A = H; R')

GLYCOLS   HO(CH₂)ₙ₋₁CH₃   Hydrogenation (15-50)
                        CO/H₂   ┌(CH₂)ₙ┐
 HO-(CH₂)ₙ-OH   ─────────────   O─────C═O   Carbonylation (40-65)   (33)

                                     HO-(CH₂)ₙ₊₁-OH   Homologation (15-25)

ORTHOFORMIC ESTERS
                        CO/H₂   CH₄ + RH   Hydrogenation (15-30)
    HC(OR)₃   ─────────────   RCOOA   Carbonylation (40-60)   (34)
                                     RCH₂OH   Homologation (10-25)

                        (A = H, R)

Scheme 4. Ruthenium carbonyl iodide catalyzed hydrocarbonylation reactions.

With the aim to direct the process toward one target product, appropriate promoters favouring the formation of the selective catalytic species are almost always added.

The ruthenium compounds involved in the catalysis are protonic hydrido iodocarbonyl and carbonyl species which can be generated in situ starting from various ruthenium precursors and an iodide promoter (CH₃I, HI, NaI etc.).

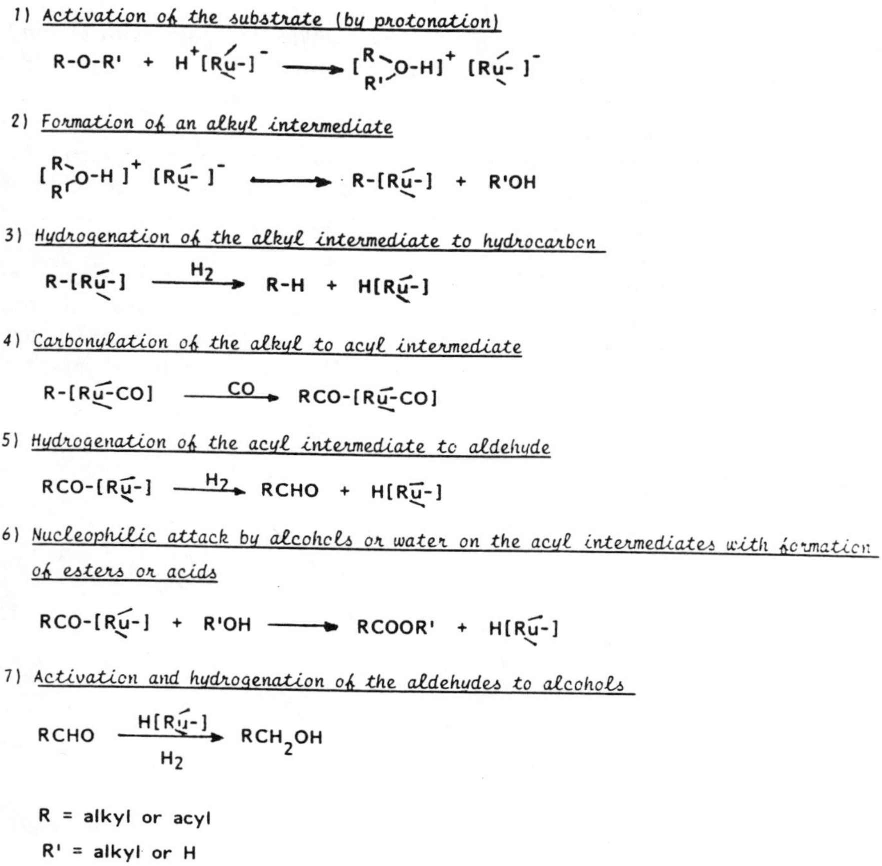

1) *Activation of the substrate (by protonation)*

$$R-O-R' + H^+[R\underset{\sim}{u}-]^- \longrightarrow [\underset{R'}{\overset{R}{>}}O-H]^+ [R\underset{\sim}{u}-]^-$$

2) *Formation of an alkyl intermediate*

$$[\underset{R'}{\overset{R}{>}}O-H]^+ [R\underset{\sim}{u}-]^- \longrightarrow R-[R\underset{\sim}{u}-] + R'OH$$

3) *Hydrogenation of the alkyl intermediate to hydrocarbon*

$$R-[R\underset{\sim}{u}-] \xrightarrow{H_2} R-H + H[R\underset{\sim}{u}-]$$

4) *Carbonylation of the alkyl to acyl intermediate*

$$R-[R\underset{\sim}{u}-CO] \xrightarrow{CO} RCO-[R\underset{\sim}{u}-CO]$$

5) *Hydrogenation of the acyl intermediate to aldehyde*

$$RCO-[R\underset{\sim}{u}-] \xrightarrow{H_2} RCHO + H[R\underset{\sim}{u}-]$$

6) *Nucleophilic attack by alcohols or water on the acyl intermediates with formation of esters or acids*

$$RCO-[R\underset{\sim}{u}-] + R'OH \longrightarrow RCOOR' + H[R\underset{\sim}{u}-]$$

7) *Activation and hydrogenation of the aldehydes to alcohols*

$$RCHO \xrightarrow[H_2]{H[R\underset{\sim}{u}-]} RCH_2OH$$

R = alkyl or acyl

R' = alkyl or H

Scheme 5. Paths involved in the hydrocarbonylation reactions catalyzed by ruthenium carbonyl iodide catalysts.

The concentration of the different species depends on the I/Ru ratio, temperature, PCO, PH_2 and type of solvent.

The triiodotricarbonyl ruthenium hydride is a strong acid hydride: it protonates weak bases like aniline giving the anilinium salt, forms tight, contact or solvent separated ion pairs with alkaline and bivalent cations depending on the reaction medium [8] and largely predominates over other species under the typical hydrocarbonylation reaction conditions (T:200°C,

I/Ru: 10, P: 10–15 MPa, CO/H_2: 0.5–2).

The observed ability of forming tight and contact ion pairs with H^+ and monovalent cations is determinant for the catalysis in view of the possible interactions of the cations with the reactive groups bound to the metal centre (CO, olefin, acyl, etc.) [28].

Looking at the roles which the metal hydrido carbonyl species and the added promoters play in the different steps of the process, it is possible to rationalize the observed changes and trends in the activity and selectivity of the catalytic systems.

The first important role of the protonic hydrido species is the protonation of the substrate (Scheme 6, path A) which appears to be necessary for the activation of the reagents and the formation of the metal alkyl intermediates. In this step, the protonic hydride may be replaced by an external protonic inorganic or carboxylic acid.

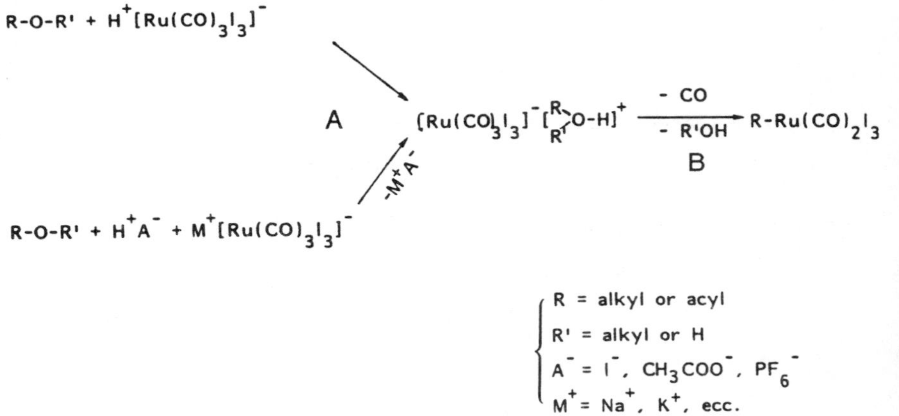

$R-O-R' + H^+[Ru(CO)_3I_3]^-$

A

$[Ru(CO)_3I_3]^- [\overset{R}{\underset{R'}{>}}O-H]^+ \xrightarrow[- R'OH]{- CO} R-Ru(CO)_2I_3$

B

$R-O-R' + H^+A^- + M^+[Ru(CO)_3I_3]^-$

$$\begin{cases} R = \text{alkyl or acyl} \\ R' = \text{alkyl or H} \\ A^- = I^-, CH_3COO^-, PF_6^- \\ M^+ = Na^+, K^+, \text{ecc.} \end{cases}$$

Scheme 6. Role of $H^+[Ru(CO)_3I_3]^-$: activation of the substrate and formation of metal-alkyl or -acyl intermediates.

Once protonated the substrate may interact with the carbonylate of a salified acid hydride forming the alkyl or acyl intermediate (Scheme 6, path B) through a direct ion-pair interaction as suggested by Wender some years ago [36].

Actually the results (Table IV) demonstrate that in absence of any proton supplier the oxygenated substrates do not undergo metal catalyzed reactions: H^- hydrides are unable to act as catalysts.

TABLE IV

Hydrocarbonylation of oxygenated substrates in the presence of different ruthenium catalytic systems[a]

Catalytic system	Oxygenated substrate (mmol 25)	Conv. (%)	Time (h)	Selectivity % to products of				Ref.
				Carbonylation	Homologation	Hydrogenation	Etherification / + esterification	
H+[Ru(CO)3I3]−	Ethanol (480)	73.0	8	9	8	13	70	35
[Et3NH]+[HRu3(CO)11]−[b]	Ethanol (480)	6.0	8	−	−	−	100	35
H+[Ru(CO)3I3]−[c,d]	Dimethyl ether (90)	69.0	28.5	32.2	28.8	28.2	10.8	31
Na+[Ru(CO)3I3]−[c,d]	Dimethyl ether (90)	no react.	28.5	−	−	−	−	31
Na+[Ru(CO)3I3]−[c,f]	Dimethyl ehter (102)	50.1	13.0	46.5	12.0	6.0	35.5	31
	Acetic acid (430)	15.0						

[a] Reaction conditions: T: 200°C; P: 14 MPa; CO/H2: 1.

[b] [Ru]: 1.1×10^{-2} mol dm^{-3}

[c] [Ru]: 1.5×10^{-2} mol dm^{-3}.

[d] Solvent: dioxane (25 ml).

[e] Solvent: toluene (25 ml).

[f] CO/H2:2.

After the activation of the substrate, the protons, coming from the acid hydride or from external acids, can be involved in the methyl migration – carbonyl insertion step, accelerating the transformation of the alkyl into acyl intermediate and decreasing the life time of the alkyl intermediate responsible for the formation of undesired hydrocarbons (Scheme 7).

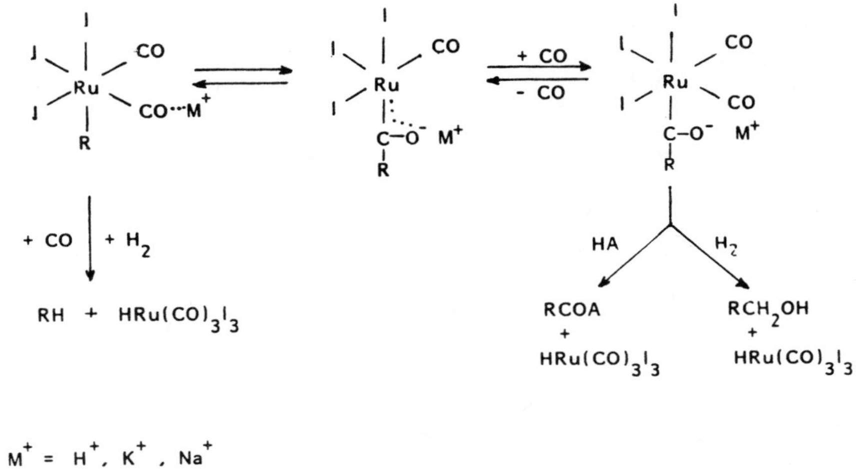

Scheme 7. Role of $H^+[Ru(CO)_3I_3]^-$: acceleration of the methyl migration-carbonyl insertion step.

This very important kinetic effect well documented for some non-catalytic metal alkyl-acyl interconversions [37] causes in this process a significant decrease of the selectivity towards the hydrocarbons favouring the formation of valuable carbonylation and homologation products.

A second effect, related to the polarization of the acyl intermediate by the H^+, favours the nucleophilic attack on the acyl intermediate by water or alcohols thus increasing the formation of esters and acids coming from simple carbonylation reactions.

A clear example of this is shown in the methyl acetate homologation (Scheme 8) [32].

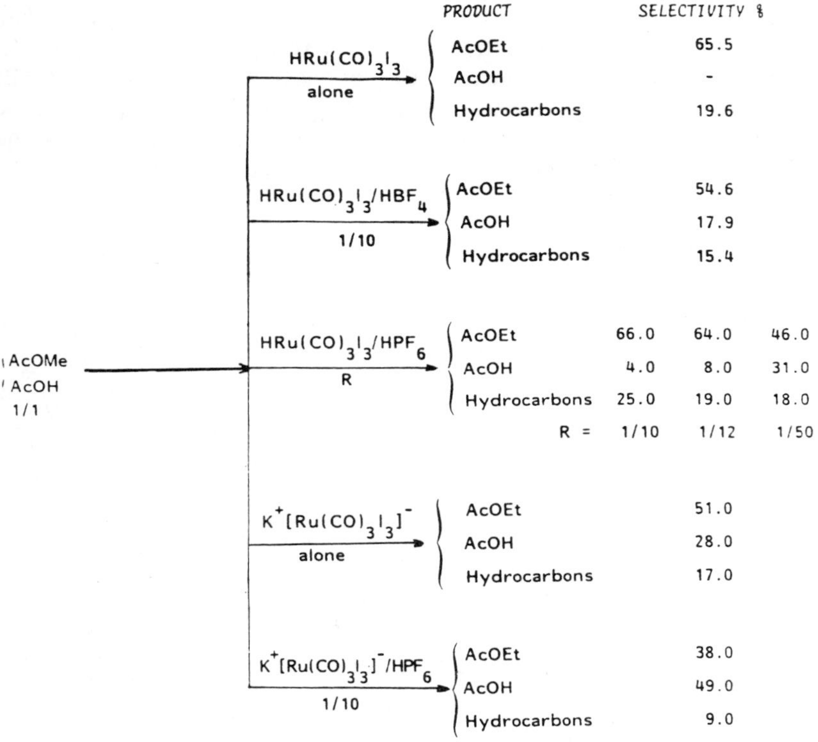

	PRODUCT	SELECTIVITY %		
$HRu(CO)_3I_3$ alone	AcOEt	65.5		
	AcOH	-		
	Hydrocarbons	19.6		
$HRu(CO)_3I_3/HBF_4$ 1/10	AcOEt	54.6		
	AcOH	17.9		
	Hydrocarbons	15.4		
$HRu(CO)_3I_3/HPF_6$ R	AcOEt	66.0	64.0	46.0
	AcOH	4.0	8.0	31.0
	Hydrocarbons	25.0	19.0	18.0
	R =	1/10	1/12	1/50
$K^+[Ru(CO)_3I_3]^-$ alone	AcOEt	51.0		
	AcOH	28.0		
	Hydrocarbons	17.0		
$K^+[Ru(CO)_3I_3]^-/HPF_6$ 1/10	AcOEt	38.0		
	AcOH	49.0		
	Hydrocarbons	9.0		

AcOMe / AcOH 1/1

Reaction conditions : T : 200 °C; P : 12 MPa; H_2/CO : 2; I/Ru : 10; Conv.% : 35-40.

Scheme 8. Effect of strong protonic acids on the methyl acetate homologation.

The addition of non complexing protonic acids like tetrafluoboric or hexafluorophosphoric acid results in an increase of the selectivity to acetic acid (product of simple carbonylation) and in the corresponding decrease of the selectivity to hydrocarbons and to ethyl acetate (homologation product). This effect augments as the acid/Ru ratio increases up to a value of about 50; at higher values the catalysts are progressively deactivated.

Alkali cations produce an analogous effect as H^+: they accelerate the carbonyl insertion step and favour the nucleophilic attack by water and alcohols; more carbonylation and less hydrogenation and homologation products are obtained. The cations efficacy is moreover enhanced when they are either less solvatable (potassium and cesium are more efficient than lithium) or when they are selectively complexed by specific crown ethers. Although these crown ethers surround the cations as an overturned umbrella [8], the latter are equally able to interact in a perpendicular plain with the coordinated reactive groups accelerating the reaction and increasing the selectivity to carbonylation and homologation products (Figure 1).

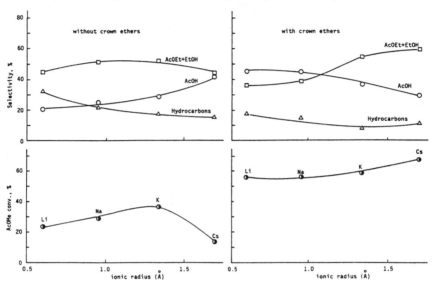

Figure 1. Hydrocarbonylation of methyl acetate in acetic acid solution with $M^+[Ru(CO)_3I_3]^-$ catalysts (M = alkaline metal) with and without crown ethers.

In an analogous manner, *i.e.* by an interaction of the cations with CO and acyl groups, some cocatalytic effects claimed for bimetallic systems and for Lewis acids promoters can be explained. In fact the addition to the ruthenium iodocarbonyl systems of a cobalt cocatalyst, $Co_2(CO)_8$ or CoI_2 [7] or of a strong Lewis acid, AlI_3, TiI_4, [32] results in a significant improvement of the reaction rate with respect to that observed with monometallic systems and in an increase of the selectivity toward simple carbonylation products with a decrease in the formation of hydrocarbons. The reasons of this behaviour cannot be related to the intervention of bimetallic cluster species since the unique metal carbonyl species detected by I.R. in the catalytic solutions are the ion-pairs between $[Ru(CO)_3I_3]^-$ and the cations of the cocatalysts added (Co^{2+}, Al^{3+}, Ti^{+4} etc.). Really this is true at least when the cocatalyst is not a carbonyl derivative or when it is added in a stoichiometric amount with respect to Ru.

Such a type of effect on the catalysis has been repeatedly evidenced in our research work [8, 29] and clearly appears looking at the hydrocarbonylation of ethanol (Figure 2) where the addition of AlI_3 instead of CH_3I significantly increases the selectivity toward the simple carbonylation products, namely ethyl propionate.

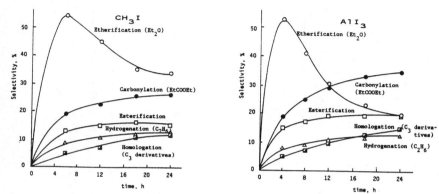

Figure 2. Ethanol hydrocarbonylation with $[Ru(CO)_3I_3]^-/CH_3I$ and $[Ru(CO)_3I_3]^-/AlI_3$ catalytic systems.

Less discussed and evident is the role played in these processes by the anionic ruthenium H− hydrido carbonyl species without iodide ligands, namely $[HRu(CO)_4]$ and $[HRu_3(CO)_{11}]^-$.

They are involved mainly in the hydrogenation reaction of the alkyl and acyl intermediates, favouring the production of hydrocarbons and homologation products (Scheme 9).

A number of experimental results confirm this role. Firstly, these H− hydrides have been demonstrated active at room temperature in the stoichiometric reduction to acetaldehyde and ethanol of $CH_3COCo(CO)_3(PPh_3)$ used as model acyl reagent and in the catalytic hydrogenation of acetaldehyde with syngas at 180°C [38].

Accordingly any basic promoter, for instance Broensted bases, or heterogeneous basic oxides such as hydroxylated MgO, Al_2O_3, which acting as iodide and proton acceptors, favour the formation in solution of the H− hydrides, depress the reaction rate and increase the selectivity toward the hydrocarbons and homologation products [39] (Table V).

In contrast, acid oxides as Nb_2O_5 cause an opposite effect favouring the formation of the acid hydride $H^+[Ru(CO)_3I_3]^-$ (Table V).

TABLE V

Ethanol hydrocarbonylation in the presence of different promoters[a]

Promoter system	Ruthenium species in solution[b]	Conversion %	Selectivity, % Homologation	Carbonylation	Hydrogenation
CH$_3$I	H$^+$[Ru(CO)$_3$I$_3$]$^-$	71	27	29	44
CH$_3$I MgO$_{25}$'[c]	[HRu$_3$(CO)$_{11}$]$^-$, [Ru(CO)$_3$I$_3$]$^-$, Ru$_3$(CO)$_{12}$	13	29	5	66
CH$_3$I Al$_2$O$_3$	[Ru(CO)$_3$I$_3$]$^-$, HRu$_3$(CO)$_{11}$]$^-$, Ru$_3$(CO)$_{12}$	29	33	15	52
CH$_3$I TiO$_2$	[Ru(CO)$_3$I$_3$]$^-$	68	25	34	41
CH$_3$I Nb$_2$O$_5$	[Ru(CO)$_3$I$_3$]$^-$	61	35	37	28

[a] Reaction conditions: Ru(acac)$_3$; 0.36 mmol; EtOH: 480 mmol; T: 200°C; time: 8 h; P: 14 MPa; CO/H$_2$: 1; I/Ru: 10; oxide: 1 g.

[b] Ruthenium species detected under reaction condition in order of decreasing concentration

[c] MgO completely hydroxylated.

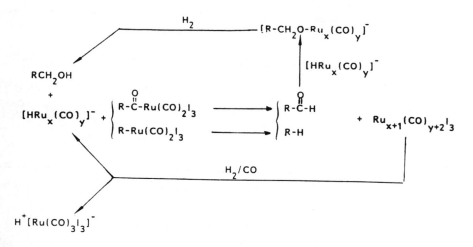

Scheme 9. Role of the anionic ruthenium H^- hydridocarbonyl species $[HRu_x(CO)_y]^-$ in the homologation of oxygenated substrates.

3. Hydrogenation of carbon monoxide

Another catalytic reaction, where the role of H^+ and H^- hydrides and the ion-pairing effects greatly influence the rate and selectivity of the process, is the hydrogenation of CO to ethylene glycol catalyzed by Rh or Ru carbonyl species.

The most important catalytic systems proposed in the literature (metal carbonyl precursors, ligands, promoters) and the carbonyl and hydrido carbonyl species identified in the catalytic solutions are indicated in Scheme 10.

The ruthenium systems active in these processes are practically the same described in the hydrocarbonylation of the oxygenated substrates and the basic concepts and effects are practically those previously discussed. The role and the cooperative effects of the H^+ and H^- hydrido carbonyl and hydrido iodocarbonyl ruthenium species have been extensively reviewed [11]. Here only the processes catalyzed by rhodium carbonyl species firstly studied by Union Carbide and more recently in Japan under the C_1 chemistry research program [40] will be discussed.

Important effects on the reaction rate and on the selectivity of the process have been observed by adding amines, alkali salts, acids, phosphines and solvents with appropriate solvating properties and the determinant intervention of H^+ and H^- hydrido carbonyl species have been recognized (Scheme 10).

Methanol, ethylene glycol and methyl formate, the main reaction products

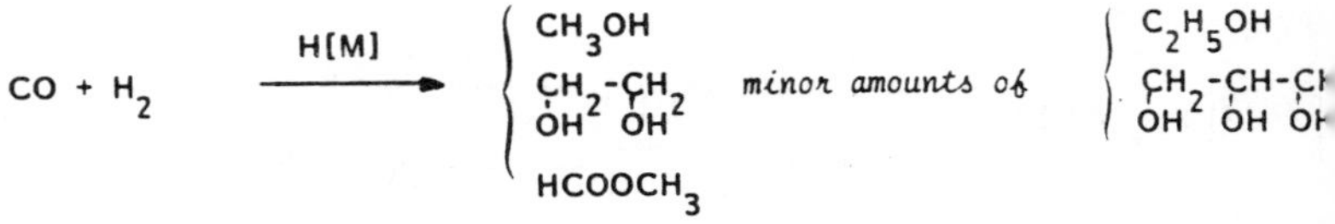

$$CO + H_2 \xrightarrow{H[M]} \begin{cases} CH_3OH \\ \underset{OH}{CH_2}\text{-}\underset{OH}{CH_2} \\ HCOOCH_3 \end{cases} \text{minor amounts of} \begin{cases} C_2H_5OH \\ \underset{OH}{CH_2}\text{-}\underset{OH}{CH}\text{-}CH \\ \quad\quad OH \end{cases}$$

Catalytic system	Metal species detected	Reference
Rhodium catalysts		
$Rh_4(CO)_{12}$/phosphine (L)	$HRh(CO)_3L + [Rh(CO)_4]^-$	(21) (41)
$Rh_4(CO)_{12}$/phosphine/HX	$HRh(CO)_3L + RhX(CO)L_2$	(42)(43)
$Rh(CO)_2(acac)$/amine	$[Rh_5(CO)_{15}]^- + [Rh(CO)_4]^-$	(44)(45)
$Rh(CO)_2(acac)$/imidazole (NMI = N-methyl imidazole)	$[Rh(CO)_4]^- + HRh(CO)_4 + HRh(CO)_3(NMI)$	(46)
$Rh_4(CO)_{12}$/amide solvent	$[Solv.H]_2[Rh_6(CO)_{15}] + [Rh(CO)_4]^-$	(47)
Ruthenium catalysts		
$Ru_3(CO)_{12}/I^-$ (or HI)	$[HRu_3(CO)_{11}]^- + [Ru(CO)_3I_3]^-$	(11)(12)
$Ru_3(CO)_{12}$/imidazole	$[imidazole.H]^+[HRu_3(CO)_{11}]^- + Ru(CO)_5 + + Ru(CO)_4(imidazole)$	(48)
$Ru_3(CO)_{12}$/PPNCl	$[HRu_3(CO)_{11}]^- + [Ru(CO)_3Cl_3]^- + + [Ru_6(CO)_{18}]^{2-}$	(10)(49)

Scheme 10. Hydrogenation of CO to methanol and ethylene glycol.

are formed through the steps indicated in Scheme 11. Ethanol and glycerol, by-products of the reaction, are formed by successive activation of methanol and ethylene glycol respectively, through steps analogous to those previously illustrated in the hydrocarbonylation of oxygenated substrates.

1) Coordination of CO on a metal centre

$$CO + [M_\leq]^- \longrightarrow [M_\leq CO]^-$$

2) Activation of CO by protonation

$$[M_\leq CO]^- + H^+ \longrightarrow H-\overset{O}{\overset{\|}{C}}-[M_\leq]$$

3) Hydrogenation of formyl group to formaldehyde

$$H-\overset{O}{\overset{\|}{C}}-[M_\leq] + H_2 \longrightarrow H-C\overset{\nearrow O}{\underset{\searrow H}{}} + H[M_\leq]$$

4) Activation of formaldehyde for carbonylation or hydro-genation

$$H-C\overset{\nearrow O}{\underset{\searrow H}{}} + H[M_\leq] \longrightarrow \begin{cases} CH_3O-[M_\leq] \\ OH-CH_2-[M_\leq] \end{cases}$$

5) Carbonylation of hydroxy-methyl intermediate

$$HO-CH_2-[M_\leq CO] \overset{CO}{\longrightarrow} HO-CH_2CO-[M_\leq CO]$$

6) Hydrogenation of the hydroxyacetyl or methoxy intermediate

$$\begin{cases} HO-CH_2CO-[M_\leq CO] \overset{H_2}{\longrightarrow} \underset{OH}{CH_2}-C\overset{\nearrow O}{\underset{\searrow H}{}} + H[M_\leq CO] \\ \qquad\qquad\qquad\qquad\qquad\qquad \textit{Glycolaldehyde} \\ CH_3O-[M_\leq] \overset{H_2}{\longrightarrow} CH_3OH + H[M_\leq] \\ \qquad\qquad\qquad\qquad \textit{Methanol} \end{cases}$$

7) Hydrogenation of glycolaldehyde

$$\underset{OH}{CH_2}-C\overset{\nearrow O}{\underset{\searrow H}{}} + H_2 \overset{H[M_\leq]}{\longrightarrow} \underset{OH}{CH_2}-\underset{OH}{CH_2}$$

$$\textit{Ethylene glycol}$$

8) Nucleophilic attack of methanol on the formyl intermediate

$$H-\overset{O}{\overset{\|}{C}}-[M_\leq] \overset{CH_3OH}{\longrightarrow} H-C\overset{\nearrow O}{\underset{\searrow CH_3}{}} + H[M_\leq]$$

$$\textit{Methyl formate}$$

Scheme 11. Steps involved in CO hydrogenation to ethylene glycol in homogeneous phase.

According to Ohgomori the activation of CO proceeds through a protonation step on a coordinated carbonyl group (Scheme 12) [47]. Either mononuclear or polynuclear protonic hydrido carbonyl rhodium species seem to be involved in this step which leads to the formation of formaldehyde.

The intermediacy of formaldehyde has been ascertained and its interaction with H^+ and H^- hydrides leading to the hydroxymethyl- or methoxy-intermediates, determines the formation of ethylene glycol and methanol respectively (Scheme 13).

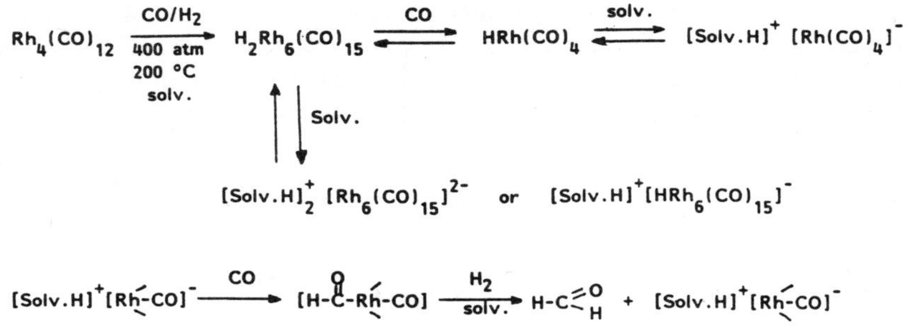

Scheme 12. Formation of formaldehyde from syngas.

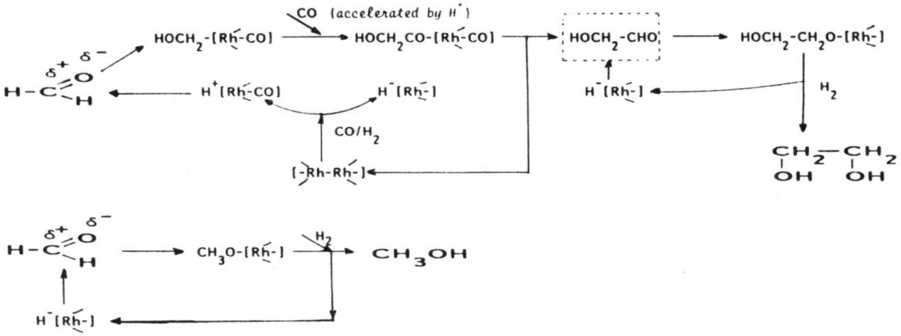

Scheme 13. Formation of methanol or ethylene glycol from formaldehyde.

Further interventions of H^+ or H^- hydrides may also affect the successive steps of the reaction. Especially the H^+ hydrides can promote the carbonyl insertion into the alkyl-metal bond, necessary for the formation of ethylene glycol, whereas the H^- hydrides favour the hydrogenation of the acyl intermediate to alcohol.

In the reaction between the H^- hydrides and the acyl intermediates polynuclear species may be formed and successively declusterized by CO/H_2 with regeneration of the active H^+ and H^- hydrides (Scheme 13).

In this view the improvement of the reaction rate and the selectivity towards ethylene glycol by free carboxylic acids [43] or by addition of salts of little solvatable cations (*i.e.* cesium benzoate, bis(triphenylphosphine)iminium acetate) [50] may be related to the favoured formation of the H^+ hydridocarbonyl species necessary for the activation of carbon monoxide and to the acceleration of the carbonyl insertion into the hydroxymethyl-rhodium bond.

The choice of the solvent is also of great importance in dependence of the type and solubility of the salt added since they can complex the counterions of the anionic rhodium species affecting their reactivity. Important classes of solvents are amides, crown-ethers and polyethers able to preferentially stabilize, perhaps in form of ion-pairs, the active rhodium catalytic intermediates [47].

Important effects on the catalysis are also played by donor ligands, like amines and phosphines [11, 21]. Both the basicity and the steric characteristics of these ligands appear to be determinant for the kinetic and selectivity of the process. It is surprising for instance that PPh_3 and PBu_3 completely inhibit the formation of ethylene glycol, whereas trialkylphosphines with a large cone angle such as $^iPr_3P$ and $^tBu_3P$ enhance the catalytic activity and the selectivity. Moreover the combination of the inhibiting phosphine with amines lead to catalytically active systems.

It is likely that the role of these ligands must be related to the stabilization of the H^+ and H^- hydrido species involved either in the activation of CO and in the carbonylation steps or in the hydrogenation of the acyl intermediates.

Another example of the role of H^+ and H^- rhodium carbonyl species in hydrogenation and carbonylation steps have been evidenced in our paper on the hydrocarbonylation reaction of ethyl orthoformate [51]. In the presence of a H^- rhodium carbonyl system $HRh_5(CO)_{15}$, the unique activation observed leads to diethoxymethane, which implies the H⁻Rh addition to the C-OR bond as illustrated in Scheme 14. In the presence of this system, absolutely no carbonylation reaction takes place.

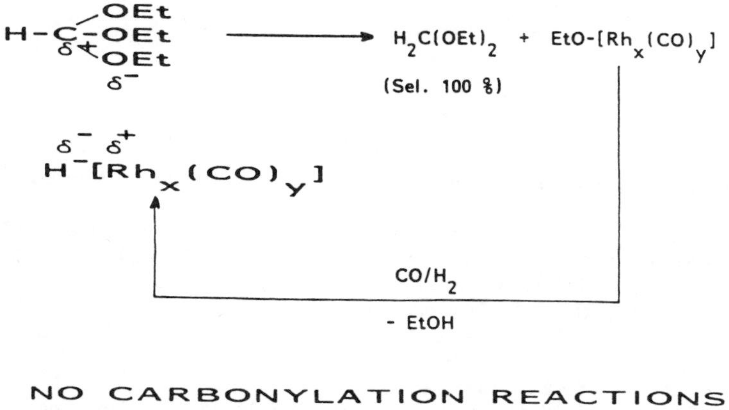

NO CARBONYLATION REACTIONS

Scheme 14. Reaction of ethyl orthoformate with CO/H_2 in the presence of H⁻[Rh] carbonyl catalysts.

On the contrary, when a H^+ hydrido halocarbonyl rhodium precursor is used, a different interaction with the substrate may take place, leading to the formation of carbonylation and hydrocarbonylation products of the ethyl group (for example ethyl propionate and diethylacetal of propionaldehyde) (Scheme 15).

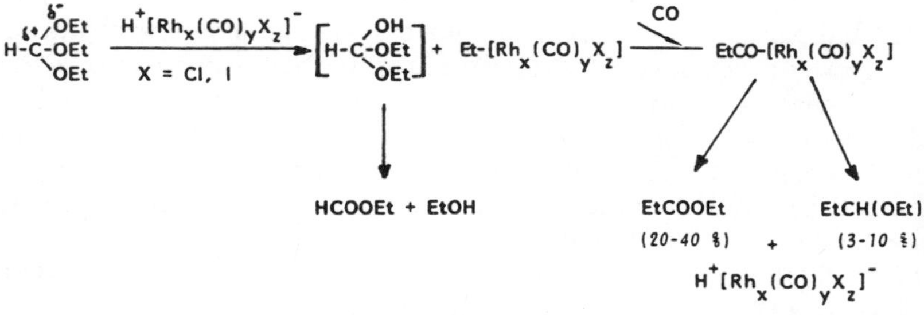

Scheme 15. Carbonylation and hydrocarbonylation of ethyl orthoformate in the presence of iodo- and chloro-carbonyl rhodium derivatives.

Acknowledgements

This work was carried out under the research program "Chimica Fine II" CNR (Rome).

Department of Chemistry and Industrial Chemistry
University of Pisa
Via Risorgimento 35, 56126 Pisa (Italy)

References

1. I. Wender and P. Pino, "Organic Syntheses via metal Carbonyls", J. Wiley & Sons, New York, **1977**.
2. J. Falbe, "New Syntheses with Carbon Monoxide", Springer-Verlag Berlin, **1980**.
3. M. Hidai, A. Fukuoka, Y. Koyasu and Y. Uchida, *J. Mol. Catal.*, **1986**, *35*, 29.
4. T.W. Dekleva and D. Forster, *Adv. Catal.* **1986**, *34*, 85.
5. G. Luft and M. Schrod, *J. Mol. Catal.* **1983**, *20*, 175.
6. G. Braca and G. Sbrana, "Aspects of Homogeneous Catalysis", R. Ugo (Ed) D. Reidel Publ. Co., Dordrecht, **1984**.
7. G. Jenner, *Applied Catal.*, **1989**, *50*, 99.
8. A.M. Raspolli Galletti, G. Braca, G. Sbrana and F. Marchetti, *J. Mol. Catal.* **1985**, *32*, 291.
9. J. Knifton, R.A. Grigsby and J.J. Lin, *Organometallics*, **1984**, *3*, 62.

10. Y. Kiso, M. Tanaka, H. Nakamura, T. Yamasaki and K. Saeki, *J. Organomet. Chem.*, **1986**, *312*, 357.
11. B.D. Dombek, *Adv. Catal.*, **1983**, *32*, 325.
12. B.K. Warren and B.D. Dombek, *J. Catal.*, **1983**, *79*, 334.
13. M. Marchionna and G. Longoni, *Organometallics*, **1987**, *6*, 606.
14. R.G. Pearson, *Chem. Rev.*, **1985**, *85*, 41.
15. E.J. Moore, J.M. Sullivan and J.R. Norton, *J. Am. Chem. Soc.*, **1986**, *108*, 2257.
16. J.L. Vidal and W.E. Walker, *Inorg. Chem.*, **1981**, *20*, 249.
17. R. Lazzaroni, P. Pertici, S. Bertozzi and G. Fabrizi, *J. Mol. Catal.*, **1990**, *58*, 75.
18. R. Lazzaroni, G. Uccello Barretta, M. Benetti, *Organometallics*, **1989**, *9*, 2323.
19. F. Piacenti, M. Bianchi, G. Bimbi, P. Frediani, U. Matteoli, *Coord. Chem. Rev.*, **1975**, *16*, 9.
20. D. Evans, J.A. Osborn, G. Wilkinson, *J. Chem. Soc.*, **1986**, 3133.
21. Y. Ohgomori, *Sekiyu Gakkaishi*, **1988**, *31(4)*, 263.
22. G. Braca, G. Sbrana, F. Piacenti and P. Pino, *Chim. Ind. (Milan)*, **1970**, *52*, 1091.
23. R.A. Sanchez-Delgado, J.S. Bradley and G. Wilkinson, *J. Chem. Soc. Dalton*, **1976**, 399.
24. T. Hayashi, Z.H. Gu, T. Sakakura and M. Tanaka, *J. Organomet. Chem.*, **1988**, *352*, 373.
25. T. Sakakura, T. Kobayashi and M. Tanaka, *Cl Mol. Chem.*, **1985**, *1*, 219.
26. H.W. Walker and P.C. Ford, *J. Organomet. Chem.*, **1981**, *214*, C43.
27. J.C. Bricker, C.C. Nagel and S.G. Shore, *J. Am. Chem. Soc.*, **1982**, *104*, 1444.
28. M.Y. Darensbourg "Progress in Inorganic Chemistry" **1985**, *33*, 221.
29. G. Braca, A.M. Raspolli Galletti and G. Sbrana "Industrial Chemicals via C1 Processes" *ACS Symposium Series*, **1987**, *329*, 220.
30. G. Braca, L. Paladini, G. Sbrana, G. Valentini, G. Andrich and G. Gregorio, *Ind. Eng. Chem. Prod. Res. Dev.*, **1981**, *20*, 115.
31. G. Braca, G. Sbrana, G. Valentini, G. Andrich and G. Gregorio "Fundamental Research in Homogeneous Catalysis" M. Tsutsui Ed., Vol. 3, p. 221 **1979**.
32. G. Braca, A.M. Raspolli Galletti, G. Sbrana and F. Zanni, *J. Mol. Catal.*, **1986**, *34*, 183.
33. G. Braca, G. Sbrana, A.M. Raspolli Galletti and S. Berti, *J. Organomet. Chem.*, **1988**, *342*, 245.
34. G. Braca, A.M. Raspolli Galletti, G. Sbrana and R. Lazzaroni, *J. Organomet. Chem.*, **1988**, *342*, 245.
35. G. Braca, A.M. Raspolli Galletti, G. Sbrana and E. Trabuco *J. Mol. Catal.* **1989**, *55*, 184.
36. I. Wender, *Catal. Rev. Sci. Eng.*, **1976**, *14*, 97.
37. S. Beda Butts, T.G. Richmond and D.F. Shriver, *Inorg. Chem.*, **1981**, *20*, 278.
38. M. Tanaka, T. Sakakura, T. Hayashi and T. Kobayashi, *Chem. Lett.*, **1986**, 39.
39. G. Braca, G. Sbrana and A.M. Raspolli Galletti, "New Science in Homogeneous Transition Metal Catalyzed Reactions", ACS Symposium Series, in press.
40. "Progress in C1 Chemistry in Japan", The Research Association for C1 Chemistry, Elsevier Tokyo **1989**.
41. M. Tamura, M. Ishino, T. Deguchi and S. Nakamura, *J. Organomet. Chem.*, **1986**, *312*, C75.
42. Y. Ohgomori, S. Yoshida and Y. Watanabe, *J. Chem. Soc., Dalton*, **1987**, 2969.
43. Y. Ohgomori, S. Mori, S. Yoshida and Y. Watanabe, *J. Mol. Catal.*, **1987**, *43*, 127.
44. J.L. Vidal and W.E. Walker, *Inorg. Chem.*, **1980**, *19*, 896.

45. T. Masuda, K. Murata and A. Matsuda, *Bull. Chem. Soc. Jpn*, **1988**, *61*, 2865.
46. Y. Kiso, M. Tanaka, T. Hayashi and K. Saeki, *J. Organomet. Chem.*, **1987**, *322*, C32–C36.
47. Y. Ohgomori, S. Mori, S. Yoshida and Y. Watanabe, *J. Mol. Catal.*, **1987**, *40*, 223.
48. Y. Kiso, K. Saeki, T. Hayashi, M. Tanaka and Y. Matsunaga, *J. Organomet. Chem.*, **1987**, *335*, C27.
49. S. Yoshida, S. Mori, H. Kinoshita and Y. Watanabe, *J. Mol. Catal.*, **1987**, *42*, 215.
50. W.E. Walker, D.R. Bryant and E.S. Brown (Union Carbide), U.S. Pat. 3 952 039 **1976**.
51. A.M. Raspolli Galletti, G. Braca and G. Sbrana, *J. Organomet. Chem.*, **1988**, *356*, 221.
52. A. Fulford and P.M. Maitlis, *J. Organomet. Chem.*, **1989**, *366*, C20–C22.

S. CENINI AND C. CROTTI

CATALYTIC CARBONYLATIONS OF
NITROGEN CONTAINING ORGANIC COMPOUNDS *

1. Introduction

The development of effective methods for promoting the carbon-nitrogen bond formation has long been a challenge in catalysis. However the addition of amines to olefins to yield alkylamines, is surprisingly catalyzed more efficiently by alkali metal amides, rather than by transition metal complexes [1]. On the other hand different routes to the metal promoted C-N bond formation lead to a different type of products. These catalytic syntheses involve the interaction of carbon monoxide with organic azides, RN_3 [2–4], with aromatic nitro compounds, $ArNO_2$ [5], and with amines in the presence of molecular oxygen [6–8] (Scheme 1).

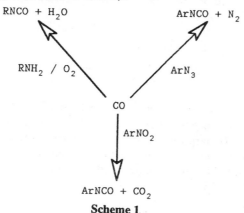

Scheme 1

* In this review, we have considered only the carbonylations of organic azides, organic nitro compounds and the oxidative carbonylation of amines, since in all these cases very important products such as isocyanates and carbamates can be obtained.

A. F. Noels et al. (Eds.), Metal promoted selectivity in organic synthesis, 311–328.
© 1991 Kluwer Academic Publishers. Printed in the Netherlands.

In general, what we expect from the use of metals in organic syntheses can be summarized as follows: (i) milder experimental conditions, (ii) better selectivities, (iii) new reaction pathways, (iv) labile organic intermediates trapped by the metal centre and moreover, (v) transition metals often are the architects of a catalytic cycle, which is of course a very important point if a large scale application of these reactions is to be considered. These points will find several examples in the course of the following discussion.

The carbonylation reactions above reported give organic isocyanates (or the corresponding carbamates, $ArNHCO_2R'$, if alcohol is present in the reaction medium), which are important final products and/or intermediates in the synthesis of pesticides and polyurethanes. The classical method for carrying out the synthesis of isocyanates is *via* catalytic hydrogenation of nitro compounds and subsequent reaction with phosgene:

$$RNO_2 \longrightarrow RNH_2 \xrightarrow{\;COCl_2\;} RNCO \xrightarrow{\;R'OH\;} RNHCO_2R' \qquad (1)$$

An enormous effort has been made on finding new, phosgene-free routes to isocyanates. Phosgene, produced by reaction of CO with chlorine, is an energy-intensive material, very poisonous and corrosive since it is a hydrolyzable chlorine derivative. Recently it has been pointed out that triphosgene, obtained by reaction of chlorine with dimethylcarbonate [9]:

$$O = C \begin{array}{c} OCH_3 \\ \\ OCH_3 \end{array} \xrightarrow[CCl_4]{h\nu\;/\;Cl_2} O = C \begin{array}{c} OCCl_3 \\ \\ OCCl_3 \end{array} + 6\;HCl \qquad (2)$$

is a stable solid, which does not create problems for transport and storage, and which can replace phosgene in its reactions, in particular with amines. However its use for large scale syntheses appears rather improbable. It is interesting to point out here that dimethylcarbonate, now produced *via* the reaction of methanol with oxygen and carbon monoxide, catalyzed by copper(I) chloride [10]:

$$2\;CH_3OH + CO + 1/2\;O_2 \xrightarrow{\;CuCl\;} (CH_3O)_2CO + H_2O \qquad (3)$$

presents much more advantages when used in the place of phosgene [11].

2. Carbonylation of organic azides

Drastic conditions must be used for the reactions of organic azides with CO to give the corresponding isocyanates ($T \approx 80\text{--}160°C$ and $P_{CO} \approx 150\text{--}300$ atm), and the selectivity is rather poor (ca.60%) [12, 13]. On the other hand phenyl azide readily reacts with $Os_3(CO)_{11}(MeCN)$, to give a metallacycle derived by the attack on coordinated carbon monoxide [14]. This metallacycle is stabilized in this case by coordination of one nitrogen atom to another osmium atom of the cluster:

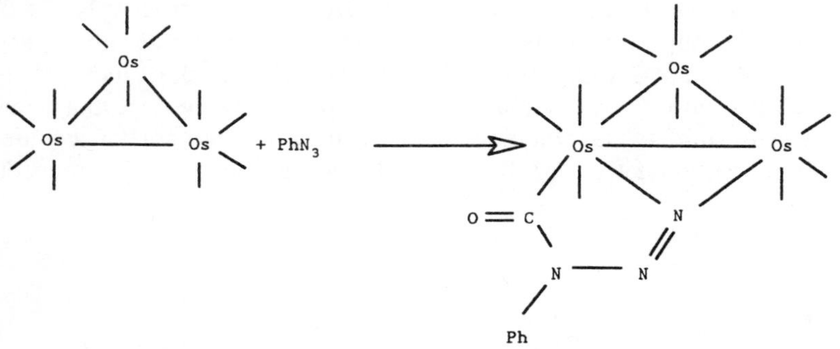

$$(4)$$

With mononuclear transition metal carbonyl complexes, the metallacycle readily loses molecular nitrogen giving an isocyanate complex. In the presence of carbon monoxide at atmospheric pressure and at 40–45°C, a catalytic cycle takes place by using *trans*-RhCl(CO)(PPh$_3$)$_2$ as catalyst [2] (Scheme 2):

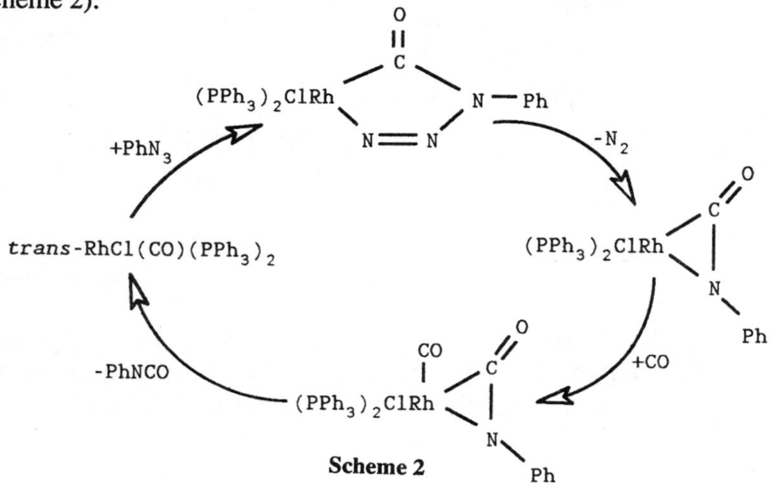

Scheme 2

A turnover of ca. 60–65 moles converted/mol cat. per hour was observed, with a selectivity in the formation of PhNCO higher than 98%. Similar results were obtained when the catalyst was heterogenized by reaction with a polymeric phosphine. Its activity and selectivity in the catalytic carbonylation of aromatic azides was found comparable to that of the homogeneous system. The heterogenized catalyst maintains its high activity even in the presence of amines, giving the corresponding ureas, or ethanol giving the carbamate directly. By carbonylation under similar conditions of 2,3-diazidonaphtalene, 2-azido-3-isocyanato-naphtalene has been obtained, even under more forcing conditions [3]. The selectivity in the carbonylation of only one azido group is probably due to an intramolecular dipolar interaction between the $-N_3$ and $-NCO$ groups, which hinders the attack of the azide to coordinated carbon monoxide. This view is supported by the fact that when the isocyanato moiety is converted into a urea or carbamate residue, the catalytic carbonylation of the second $-N_3$ group readily proceeds (Scheme 3).

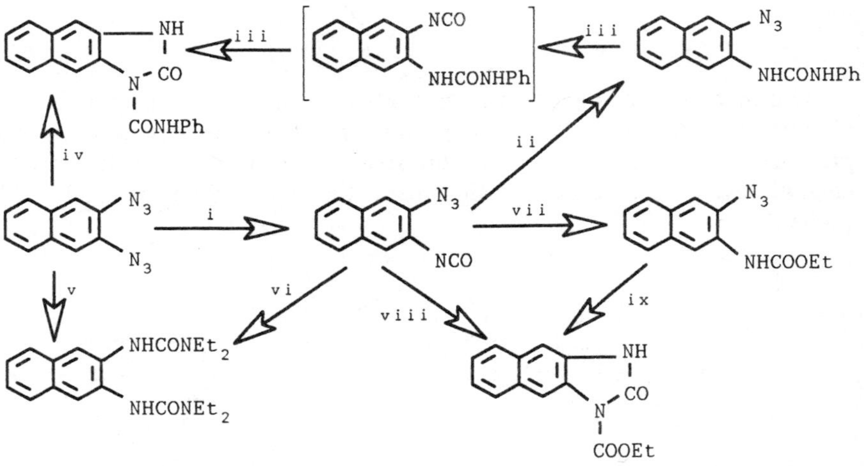

Scheme 3

(i) homogeneous catalyst (hc) or supported catalyst (sc) (see text), CO; (ii) PhNH$_2$; (iii) CO, (hc); (iv) CO, PhNH$_2$ (hc); (v) CO,Et$_2$NH, (hc); (vi) CO, Et$_2$NH, (hc); (vii) EtOH; (viii) CO,EtOH, (hc); (ix) CO,EtOH, (hc).

New heterocyclic organic derivatives, which appear to be difficult to prepare by conventional routes, were thus obtained. More generally, heterocyclic organic derivatives can be obtained by carbonylation of aromatic azides bearing a substituent in the *ortho*-position, able to attack the intermediate isocyanate function formed during the carbonylation reaction [4].

On considering the very mild reaction conditions, the total selectivity and the easy removal of the catalyst, the carbonylation of organic azides represents the simplest way to prepare on a laboratory scale a solution containing pure organic isocyanates, which can then be used for a variety of reactions.

3. Carbonylation of aromatic nitro compounds

In recent years, the catalytic carbonylation of aromatic nitro compounds has become a very intense field of research [5]. This is due to the fact that a series of industrially important compounds can be obtained from nitro compounds and carbon monoxide in a single step. By this way in fact, amines, azo derivatives, Schiff bases, heterocycles, isocyanates, carbamates, ureas, N-substituted amides and imides, have been selectively obtained.

The reaction of carbon monoxide with nitro compounds to give the isocyanates is a thermodynamically highly favorable process. In the case of nitrobenzene a $\Delta H^\circ = -128.8$ KCal/mol can be calculated, assuming $\Delta H^\circ_f = -16.1$ KCal/mol for $PhNCO_{(l)}$ [5].

$$PhNO_{2\ (l)} + 3\ CO_{\ (g)} \longrightarrow PhNCO_{\ (l)} + 2\ CO_{2\ (g)} \qquad (5)$$

$$\Delta H^\circ = -128.8\ Kcal/mol$$

The unfavorable $T \times \Delta S^\circ$ term only marginally modifies this value. Even the reaction of PhNCO with methanol is exothermic [15]:

$$PhNCO + CH_3OH \rightleftharpoons PhNHCO_2Me \qquad (6)$$

$$K\ (1/mol):\ 1.34\ (15°C);\ 0.90\ (25°C);\ 0.55\ (35°C)$$

Thus the direct carbonylation of a nitro compound in the presence of alcohol to give the carbamate is an even more exothermic reaction. However in the absence of a catalyst these reactions do not occur.

The mechanism generally accepted for the reaction catalyzed in the homogeneous phase is based on the following steps (Scheme 4):

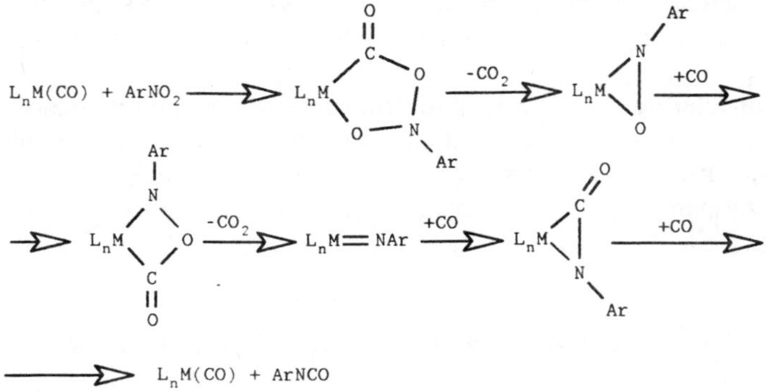

Scheme 4

The nitro compound is at first reduced to a nitroso derivative by an attack of the two oxygen atoms of the nitro group respectively on the metal and on the carbon atom of the carbonyl, giving a metallacycle that undergoes a decarboxylation, thus leaving the nitroso group η^2-bound to the metal. The hypothetical metallacycle seems to be confirmed by analogy with similar intermediates postulated in the deoxygenation of amine N-oxides [16] (**7**) or isolated in the deoxygenation of nitrile N-oxides [17](**8**).

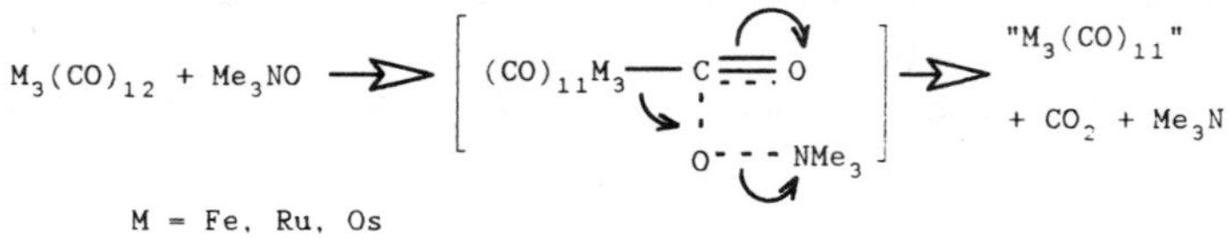

$$M_3(CO)_{12} + Me_3NO \longrightarrow \left[(CO)_{11}M_3\!-\!C \rightleftharpoons O \atop O^- - - NMe_3 \right] \longrightarrow "M_3(CO)_{11}" + CO_2 + Me_3N \qquad (7)$$

M = Fe, Ru, Os

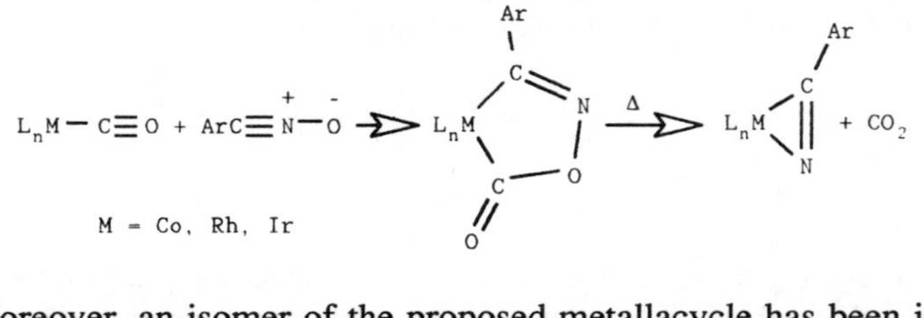

M = Co, Rh, Ir

Moreover, an isomer of the proposed metallacycle has been isolated, and it loses carbon dioxide very readily [18].

The subsequent reaction with CO gives the key intermediate, that is the nitrene complex $L_nM = NAr$, which is then carbonylated to the isocyanate derivative. Some recent results obtained in our laboratories seem however to suggest that a N-σ bound nitroso derivative should be the proper intermediate in these reactions [19].

A recent work by Gladfelter *et al.* [20] has reported an *in situ* FT-IR spectroscopy study at elevated CO pressure of the carbonylation of nitroaromatics catalyzed by $Ru(DPPE)(CO)_3$ (DPPE = 1,2-bis(diphenylphosphino)ethane). Evidence has been found by EPR spectroscopy for single-electron transfer in the reaction of $Ru(DPPE)(CO)_3$ with p-$ClC_6H_4NO_2$, since a signal due to the radical anion of p-nitrosochlorobenzene was detected. Precedent exists for the single-electron oxidation of metals by nitroaromatics, especially in the work on the reaction of $Ni(PR_3)_4$ with nitrobenzene to give $Ni(PR_3)_2(PhNO)$, PR_3 and OPR_3 [21]. In any case, these results are not in conflict with the intermediate proposed in Scheme 4, which can be formed after an initial one electron oxidation of the metal by $ArNO_2$.

The formation of a nitrene complex by reaction of nitro compounds with transition metal carbonyls is a well documented reaction [5].

For the reactions carried out in alcohol, it is in general considered that the carbamate is formed by reaction of ArNCO with ROH outside the coordination sphere of the metal. However some results obtained by us suggest that another mechanism could be involved [22]:

$$L_nM = NAr \xrightarrow{CO/ROH} L_nM \overset{NHAr}{\underset{COOR}{<}} \xrightarrow{CO} L_nM(CO) + ArNHCO_2R \qquad (9)$$

With catalysts working in the heterogeneous phase another mechanism can be considered (Scheme 5):

$$M + RNO_2 \longrightarrow M = O + RNO$$

$$M + RNO \longrightarrow M = O + "RN" \longrightarrow products$$

$$M = O + CO \longrightarrow M + CO_2$$

Scheme 5

A model reaction for the deoxygenation of the nitro group by the direct action of the metal can be [23]:

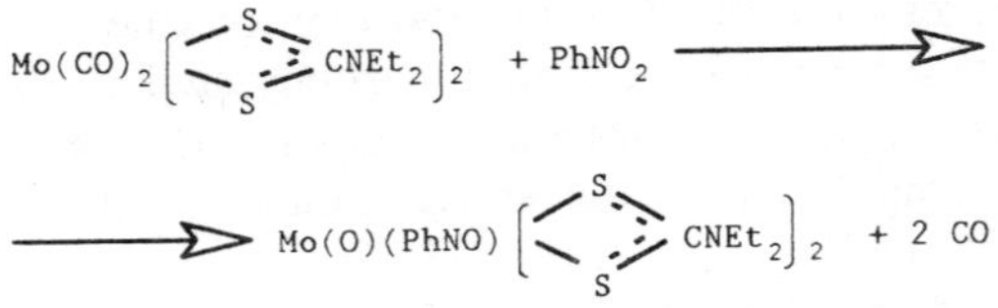

$$Mo(CO)_2 \left[\underset{S}{\overset{S}{<}} CNEt_2 \right]_2 + PhNO_2 \longrightarrow$$

$$(10)$$

$$\longrightarrow Mo(O)(PhNO) \left[\underset{S}{\overset{S}{<}} CNEt_2 \right]_2 + 2\ CO$$

Rhodium on alumina in the presence of pyridine, at 240°C and 200 atm of CO, gives PhNCO in 70% yield from $PhNO_2$, with 100% conversion. Based on i.r. studies, the redox mechanism reported in Scheme 6 has been suggested [24].

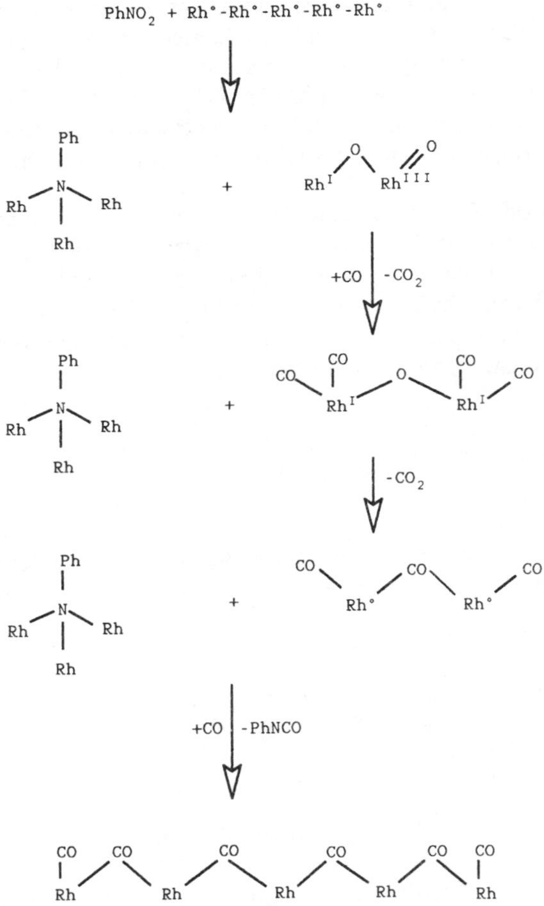

$$PhNO_2 + Rh°-Rh°-Rh°-Rh°-Rh°$$

Scheme 6

This mechanism appears reasonable, but leaves unsolved the problem related to the function of the co-catalyst, which, moreover, in the case of rhodium can be a base but even a Lewis acid.

Among the various homogeneous catalysts able to catalyze the carbonylation of aromatic nitro compounds to isocyanate, of particular interest are the few working at atmospheric pressure. Of these, $[Rh(CO)_2Cl]_2$ in the presence of $MoCl_5$ [25] and $Pd_2Cl_6^{2-}$ or $[Pd(CO)Cl_3]^-$ with vanadium oxychlorides as co-catalysts [26], have been particularly studied. For the Rh-Mo catalytic system, a Lewis acid-nitrogroup σ interaction has been proposed. This interaction should favor a side-on coordination of the nitro group to rhodium, allowing an easy deoxygenation by CO to give the nitroso derivative.

As we have already mentioned, the syntheses of carbamates by reductive carbonylation of nitro compounds is usually considered to proceed *via* the formation of the corresponding isocyanate, followed by reaction with alcohol outside the coordination sphere of the metal. However we have found two catalytic systems in which the alcohol should participate to the catalytic cycle, since when it was absent practically no isocyanate was obtained [22, 27].

A variety of aromatic nitro compounds can be carbonylated in toluene/methanol at 170°C and 60 atm of carbon monoxide to give the corresponding carbamates in high yields and very high selectivity (up to 95%), by using $Ru_3(CO)_{12}$ as catalyst with $NEt_4^+Cl^-$ as co-catalyst [22]. An interesting effect was observed when the temperature inside the autoclave was followed during the reaction. It has been found that most part of the reaction occurs in about 30 minutes, during which the temperature was notably higher with respect to the temperature registered during a blank experiment. This of course is due to the exothermic reaction, and emphasizes the importance of a diluting solvent when these reactions are carried out. On considering only the time where the exothermic reaction occurs, we could calculate that about 400 moles of $PhNO_2$ are converted per mole of catalyst per hour. On the basis of what is known in literature and by the fact that the nitrene complex, $Ru_3(CO)_{10}(\mu_3-NPh)$, is a catalyst as good as $Ru_3(CO)_{12}$ in this reaction, we considered the following mechanism as the most likely (Scheme 7).

Support to the above mechanism came recently from the observation that methyl-N-phenylcarbamate is produced from nitrene-methoxycarbonyl coupling, when a methoxycarbonyl iron nitrene cluster was heated in methanol solution [28].

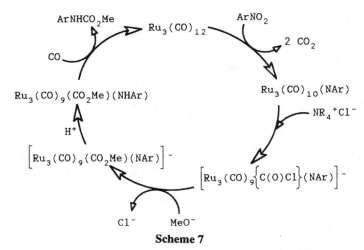

Scheme 7

Recently, a comprehensive work by Geoffroy [29] summarized all the efforts to clarify the halide promotion of nitroaromatic carbonylation catalysis; although a mechanism like the one reported on Scheme 7 still is considered among the most likely, the authors point out the importance that other pathways can assume, particularly in some reaction conditions (*i.e.* different solvent and/or temperature). Indeed, it is outlined that (i) halides promote the formation of imido (or nitrene) ligands from nitrosobenzene (but not from nitrobenzene), probably as a consequence of the promotion of substitution of PhNO for CO; (ii) halides catalyze the carbonylation of nitrene complexes to isocyanates at very mild conditions (atmospheric CO pressure and 22°C) (Scheme 8):

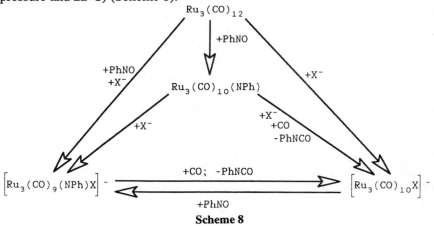

Scheme 8

To explain the easier carbonylation of the imido complex, Geoffroy suggests that halides weaken the Ru–N bonds by assuming a bridging position and displacing a N–Ru bond to form a *di*bridging imido ligand which then undergoes nucleophilic attack on a coordinated CO (Scheme 9):

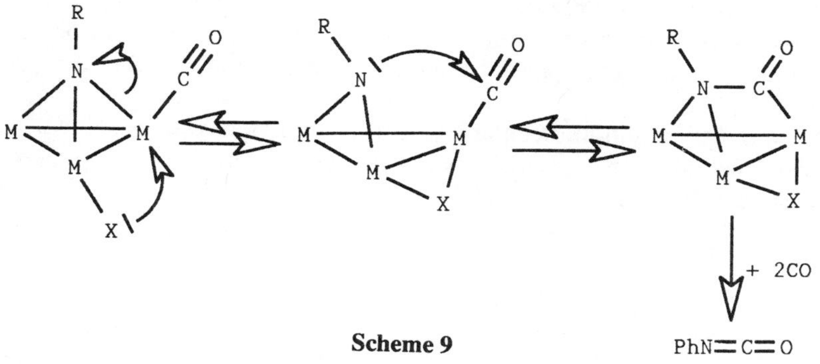

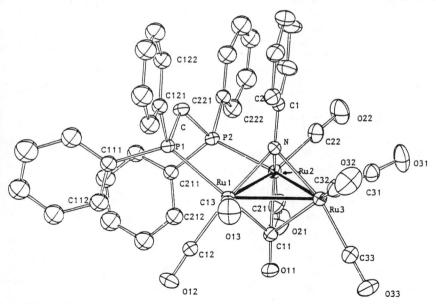

Scheme 9

$PhN = C = O$

For the other catalytic systems discovered by us [27], and which use $Ru_3(CO)_{12}$ or $Rh_6(CO)_{16}$ as catalysts with chelating ligands such as DPPM, DPPE, DPPP and bipyridine as co-catalysts, the mechanism of the reaction is

Figure 1. (Reprinted by Kinol permission from *J. of Organometallic Chemistry*, Elsevier Sequoia S.A., Lausanne).

still obscure. We have observed that a complex such as $Ru_3(\mu_3-NPh)$ $(\mu_3-CO)(\mu-DPPM)(CO)_7$ (Figure 1) (likewise to the unsubstituted nitrene complex, $Ru_3(\mu_3-NPh)(\mu_3-CO)(CO)_9$ [22, 30]) is very robust, and gives only small amounts of $PhNHCO_2Me$ and $Ru_3(CO)_{10}(\mu-DPPM)$ when heated at 170°C under CO pressure in the presence of methanol [31]. This result does not clarify the role of the co-catalyst, the chelating diphosphine, and thus other complexes, possibly mononuclear, could be the true active catalytic species.

An interesting intramolecular cyclization reaction was observed, when *ortho*-nitrostyrenes were carbonylated in the presence of metal carbonyls such as $Fe(CO)_5$, $Ru_3(CO)_{12}$ and $Rh_6(CO)_{16}$ [32].

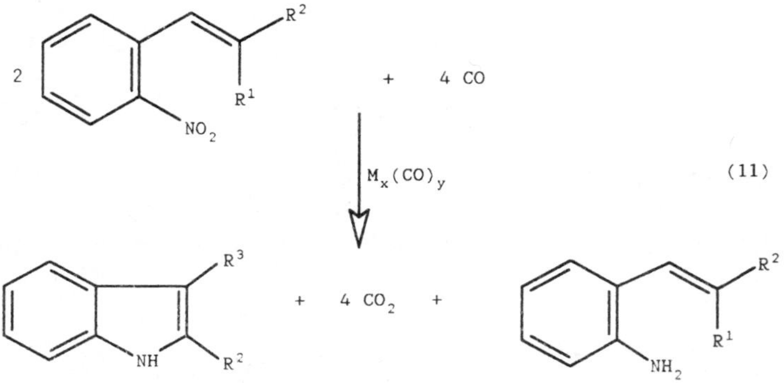

$$M_x(CO)_y = Fe(CO)_5;\ Ru_3(CO)_{12};\ Rh_6(CO)_{16}$$

a: $R^1 = R^2 = R^3 = H$
b: $R^1 = R^3 = H;\ R^2 = Me$
c: $R^1 = R^3 = H;\ R^2 = CO_2Me$
d: $R^1 = R^3 = H;\ R^2 = Ph$
e: $R^1 = R^3 = H;\ R^2 = 2\text{-}C_5H_5N$
f: $R^1 = R^2 = R^3 = Me$
g: $R^1 = R^3 = Ph;\ R^2 = Me$

Reaction products are mainly the indoles, derived by attack of an intermediate nitrene to the β-carbon of the styrene double bond. With substrates having a β,β-disubstituted styrene double bond, a 1,2-shift of one of the two substituents occurs during the reaction.

A chemometric optimization of the ruthenium carbonyl catalyzed cyclization of 2-nitrostilbene to 2-phenylindole has been carried out [33]. It has been found that the conversion % of the reaction is almost independent on the amount of the substrate; moreover it decreases by increasing the CO pressure. These results suggest that the rate determining step should be the dissociation of carbon monoxide from the ruthenium carbonyl. Evidences have also been found for the participation of the ruthenium cluster, but also of the monomeric $Ru(CO)_5$, to the catalytic cycle.

Other heterocycles can be obtained by carbonylation of *ortho*-substituted nitro compounds such as 2-substituted benzimidazole from N-benzylidene-2-nitro-aniline [34] (12) or carbazole from 2-nitrobiphenyl [35] (13):

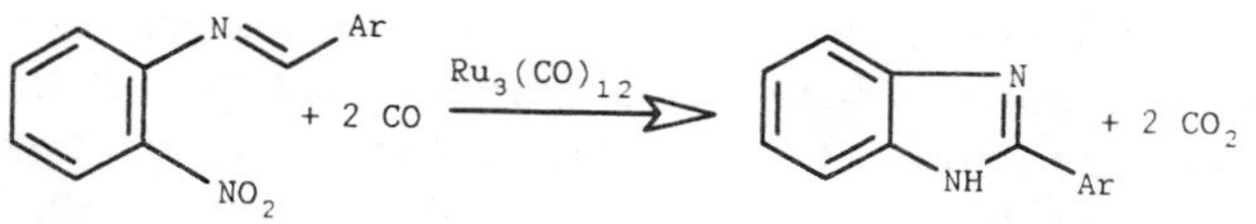

$$+ 2\ CO \xrightarrow{Ru_3(CO)_{12}} + 2\ CO_2 \qquad (12)$$

$$+ 2\ CO \xrightarrow{Ru_3(CO)_{12}} + 2\ CO_2 \qquad (13)$$

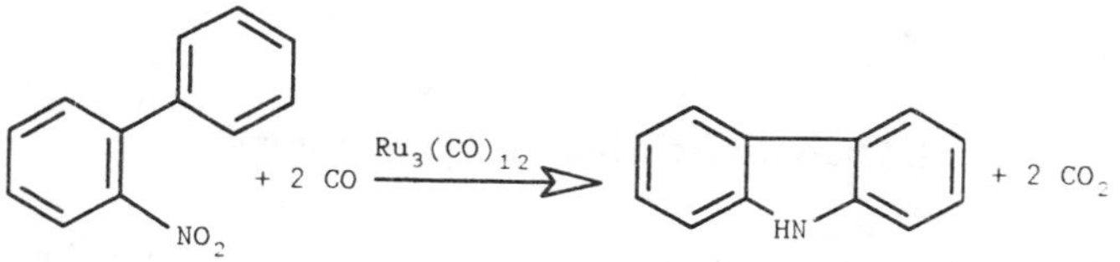

On the other hand, the cyclization can fail for the occurrence of undesired side-reactions, like (14) [36], or because the intermediate nitrene finds other favorite reaction pathways, such as (15) [34]:

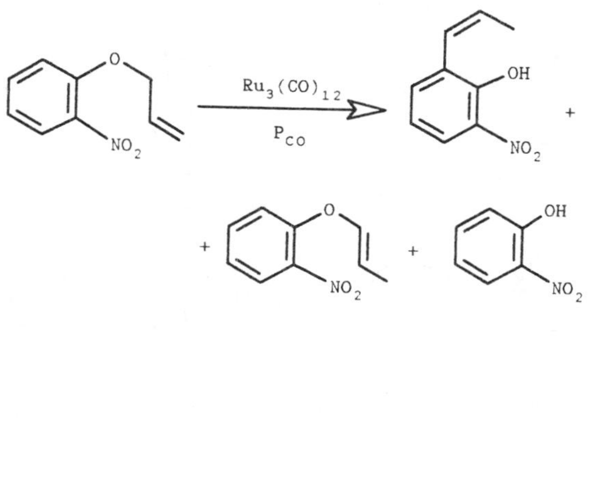

$$\xrightarrow[P_{CO}]{Ru_3(CO)_{12}} \qquad + \qquad (14)$$

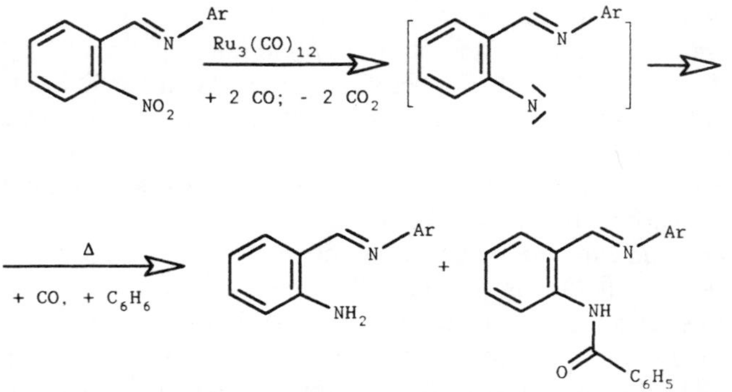
Ar
N
NO2
Ru3(CO)12
+ 2 CO; - 2 CO2
Ar
N
N
Δ
+ CO, + C6H6
Ar
N
NH2
+
Ar
N
NH
O
C6H5

(15)

Although the homogeneous catalytic systems above described (particularly the $Ru_3(CO)_{12}$–$NEt_4^+Cl^-$ catalyst) are very efficient and selective, their use for large scale preparations creates the serious problem of the catalyst recycle, which appears to be rather difficult to overcome. For this reason we are now studying the activity of heterogeneous catalysts,where a transition metal is used in the presence of an appropriate co-catalyst (the metal itself is not active in this reaction).

4. Oxidative carbonylation of amines

The catalytic oxidative carbonylation of aniline has been used for the synthesis of MDI [6] (methylenediphenyldiisocyanate, one of the most important monomer for the production of polyurethanes together with TDI, 2,4-toluenediisocyanate) (Scheme 10):

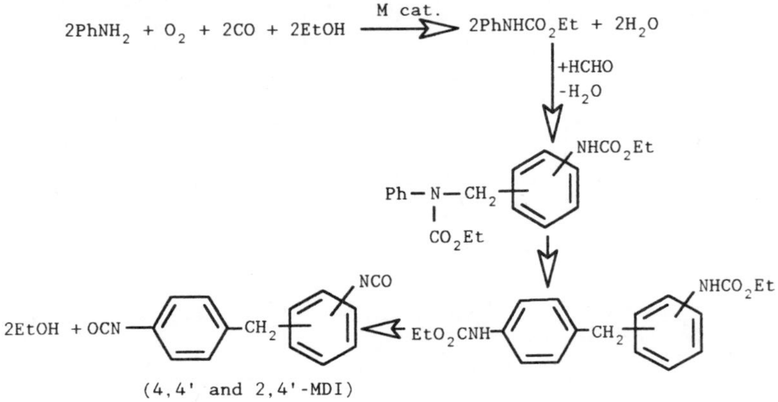
M cat.
2PhNH2 + O2 + 2CO + 2EtOH
2PhNHCO2Et + 2H2O
+HCHO
-H2O
NHCO2Et
Ph—N—CH2
CO2Et
NCO
NHCO2Et
2EtOH + OCN
CH2
EtO2CNH
CH2
(4,4' and 2,4'-MDI)

Scheme 10

The catalyst for the oxidative carbonylation to give $PhNHCO_2Et$ is metallic palladium with iodide compounds as promoters. The selectivity is superior to 97%, with a turnover of about 260 moles converted/mol cat. per hour. Other platinum group metals, particularly rhodium, were active, although less efficiently, as catalysts [37]. The reaction is conducted with 80 atm of CO and 6 atm of O_2, at 160–170°C, and can also be applied to aliphatic amines. Diphenylurea is intermediately formed in the oxidative carbonylation of aniline. More recently it has been reported that aromatic amines react at room temperature and atmospheric pressure with carbon monoxide, dioxygen, alcohols and hydrochloric acid, with palladium chloride as the catalyst and copper(II) chloride as re-oxidant, to give carbamate ester in fair to quantitative yields [38]. This homogeneously catalyzed reaction is therefore superior to the heterogeneously catalyzed reaction described above. Palladium complexes such as $PdCl_2(PhNH_2)_2$ have also been used as catalysts in this reaction in slightly more drastic conditions [39]:

$$PdCl_2(PhNH_2)_2 + 2PhNH_2 + CO \longrightarrow CO(NHPh)_2 + 2PhNH_3^+Cl^- + Pd \quad (16)$$

$$CO(NHPh)_2 + MeOH \rightleftharpoons PhNHCO_2Me + PhNH_2 \quad (17)$$

$$Pd + 2PhNH_3^+Cl^- + \tfrac{1}{2}O_2 \longrightarrow PdCl_2(PhNH_2)_2 + H_2O \quad (18)$$

Thus even in this case the reaction proceeds *via* the intermediate formation of the urea (the real catalytic reaction), followed by alcoholysis of the urea to give carbamate. The intimate mechanism by which the urea is formed is not well established. A possible reaction scheme in the case of the heterogeneous reaction could be the one reported in Scheme 11 [37]:

A different mechanism has been observed when Co^{II}(salen) has been used as catalyst in this reaction [7, 8]. In fact in this case the isocyanate corresponding to the amine is the primary product of the reaction, which can react with alcohol, giving the carbamate, or with the excess of amine, giving the urea. However this catalyst was not particularly efficient, since the salen ligand was gradually oxidized during the catalytic reactions.

Very recently it has been reported that the complex $Ru(saloph)Cl_2$ (saloph = bis(salicylaldehyde-*o*-phenylenediimine) catalyses the oxidative carbonylation of cyclohexylamine in ethanol to cyclohexylurethane selectively, at 160°C and CO + O_2 (1:0.5) pressure of 21 bar [40]. The rate of the reaction is first order with respect to catalyst, amine and carbon monoxide, and one-half order with respect to dioxygen. The activation energy has been

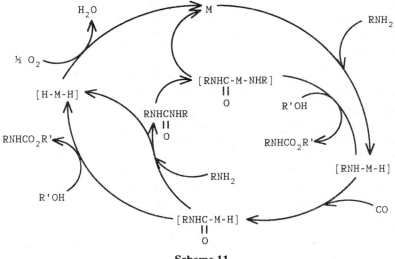

Scheme 11

evaluated as 21.0 Kcal mol^{-1}. A probable mechanism, based on kinetic observations involving a ruthenium(V)-oxo species as intermediate, has been proposed for the oxidative carbonylation of cyclohexylamine. The reaction seems to proceed *via* the formation of cyclohexylisocyanate within the coordination sphere of the metal, with subsequent attack by ethanol to give the carbamate [40].

5. Conclusions

A variety of important organic products can now be selectively obtained, particularly by the carbonylation of nitro compounds catalyzed by transition metal complexes. These reactions generally proceed *via* the intermediate formation of nitrene (or imido) complexes, which are thus key intermediates. It is interesting to note that carbene complexes (A) are next-of-kind to nitrene complexes (B):

$$L_nM=C\overset{R}{\underset{R'}{\big\langle}} \qquad\qquad L_nM=N\overset{}{\underset{R}{\big\backslash}} \quad\longleftrightarrow\quad L_{(n-1)}M\equiv NR \;+\; L$$

(A)   (B)

Structures (A) and (B)

The most important difference is the lone pair on nitrogen, which can be involved in a triple bond with the metal or can bridge a triangle of transition metal atoms [41]. However the addition of an appropriate ligand can induce the transformation into a bent, doubly bound nitrene species, which must be more reactive.

Dipartimento di Chimica Inorganica e Metallorganica and C.N.R. Center
Via Venezian 21
20133 Milano (Italy)

References

1. Pez, G.P. and Galle, J.E. *Pure Appl. Chem.* **1985**, *57*,1917.
2. La Monica, G.; Monti, C. and Cenini, S. *J. Molecular Cat.* **1984**, *18*, 93.
3. La Monica, G.; Monti, C.; Cenini, S. and Rindone, B. *J. Molecular Cat.* **1984**, *23*, 89.
4. La Monica, G.; Ardizzoia, G.; Maddinelli, G. and Tollari, S. *J. Molecular Cat.* **1986**, *38*, 327.
5. Cenini, S.; Pizzotti,M. and Crotti, C. *Metal catalysed deoxygenation reactions by carbon monoxide of nitroso and nitro compounds*, in *Aspects of Homogeneous Catalysis*; Vol. VI, R. Ugo, D.Reidel Publishing Company, 1988, and references cited therein.
6. Fukuoka, S.; Chono, M. and Khono, M. *Chemtech* **1984**, 670.
7. (a) Benedini, F.; Nali, M.; Rindone, B.; Tollari, S.; Cenini, S.; La Monica, G. and Porta, F. *J. Molecular Cat.* **1986**, *34*, 155; (b) Maddinelli, G.; Nali, M.; Rindone, B.; Tollari, S.; Cenini, S.; La Monica, G. and Porta, F. *J. Molecular Cat.* **1987**, *39*, 71.
8. Bassoli, A.; Rindone, B.; Tollari, S. and Chioccara, F. *J. Molecular Cat.* **1990**, *60*, 41.
9. Eckert, H. and Forster, B. *Angew. Chem., I.E.* **1987**, *26*, 894.
10. Romano, U.; Tesei, R.; Mauri, M. M. and Rebora, P. *Ind. Eng. Chem. Prod. Res. Dev.* **1980**, *19*, 396.
11. Mauri, M.M.; Romano,U. and Rivetti, F. *Ing. Chim. Ital.***1985**, *21*, 6.
12. Ribaldone, G.; Caprara, G. and Borsotti, G. *Chim. Ind. (Milan)* **1965**, *50*, 1200.
13. Weigert, F.J. *J. Org. Chem.* **1973**, *38*, 1316.
14. Burgess, K.; Johnson, B.F.G.; Lewis, J. and Raithby, P.R. *J. Chem. Soc., Dalton Trans.* **1982**, 2119.
15. Nesterov, O.V.; Zabrodin, V.B.; Chirkov, Yu.N. and Entelis, S. *Kinetica Katal.* **1974**, *15*, 1341.
16. Shen, J.; Shi, Y.; Gao, Y.; Shi, Q. and Basolo, F. *J. Am. Chem. Soc.* **1988**, *110*, 2414.
17. Chetcuti, P.A.; Walker, J.A.; Knobler, C.B. and Hawthorne, M.F. *Organometallics* **1988**, *7*, 641.
18. Cenini, S.; Porta, F ; Pizzotti, M. and Crotti, C. *J. Chem. Soc., Dalton Trans.* **1985**, 163.
19. Pizzotti, M.; Porta, F.; Cenini, S.; Demartin, F. and Masciocchi, N. *J. Organometal. Chem.* **1987**, *330*, 265.
20. Kunin, A.J.; Noirot, M.D. and Gladfelter, W.L. *J. Am. Chem. Soc.* **1989,**, *111*, 2739.
21. Berman, R.S. and Kochi, J.K. *Inorg. Chem.* 1980, *19*, 248.

22. Cenini, S.; Crotti, C.; Pizzotti, M. and Porta, F. *J. Org. Chem.* **1988**, *53*, 243.
23. Maatta, E.A. and Wentworth, R.A.D. *Inorg. Chem.* **1980**, *19*, 2597.
24. Elleuch, B.; Ben Taarit, Y.; Basset, J.M. and Kervennal, J. *Angew. Chem., I.E.* **1982**, *21*, 687.
25. Unverferth, K.; Hontsch, R. and Schwetlich, K. *J. Prakt. Chem.* **1979**, *321*, 86 and 928.
26. Tietz, H. and Schwetlich, K. *Z. Chem.* **1985**, *25*, 147.
27. Cenini, S.; Pizzotti, M.; Crotti, C.; Ragaini, F. and Porta, F. *J. Molecular Cat.* **1988**, *49*, 59.
28. Williams, G.D.; Whittle, R.R.; Geoffroy, G.L. and Rheingold, A.L. *J. Am. Chem. Soc.* **1987**, *109*, 3936.
29. Han, S.H. and Geoffroy, G.L. *Polyhedron* **1988**, *7*, 2331.
30. Smejia, J.A.; Gozum, J.E. and Gladfelter, W.L. *Organometallics* **1987**, *6*, 1311.
31. Pizzotti, M.; Porta, F.; Cenini, S. and Demartin, F. *J. Organometallic Chem.* **1988**, *356*, 105.
32. Crotti, C.; Cenini, S.; Rindone, B.; Tollari, S. and Demartin, F. *J. Chem. Soc., Chem. Commun.* **1986**, 784.
33. Crotti, C.; Cenini, S.; Todeschini, R. and Tollari, S. *J. Chem. Soc., Faraday Trans.*, in press.
34. Crotti, C.; Cenini, S.; Tollari, S. and Bassoli, A. *VI°* International Symposium on Homogeneous Catalysis , Vancouver (Canada), 1988; manuscript in preparation.
35. Crotti, C.; Cenini, S.; Bassoli, A.; Rindone, B. and Demartin, F., submitted for publication.
36. Bassoli, A.; Rindone, B.; Tollari, S.; Cenini, S. and Crotti, C. *J.Molecular Cat.* **1989**, *52*, 445.
37. Fukuoka, S.; Chono, M. and Khono, M. *J. Org. Chem.* **1984**, *49*, 1458.
38. Alper, H. and Hartstock, F.W. *J. Chem. Soc., Chem. Commun.* **1985**, 1141.
39. (a) Giannoccaro, P. *Inorg. Chim. Acta* **1987**, *336*, 271; (b) *ibidem* **1988**, *142*, 81.
40. Taqui Khan, M.M.; Halligudi, S.B. and Sumita, N. *J. Molecular Cat.* **1990**, *59*, 303.
41. (a) Cenini, S. and La Monica, G. *Inorg. Chim. Acta, Rev.* **1976**, *18*, 279; (b) Nugent, W.A. and Haymore, B.L. *Coord. Chem. Rev.* **1980**, *31*, 123.

D. BROWN, B.T. HEATON AND J.A. IGGO

APPLICATIONS OF SPECTROSCOPIC
MEASUREMENTS TO HOMOGENEOUS CATALYSIS

Many industrially important homogeneously catalysed reactions use rhodium carbonyl or rhodium phosphine compounds as catalysts, (*e.g.* the Celanese-Union Carbide hydroformylation [1], the production of L-DOPA by asymmetric hydrogenation [2], the BP-Monsanto carbonylation of methanol to give acetic acid [3] and the related Eastman Kodak carbonylation of methyl acetate to give acetic anhydride [4]). This is a lucky coincidence since the two most powerful spectroscopic techniques for *in situ* studies of homogeneous catalysts are infra-red (*via* metal-carbonyl and metal-hydride absorptions) and NMR.

^{103}Rh is 100% abundant with spin 1/2. Thus, the possibility of unambiguous characterisation by NMR of rhodium species in solution is enhanced due to J(^{103}Rh–X) (X = ^{1}H, ^{13}C, ^{31}P, etc.) coupling. This is extremely useful in determining both how many X groups are present per Rh atom and the ligand stereochemistry about the rhodium. Additionally, the observation of spin-spin coupling readily allows ligand inter- and intra- molecular exchange data to be obtained both *via* variable temperature measurements and from classical line-broadening effects. Exchange data may also be obtained *via* spin-saturation transfer [5] experiments using the DANTE pulse sequence for selective excitation of an exchanging nucleus [6].

Unfortunately, direct observation of ^{103}Rh NMR spectra is not easy because of the low resonance frequency of ^{103}Rh (3.16 MHz at a field such that the protons of TMS resonate at 100 MHz) and the resulting low sensitivity. As a result, direct observation of ^{103}Rh NMR spectra requires large sample sizes (15 mm o.d.) and a high field spectrometer. In practice it is much easier to collect ^{103}Rh NMR data indirectly and methods are available for compounds containing ^{103}Rh–^{1}H, ^{103}Rh–^{13}C, ^{103}Rh–^{31}P groups [7, 8, 9].

The relative merits of IR and NMR may be summarised thus:
– IR: fast timescale, high sensitivity, but crowded spectra giving low quality structural information and no quantitative information unless

329

A. F. Noels et al. (Eds.), Metal promoted selectivity in organic synthesis, 329–365.
© 1991 *Kluwer Academic Publishers. Printed in the Netherlands.*

– NMR: slow timescale, low sensitivity, but high dispersion spectra giving detailed structural information from chemical shifts and couplings and multinuclear operation, quantitative spectra. Informative about chemical exchange.

The two techniques are thus complementary.

Although multinuclear NMR measurements provide the most powerful spectroscopic technique for characterisation of species in solution and for following the course of reactions, elucidation of the complete reaction pathway is fraught with difficulties and relies heavily upon an armoury of physical techniques (X-ray structural determinations, NMR, IR, kinetic studies, isotopic labelling studies etc).

In order to illustrate the usefulness of NMR measurements in helping to elucidate the pathways of catalytic reactions, I will discuss first the rhodium-catalysed hydrogenation of alkenes and then describe the application of high pressure IR and NMR spectroscopies to the study of an olefin hydroformylation catalyst.

1. Catalytic hydrogenation of alkenes

1.1. PREAMBLE

The mechanism of action of Wilkinson's catalyst, [RhCl(PPh$_3$)$_3$], (1), has been studied in considerable detail since its discovery in the mid-1960's [10]. There is now general acceptance of the schematic mechanism shown in Scheme 1 but work is still continuing to elucidate the details of the reaction pathway.

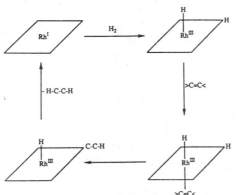

Scheme 1. Schematic mechanism for olefin hydrogenation catalysed by rhodium(I) complexes.

This is difficult because there are many steps involved and the inter-mediates in the catalytic pathway can coexist in several forms related by intra- and inter-molecular rearrangements. The mechanism which is generally accepted is shown in Scheme 2 [11] and involves bisphosphine-rhodium species which sequentially coordinate H_2 and the alkene followed by the rate-controlling step which is migration of a *cis*-hydride onto the olefin.

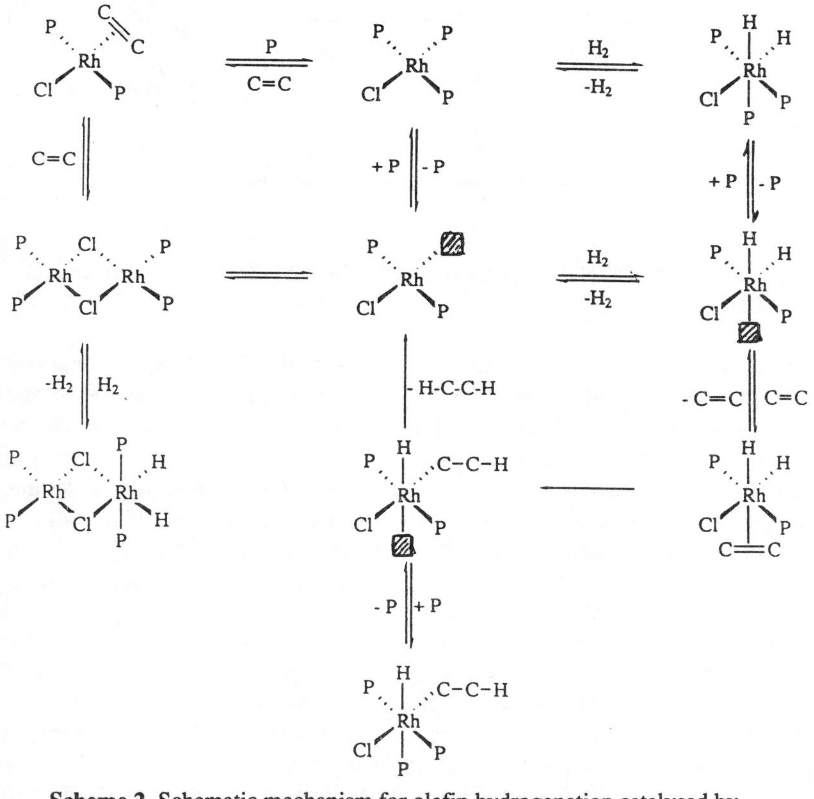

Scheme 2. Schematic mechanism for olefin hydrogenation catalysed by [RhCl(PPh₃)₃].

Despite much activity, characterisation of the intermediates associated with this catalytic pathway is lacking and the stereochemistry of [RhH₂Cl(PPh₃)₃](2) apart, the stereochemistry of the coordination sphere during the catalytic cycle has been inferred from the structural characterisa-tion/behaviour of more stable species, *e.g.* iridium complexes or complexes containing non-dissociating ligands.

I would like to concentrate on how recent NMR measurements have been used to obtain rates of ligand inter- and intra-molecular exchange in (1) and (2) [12] and how para-H_2 [13] has been used to establish the reversibility of the reaction shown below.

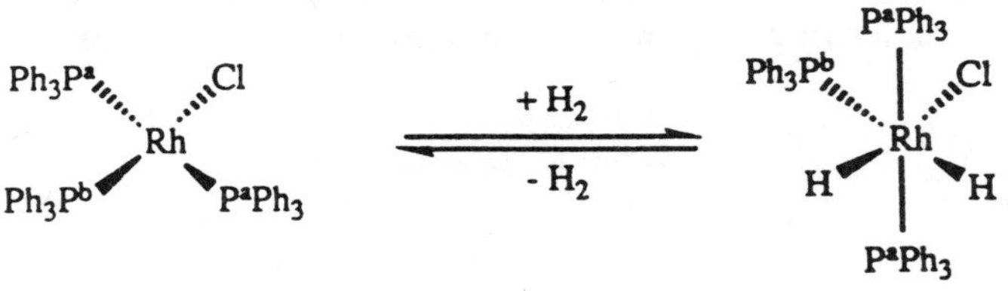

Scheme 3. Reversible uptake of H_2 by (1).

1.2. APPLICATION OF NMR SPECTROSCOPY IN DETERMINING CHEMICAL EXCHANGE

NMR studies of exchange processes have traditionally relied on analysis of the changes in the NMR spectrum of the exchanging system as the temperature of measurement is altered. At sufficiently low temperature the exchange will be frozen out and discrete resonances for each site in the exchanging system will be observed. As the temperature is raised the resonances undergoing exchange first broaden, then coalesce, until at sufficiently high temperature a single resonance is observed at the weighted average chemical shift. In this way it is possible to extract kinetic and thermodynamic parameters for the exchange.

More recently, saturation transfer techniques have been used to study exchange. These are particularly useful where several competing exchange processes occur or where the exchange occurs at such a slow rate that coalescence cannot be observed at a convenient temperature. The saturation transfer experiment relies on first exciting the nuclei at one site of the exchanging system. As these nuclei leave this site they will take their excited spin with them and hence the original excitation is carried to all sites in the exchanging system. By monitoring the progress of the excitation around the system with time after the initial excitation, rate constants for the exchange process can be obtained. Brown and coworkers have demonstrated the power of this approach in studies of homogeneous catalysis in an elegant series of papers [12 and refs. therein].

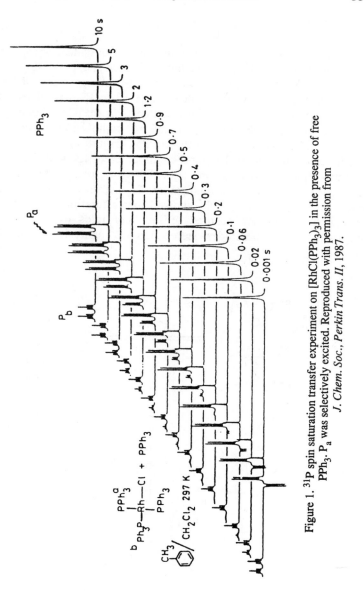

Figure 1. ^{31}P spin saturation transfer experiment on [RhCl(PPh$_3$)$_3$] in the presence of free PPh$_3$. P$_a$ was selectively excited. Reproduced with permission from *J. Chem. Soc., Perkin Trans. II*, 1987.

1.3. LIGAND EXCHANGE IN [RhCl(PPh$_3$)$_3$], (1)

The ^{31}P NMR spectrum of (1) shows two inequivalent phosphorus environments P^a and P^b in the ratio 2:1 both of which show coupling to rhodium. A spin saturation transfer experiment on (1) in the presence of free PPh$_3$ was carried out in order to establish whether P^a or P^b dissociates more rapidly [12]. Excitation of the low field doublet, *via* the DANTE pulse sequence, showed that magnetisation was transferred rapidly to the low field triplet of P$_b$ and *much more slowly to external* PPh$_3$ (Figure 1).

The equivalencing of P^a and P^b without exchange with free PPh$_3$ requires an intra-molecular rearrangement of the square-planar complex, presumably through a tetrahedral intermediate of C$_{3v}$ symmetry. In a separate experiment, excitation of free PPh$_3$ showed that saturation transfer to all the co-ordinated phosphines occured at *identical* rates (0.31 s^{-1}). This rather surprising result precludes stereochemical analysis of the dissociation process.

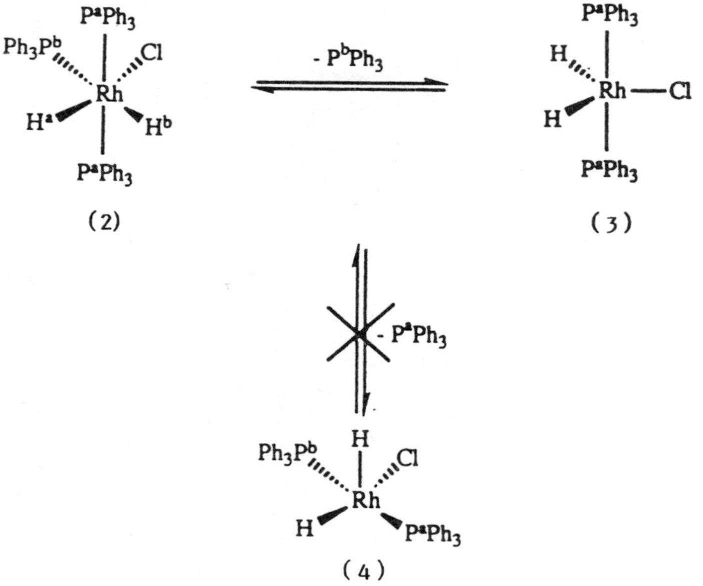

Scheme 4. Possible intermediates resulting from dissociation of PPh$_3$ from [RhH$_2$Cl(PPh$_3$)$_3$].

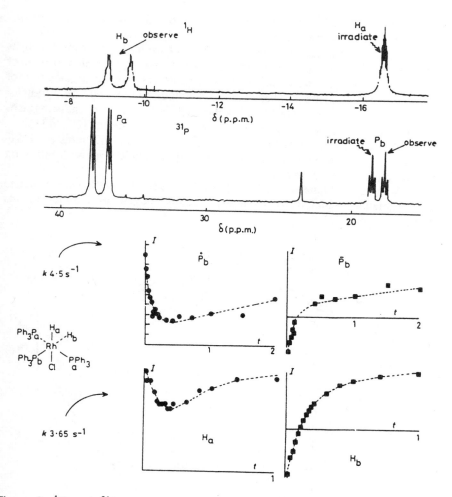

Figure 2. ^1H and ^{31}P spin saturation transfer experiments on [RhH$_2$Cl(PPh$_3$)$_3$]. The resonance indicated by a wavy arrow was excited. Reproduced with permission from *J. Chem. Soc., Perkin Trans. II, 1987.*

1.4. LIGAND EXCHANGE IN [RhH$_2$Cl(PPh$_3$)$_3$], (2)

The ^{31}P and ^1H NMR spectra of (2) at low temperature are entirely consistent with the stereochemistry shown in (2). Variable temperature studies have shown that the phosphine *trans* to H^b dissociates very readily (*ca.* 400 s^{-1} at ambient temperature) [14]. The resulting five-coordinate intermediate is of unknown stereochemistry but we do know from the NMR spectra that the fast exchange does not make P^a and P^b equivalent. Either of two five-coordinate intermediates (3 or 4) could result from dissociation of PPh_3 from (2), Scheme 4. These can be distinguished by saturation transfer experiments. If intermediate (3) is involved H^a and H^b become indistinguishable whereas intermediate (4) retains the separate identities of H^a and H^b.

Using saturation transfer techniques, Brown *et al.* find the rates of H^a/H^b exchange and P^bPh_3 dissociation to be indistinguishable, Figure 2, implying a common reaction pathway *i.e. via* intermediate (3).

Such an intermediate would be expected to be fluxional, Scheme 5, hence the intermediate which traps alkene in the catalytic cycle need not have *trans* phosphines as in (3).

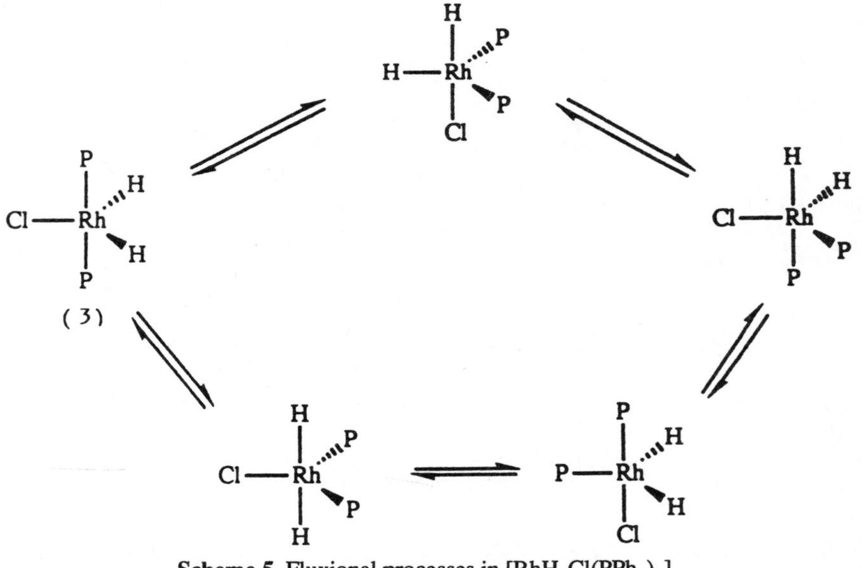

Scheme 5. Fluxional processes in $[RhH_2Cl(PPh_3)_3]$.

In fact the saturation transfer experiments on (2) show there is also exchange of:

- (a) P^a with P^b; k = 0.2 s^{-1} at 303 K and
- (b) P^a with free PPh_3.

Both these processes occur on the timescale of the catalytic hydrogenation reaction. Brown *et al.* [12] conclude that a *cis*-bisphosphinerhodium complex maybe the intermediate responsible for trapping the alkene rather than (3).

1.5. REVERSIBILITY OF H_2 UPTAKE IN OXIDATIVE-ADDITION REACTIONS DETECTED VIA PARA-HYDROGEN INDUCED POLARISATION

The reversibility of H_2 oxidative addition reactions in homogeneous hydrogenations has been investigated by Raman spectroscopy using *para*-enriched H_2 [15]. [RhCl(PPh$_3$)$_3$] was shown to react with H_2 reversibly whereas both [Rh(COD)(dppe)]$^+$ and [Rh(COD)(chiraphos)]$^+$, an asymmetric hydrogenation catalyst, reacted, in the presence of alkene, by reversible alkene coordination followed by irreversible H_2 addition [15]. Now, *para*-hydrogen will also give dramatically enhanced nuclear alignment and hence enhanced NMR absorptions and emissions in reactions in which para-H_2 is added or transferred pairwise to a substrate to yield a product in which the two transferred protons are magnetically distinct. Assuming this addition is fast relative to proton relaxation then the two transferred protons will initially reflect the nuclear spin populations of the starting H_2. As a result, some states will be overpopulated relative to the normal Boltzmann distribution giving rise to polarised absorbtion (A)/emission(E) or E/A spectra similar to a multiplet effect [13]. This is illustrated in Figure 3 [13].

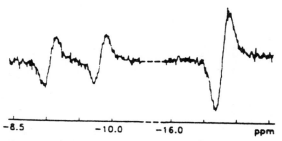

Figure 3. E/A ^{1}H NMR spectrum resulting from addition of *para*-H_2 to [RhCl(PPh$_3$)$_3$]. Reproduced with permission from *J. Am. Chem. Soc.*, 1987.

Such an effect shows that there is fast pairwise transfer of H's and, if the effect is long-lasting, that the reverse reaction is slow. This promises to be a most useful innovation for establishing mechanisms of hydrogenation since *para*-H_2 is simply prepared by storing a sample of H_2 at $-196\,°C$ over Fe_3O_4 for several hours. In favourable cases *para*-enrichment can be achieved simply by storing a hydrogen saturated solution of the hydrogenation catalyst at this temperature overnight.

2. Rhodium catalysed asymmetric hydrogenation

The catalytic hydrogenation of prochiral olefins by cationic rhodium complexes containing optically active ligands has been the basis for the commercial manufacture of L-DOPA (3,4-dihydroxphenylalanine), a drug used for treating Parkinson's disease. L-DOPA can now be obtained in > 99% enantiomeric excess *via* the asymmetric hydrogenation of a pro-chiral intermediate using a rhodium/optically active diphosphine catalyst system.

Much work [2, 11, 16, 17, 18] has been carried out in order to clarify the stereochemistry and reactivity of the intermediates involved in asymmetric catalysis. The generally accepted mechanism for hydrogenation of alkenes using non-optically active ligands with a cationic rhodium catalyst is shown in Scheme 6 [17].

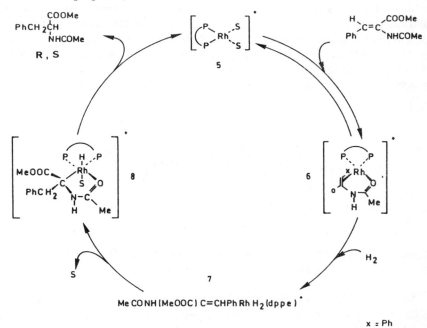

Scheme 6. Schematic mechanism for olefin hydrogenation catalysed by cationic rhodium complexes.

Species **5**, **6** and **8** have been characterised by NMR spectroscopy and **6** has been characterised crystallographically. According to Halpern [17], there is a rapid equilibration of **5** with **6** so that oxidative-addition of H_2 (**6** → **7**) is the rate limiting step in the catalytic cycle. At low temperatures ($< -40\,°C$), the conversion of **8** → **5** becomes rate limiting so that **8** accumulates and can be isolated and characterised crystallographically.

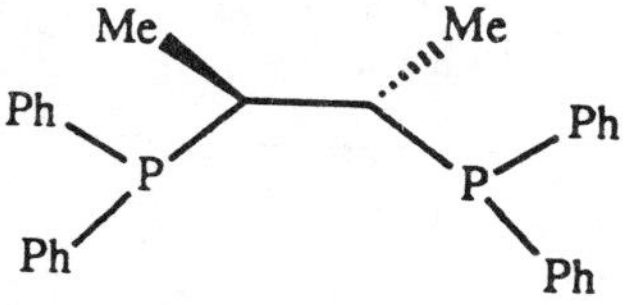

(9)    **(10)**

Extension of this mechanism to include chiral ligands, chiraphos (**9**) and dipamp (**10**), necessitates the inclusion of two diastereoisomers of (**6**) resulting in the two mechanistic pathways shown in Figure 4.

The isolation and structural characterisation of (**12**), the major isomer of [Rh(chiraphos) (EAC)]$^+$, (EAC = **11a**), is a surprising result since the minor isomer (**12′**) has to be involved in the formation of the enantiomeric product and thus must have the higher reactivity towards H_2.

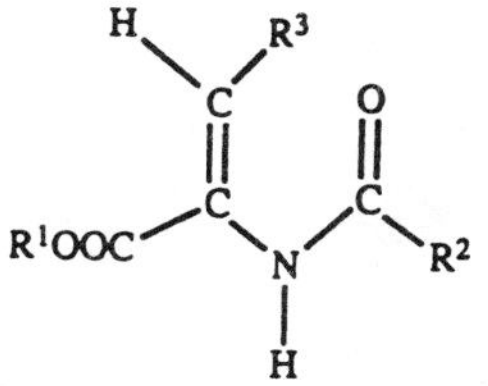

	R^1	R^2	R^3
a	Et	Me	Ph
(11) b	Me	Me	Ph
c	Me	Ph	Ph

In the case of [Rh(chiraphos)(EAC)]$^+$, only the major isomer (**12**) could be observed in solution but both the major (**12**) and the minor (**12′**) isomers have been observed for [Rh(dipamp)(MAC)]$^+$ and indeed the minor isomer has been shown to react at a much faster rate with H_2 [17].

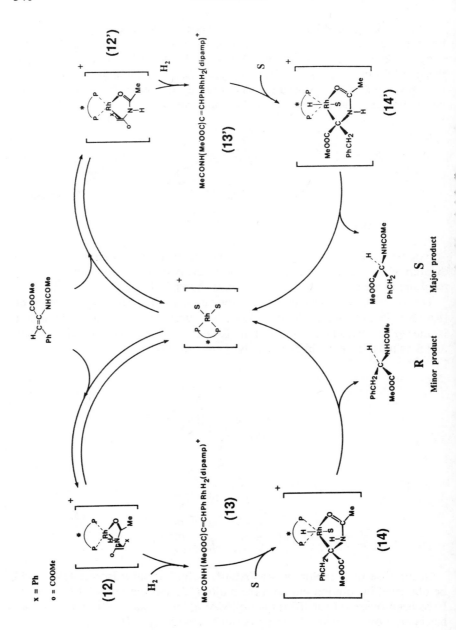

2.1. MECHANISM OF INTERCONVERSION OF THE DIASTEREOISOMERIC COMPLEXES (12) AND (12′)

Both diastereoisomers of [Rh(dipamp)(MAC)]$^+$ can be observed by ^{31}P NMR in solution. Brown [18] has shown that selective inversion of one of the P^a doublets of the major (12) diastereoisomer using the DANTE pulse sequence results in exchange of magnetisation with P^a of the minor diastereoisomer (12′) only. This implies an intramolecular process involving dissociation/recombination of the olefin but retention of the N-acyl to Rh bond. Rotation about the N-vinyl bond in the non-coordinated state equilibrates P^a in (12) with P^a in (12′). Support for this proposal comes from experiments using ^{13}C labelled (11c′). Saturation transfer ^{13}C measurements on solutions of [Rh(dipamp)(11c′)]$^+$ in the presence of free enamide (11c′) showed that the initial flux of magnetisation was from the major (12) to the minor (12′) diastereoisomer rather than to the free enamide (11c′).

2.2. STEREOCHEMISTRY OF THE HYDROGENATED INTERMEDIATES

This is much more difficult to ascertain and there is, as yet, no information on the stereochemistry of (13) and (13′). There are two diastereoisomers for each of the four possible geometric isomers that might result from *cis*-oxidative addition of di-hydrogen to the intermediates (12) and (12′). It is unclear which is first formed. It is also unclear which of the hydride ligands in (13′) is transferred first to give the intermediate (14′). It is known however that (14′) contains H *cis* to the rhodium-carbon bond. H transfer to give the hydrogenated product is thus easy.

3. Spectroscopic measurements under high pressures of gas

There are many industrially important chemical reactions carried out in solution which require high pressures of gas and/or high temperatures. Withdrawal of samples from high pressure reactions, followed by analysis under ambient conditions, obviously suffers from the disadvantage that the species identified under ambient conditions may be quite different from those that exist at high pressures and temperatures. The importance of spectroscopically following such reactions *in situ* is thus immediately apparent.

3.1. IR (HPIR)

During the last 25 years, high pressure infrared, (HPIR), spectroscopy has been the preferred technique for the spectroscopic study of homogeneous catalysis under operating conditions of temperature and pressure. It is now becoming routine in many industrial laboratories. As many of the reactions involve metal carbonyls, HPIR is quite a good method for monitoring the course of the catalytic reaction and for identifying the metal carbonyl species present. The various HPIR cells, which have been used, have been described [19, 20]. One such cell is shown in Figure 5.

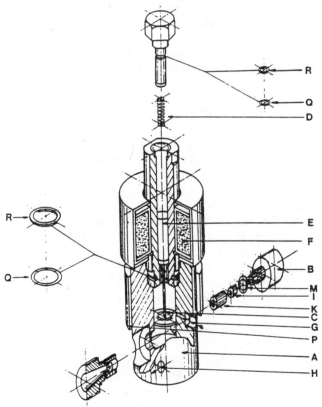

Figure 5. High pressure infrared cell. Exploded view of Type A cell: A, cell body; B, window holder; C, stirrer; D, spring; E, Hastelloy-sheathed soft iron core; F, solenoid; G, entrance port; H, exit port, I, window; K, screw cap; M, window holder 'O' ring; P, band heater; Q, cell 'O' rings; R, anti-extrusion rings.

This cell has the advantage of a flip-flop stirrer which allows efficient mixing of the gas with the solution. The information which can be obtained from HPIR spectra has been significantly enhanced through the advent of Fourier Transform infrared spectroscopy (FTIR). Accurate subtraction of background absorptions is readily performed so useful information can now be obtained in the region associated with bridging carbonyls. This important region of the spectrum is normally obscured by solvent absorptions (Figure 6).

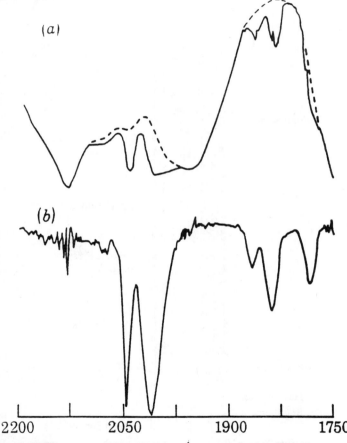

Figure 6. FTIR spectra (2200–1750 cm^{-1}) of [Rh(CO)$_2$(acac)] (0.3 mmol), 2-hydroxypyridine (0.9 mmol), caesium benzoate (0.05 mmol) in 20 cm^3 of N-methylpyrolidine/tetraglyme (1:4) measured at 300 atm. CO/H$_2$ (1:1) and 50°C. (a) Spectrum of reaction mixture (——); reference spectrum of the system in the absence of the rhodium complex (----). (b) Spectrum of the rhodium containing species after subtraction of (----) from (——).

3.2. NMR (HPNMR)

NMR measurements allow much more unambiguous structural and exchange information to be obtained than is possible with IR. We embarked upon the design and construction of an NMR cell to allow measurements under high pressures of gas *ca.* 10 years ago. Our initial HPNMR results were obtained using a modified version of a cell due to Jonas *et al.* [21]. This cell is shown in Figure 7.

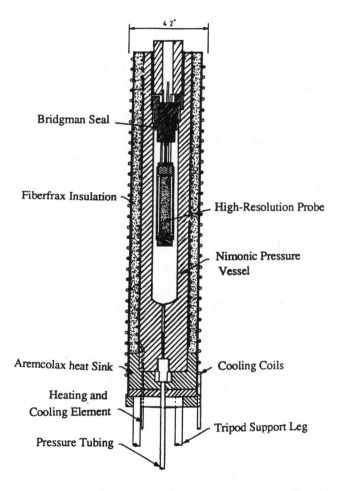

Figure 7. HPNMR cell and probe for single nucleus work.

We chose at that time to study the conversion of CO/H_2 to ethylene glycol and we were able to show [22] by ^{13}C HPNMR that:

$$3[Rh_{12}(CO)_{30}]^{2-} \xrightarrow{\text{1000 bar CO}} 6[Rh_5(CO)_{15}]^- + [Rh_6(CO)_{16}] \qquad (1)$$

At that time, we used a 4-turn Helmholtz coil of copper wire, tuned to a single frequency (^{13}C), which enabled spectra with a S/N ratio of 50:1 and $W_{1/2} \sim 25$ Hz to be obtained. Although the spectra obtained were not of high quality, this was the first report of HPNMR spectra of metal carbonyls and clearly showed that terminal and bridging CO's could be distinguished and that exchange with free CO was slow on the NMR time scale.

TABLE I

Comparison of S/N and $W_{1/2}$ for $P(OCH_2CH_3)_3$ in $CDCl_3$ (38%) in 10 mm NMR tubes and a 5 mm sapphire tube using VSP and BB probes at ambient pressure and the high pressure probe under 140 bar N_2

	VSP	BB	Sapphire/BB	HPNMR probe
S/N				
^{13}C	137	100	< 1	14
^{31}P	1000	600	100	650
$W_{1/2}$(Hz)				
^{13}C	ca. 1		6	5
^{31}P	2.5		12	4

TABLE II

Comparison of 90° pulse length (μs) for $P(OCH_2CH_3)_3$ in $CDCl_3$ (38%) in a 10 mm NMR tube using the VSP probe, a 5 mm sapphire tube in the BB probe and the high pressure probe under 140 bar N_2

	VSP	$P_{90°}$ (μ) Sapphire/BB	HPNMR probe
^1H	16.5	25	45
^{13}C	24	22.5	35
^{31}P	12.5	32.5	22

Since these first HPNMR measurements, we have modified the high pressure probe and cell to achieve:

1) Significant improvements to S/N (Table I). It is now possible to measure *e.g.* ^{31}P NMR spectra of samples approaching catalytic concentrations in Rh (*ca.* 3 mM) in a reasonable time (*ca.* 2h).

2) An order of magnitude improvement in resolution (Table I).

3) Short 90° pulse times (Table II) which allow us to perform multipulse HPNMR experiments.

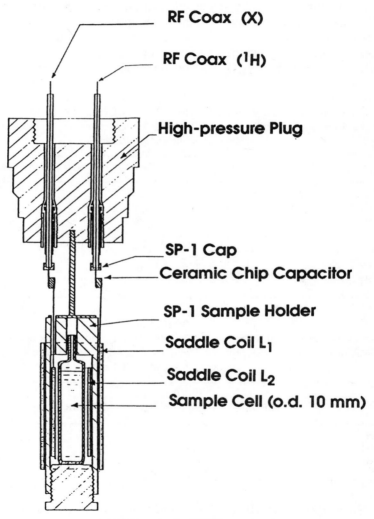

Figure 8. HPNMR probe for X[1H] spectroscopy.

4) Multinuclear (^{1}H, ^{13}C, ^{31}P, ^{59}Co etc.) operation. This allows measurement of more than one nucleus (with or without ^{1}H-decoupling) under the same *in situ* conditions without the need for depressurisation and dismantling.

5) High pressure, high temperature measurements under H_2 without the risk of H-embrittlement. This has been accomplished by the use of a Nimonic 90 pressure vessel rather than the IMI-Titanium (6A1-4V) vessel originally described [22].

These improvements have been achieved by the use of the probe shown in Figure 8 together with the impedance matching circuit shown in Figure 9 [23, 24].

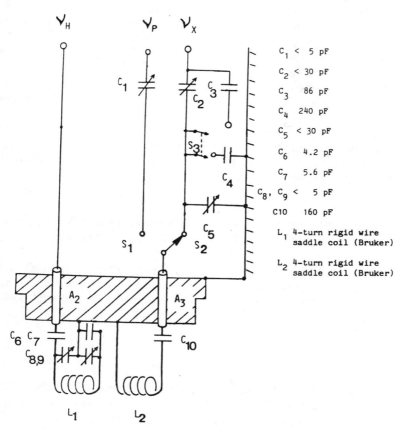

Figure 9. Impedance matching circuit for the probe shown in Fig. 8.

The probe now consists of a 10 mm (o.d.) sample cell within a 4-turn vertical, rigid wire saddle coil (Bruker) within a similar 4-turn saddle coil (Bruker). The inner 4-turn saddle coil can be tuned within the limits 82–29 MHz and the outer 4-turn saddle coil is used for either ^1H measurement or decoupling experiments. Thus, it is now realistically possible to obtain good quality, high resolution, multinuclear NMR spectra under high pressures of gas, (1–700 bar), from –80 to +190°C.

There is no facility within the above cell to allow rapid mixing of gas with the solution. As a result, the reactions are quite slow because of diffusion control of the rate. Vander Velde and Jonas [25] have recently described a high pressure probe for ^1H and ^{13}C measurements. This uses an 8 mm sample tube inside a Berylco-25 pressure vessel. The probe has good S/N and resolution with a 90° pulse length of 15 μs. However each coil is tuned to a single frequency preventing multinuclear work.

Another interesting development in HPNMR is the use of a sapphire NMR tube inside a conventional high resolution probe [26]. Although the burst pressure was found to be *ca.* 700 bar for one 5 mm tube, it is impossible to specify the safe upper limit with any accuracy since the tube's performance is critically dependent upon:

1) The tube's detailed history,
2) The epoxy seal of the sapphire tube onto the valve.

These limitations are serious for very high pressure work but this cell is useful for other measurements despite the low cell volume.

4. HPIR and HPNMR study of the reactions of $[Rh_4(CO)_{12-x}L_x]$, ($x = 1$ to 4; $L = P\{OPh\}_3$) with CO, H_2 or syngas

I now wish to illustrate the complementary use of HPIR and HPNMR in the study of homogeneous catalysis by considering some results we have obtained for rhodium-phosphite alkene hydroformylation catalysts derived from the clusters $[Rh_4(CO)_{12-x}L_x]$, ($x = 1–4$; $L = P\{OPh\}_3$).

4.1. INTRODUCTION

The remarkable catalytic activity of the cobalt sub-group carbonyls and their phosphinated derivatives has spawned a vast literature on the chemistry of these compounds. Of particular relevance to our study are Whyman's [27] and Vidal's [28] HPIR studies of $[Rh_4(CO)_{12}]$ derivatives and Horvath and Bor's HPIR studies of $[M_{4-n}M'_n(CO)_{12}]$, ($M = Co$, $M' = Rh$; $n = 1$ to 3) derivatives [29]. Whyman reports the reversible fragmentation of $[Rh_4(CO)_{12}]$ to $[Rh_2(CO)_8]$ on pressurisation with CO or $CO:H_2$ gas mixtures (Equation (2)).

$$[Rh_4(CO)_{12}] + 4CO \underset{\text{warm}}{\overset{\text{430 bar CO, cool}}{\rightleftharpoons}} 2[Rh_2(CO)_8] \qquad (2)$$

Whyman also studied the effect of phosphine substituents on the reaction (Equation (3)).

$$[Rh_4(CO)_{10}(PPh_3)_2] + 2PPh_3 + 2CO \underset{\text{1 bar } N_2}{\overset{\text{80 bar CO}}{\rightleftharpoons}} 2[Rh_2(CO)_6(PPh_3)_2] \qquad (3)$$

In neither reaction was further fragmentation to mononuclear species on pressurisation with $CO:H_2$ observed. Vidal [28] reports that much more forcing conditions are required to effect this reaction (Equation (4)).

$$[Rh_4(CO)_{12}] + 4CO + 2H_2 \underset{\text{273 to 258 K}}{\overset{\text{1542 bar CO/H}_2}{\rightleftharpoons}} 4[HRh(CO)_4] \qquad (4)$$

More recently several groups have reported the formation of hydrido species in the presence of basic solvents/ligands (Equations (5)–(7)) [30, 31, 37].

$$[Rh_4(CO)_{12}] + 4PPr^i_3 \underset{\text{493 K}}{\overset{\text{443 bar CO/H}_2}{\rightleftharpoons}} 4[HRh(CO)_3P(i\text{-}Pr)_3] \qquad (5)$$

$$[Rh(CO)_2acac] + 4CO + 2H_2 \underset{\text{473 K, NMI}}{\overset{\text{296 bar CO/H}_2}{\longrightarrow}} 4[HRh(CO)_3L'] \qquad (6)$$
$$L' = \text{N-methylimidazole}$$

$$[Rh_4(CO)_{12}] + 4L^* \underset{\text{313 K}}{\overset{\text{12 bar CO/H}_2}{\rightleftharpoons}} 4[HRh(CO)_2L^*_2] \qquad (7)$$
$$L^* = (\text{-})\text{-DIOP}$$
$$\text{or L-EPHOS}$$

Horvath and Bor report fragmentation of the mixed-metal clusters $[M_{4-n}M'_n(CO)_{10}L_2]$, (M=Co, M'=Rh, n = 1 to 3; L=PEt$_3$, CO) to the «heptacarbonyl» type dimers at moderate CO pressure, (Equation (8)) which convert to «octacarbonyl» type dimers at higher pressures, (Equation (9)).

$$[Co_2Rh_2(CO)_{10}L_2] + 2CO \overset{\text{1 bar CO, 298 K}}{\longrightarrow} 2[(OC)_4CoRh(CO)_2L] \qquad (8)$$

$$[(OC)_4CoRh(CO)_2L] + CO \underset{\text{1 bar CO, 298 K}}{\overset{\text{15 bar CO, 278 K}}{\rightleftharpoons}} [(OC)_4CoRh(CO)_3L] \qquad (9)$$
$$L = \text{CO or PEt}_3$$

Such fragmentation/recombination reactions on varying the applied pressure of CO gas and on changing the temperature are to be expected. Consider the following equilibrium (Equation (10)):

$$M_x(CO)_y + zCO \rightleftharpoons M_a(CO)_b + M_c(CO)_d \qquad (10)$$

Approximate metal-metal and metal-CO bond energies might be 371 kJ mol^{-1} and 546 kJ mol^{-1} respectively. These bond energies clearly favour fragmentation at high CO applied pressure. However, there is also an entropy term $T\Delta S$ of *ca.* 118 kJ mol^{-1} of liberated CO at 298 K favouring agglomeration. Clearly this entropy term will become more important as the

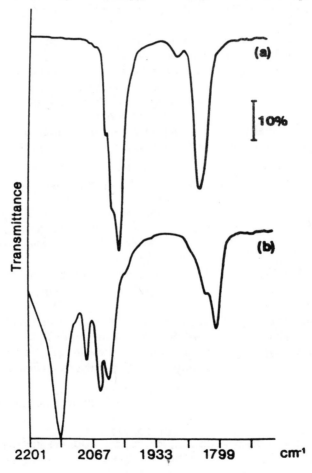

Figure 10. HPIR spectra of $[Rh_4(CO)_8 \{P(OPh)_3\}_4]$ under (a) 1 bar N_2 and (b) 400 bar CO.

temperature is increased. We can therefore say that fragmentation will be favoured at high pCO, agglomeration at high temperature.

It is well known that metal to ligand stoicheiometry affects catalytic activity. We therefore chose to carry out HPIR and HPNMR studies of the fragmentation/recombination reactions of the series of clusters $[Rh_4(CO)_{12-x}\{P(OPh)_3\}_x]$, ($x = 1$–$4$) to determine the effect of rhodium to phosphorus ligand ratio on the species present in solution under high applied pressures of CO and $CO:H_2$ gas mixtures.

4.2. REACTION OF $[Rh_4(CO)_8\{P(OPh)_3\}_4]$ WITH CO

The IR spectrum in the carbonyl stretching region of a 3 mM CH_2Cl_2 solution of $[Rh_4(CO)_8\{P(OPh)_3\}_4]$ under 400 bar applied CO pressure is shown in Figure 10. The assignment of the major absorptions is given in Table III.

TABLE III

High pressure infrared spectroscopic data ($v(CO)/cm^{-1}$) for $[Rh_4(CO)_{12-x}L_x]$ (L = $P(OPh)_3$; $x = 0$, 2, 4) and the reaction products with CO in CH_2Cl_2 solution

$Rh_4(CO)_{12}$	$Rh_4(CO)_{12}$ + 400 bar CO	$Rh_4(CO)_{10}L_2$	$Rh_4(CO)_{10}L_2$ + 400 bar CO	$Rh_4(CO)_8L_4$	$Rh_4(CO)_8L_4$ + 400 bar CO
	2110 m		2099 mw		
	2083 sh	2078 m	2077 sh		2078 s
2070 vs	2070 s				
	2060 sh		2064 vs		
		2054 vs			2052 vs
2040 s	2041 s		2041 vs	2036 w	2033 s
		2029 s		2022 sh	
				2008 vs	
1875 s	1874 ms				1884 w
	1848 sh	1850 ms	1855 m		
		1842 ms	1840 m	1836 s	
	1829 m				1825 m
			1815 m		
					1803 s
Assignments	$Rh_2(CO)_8$ + $Rh_4(CO)_{12}$		$Rh_2(CO)_7L$ + $Rh_4(CO)_{11}L$		$Rh_2(CO)_6L_2$

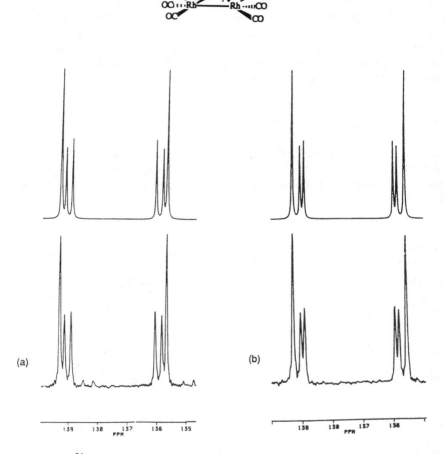

Figure 11. ^{31}P HPNMR spectra of [Rh$_4$(CO)$_8$ {P(OPh)$_3$}$_4$ at (a) 213 K, (b) 260 K. Simulated spectra are given above the experimental spectra.

The ^{31}P NMR spectrum of the same cluster (120 mM) in CDCl$_3$ under 400 bar applied CO pressure is shown in Figure 11. A similar ^{31}P NMR spectrum was obtained from a 3 mM solution of the same cluster under the same conditions of temperature and pressure. Our assignment of the resonances is given in Table IV. Both spectra are consistent

TABLE IV
^{31}P NMR spectroscopic data for the complexes

	δ(P)/p.p.m.[a]	^{1}J(RhP)/Hz[b]	^{2}J(RhP)/Hz	^{3}J(PP)/Hz
$Rh_4(CO)_{11}$ [P(OPh)$_3$][c]	100.9	205	–	
$Rh_4(CO)_{10}$ [P(OPh)$_3$]$_2$[c]	102.2, 116.6	212, 260	–	
$Rh_4(CO)_9$ [P(OPh)$_3$]$_3$[c]	102.2, 118.0	215, 263	–	167
$Rh_4(CO)_8$ [P(OPh)$_3$]$_4$[d]	103.4, 121.2	213, 267	–	121
$Rh_2(CO)_7$ [P(OPh)$_3$][d,e]	137.4	279	30	–
$Rh_2(CO)_6$ [P(OPh)$_3$]$_2$[d,f]	138.0	277.7	25.5	10.4
$Rh_2(CO)_6$ [P(OPh)$_3$]$_2$[f,g]	138.0	271.9	22.0	18.0

[a] Chemical shift of axial phosphite first.
[b] First value is ^{1}J(RhP$_{ax}$); second value is ^{1}J(RhP$_{rad}$).
[c] CD$_2$Cl$_2$, 221 K.
[d] CDCl$_3$, 213 K.
[e] 250 bar CO.
[f] 300 bar CO.
[g] CDCl$_3$, 260 K.

with the complete conversion (> 95%) of [Rh$_4$(CO)$_8${P(OPh)$_3$}$_4$] to [Rh$_2$(CO)$_6${P(OPh)$_3$}$_2$] (Equation (11)).

$$[Rh_4(CO)_8\{P(OPh)_3\}_4] \xrightarrow{\text{300-400 bar CO}} 2[Rh_2(CO)_6\{P(OPh)_3\}_2] \qquad (11)$$

Assignment of the HPIR spectrum relies on comparisons with the known IR absorptions of [Co$_2$(CO)$_6$(PEt$_3$)$_2$] [32]. By contrast the additional information obtainable from NMR spin-spin couplings allows us to unambiguously interpret the ^{31}P NMR spectrum in terms of a [Rh$_2$(CO)$_6$L$_2$] species. Thus, the spectrum is that of a four spin AA'XX' (AA' = ^{31}P; XX' = ^{103}Rh) system and can be accurately simulated using the PANIC [33] programme, Figure 11. The coupling constants derived from the simulation are given in Table IV. The values of ^{1}J(Rh–P), ^{2}J(Rh–P) and ^{3}J(P–P) are most reasonably interpreted in terms of a CO-bridged, symmetrical structure such as that shown in Figure 11. That the structure is symmetrical follows from the AA'XX' nature of the spin system. This requires that the phosphorus atoms are chemically but not magnetically equivalent, similarly the rhodium atoms. The presence of bridging carbonyl groups is clearly demonstrated by the HPIR bands at 1825 and 1803 cm^{-1}. The formation of a carbonyl bridged species in this reaction of a *phosphite* substituted cluster contrasts with the isolation of non-CO-bridged dimers from the reaction of CO with a range of tertiary *phosphine* substituted rhodium carbonyls [27]. However bridged structures have previously been proposed for both $[Rh_2(CO)_8]$ [27] and $[RhCo(CO)_8]$ [29] and found by X-ray structure determination for $[Rh_2(CO)_2(triphos)_2]$, [triphos = $MeC(CH_2PPh_2)_3$] [35].

It should be noted that under these conditions (indeed in all this work) free phosphite is not observed.

Cleavage of $[Rh_4(CO)_8\{P(OPh_3\}_4]$ to give $[Rh_2(CO)_6\{P(OPh)_3\}_2]$ is reasonable both chemically and statistically. We have previously shown that the cluster $[Rh_4(CO)_8\{P(OPh)_3\}_4]$ undergoes a fluxional process that equivalences the $[Rh(CO)_2L]$-fragments [36].

4.3. REACTION OF $[Rh_4(CO)_{11}\{P(OPh)_3\}]$ WITH CO

Again we can predict unambiguously the likely fragmentation reaction (Equation (12)).

$$[Rh_4(CO)_{11}\{P(OPh)_3\}] \xrightarrow{\text{CO}} [Rh_2(CO)_7\{P(OPh)_3\}] + [Rh_2(CO)_8] \quad (12)$$

This fragmentation is confirmed by the ^{31}P NMR spectrum, Figure 12. A simple AX_2 doublet of doublets is observed, consistent with the structure shown in Figure 12.

We have also studied the recombination reaction in Equation (13):

$$[Rh_2(CO)_7\{P(OPh)_3\}] + [Rh_2(CO)_8] \longrightarrow \begin{array}{c} [Rh_4(CO)_{12}] \\ + \\ [Rh_4(CO)_{11}\{P(OPh)_3\}] \\ + \\ [Rh_4(CO)_{10}\{P(OPh)_3\}_2] \end{array} \quad (13)$$

Several recombination reactions are possible to give Rh_4 clusters containing any of zero, one and two phosphite ligands. The last two are observed by ^{31}P NMR confirming the identity of the fragmented species as the mono-substituted dimer $[Rh_2(CO)_7\{P(OPh)_3\}]$. $[Rh_4(CO)_{12}]$ of course cannot be observed by ^{31}P NMR. Although it could be observed by ^{13}C NMR this would require the use of isotopically enriched materials. Clearly a better solution would be to measure the IR spectrum of the recombined solution. IR bands characteristic of $[Rh_4(CO)_{12}]$ are observed demonstrating the complementary nature of the two techniques HPNMR and IR.

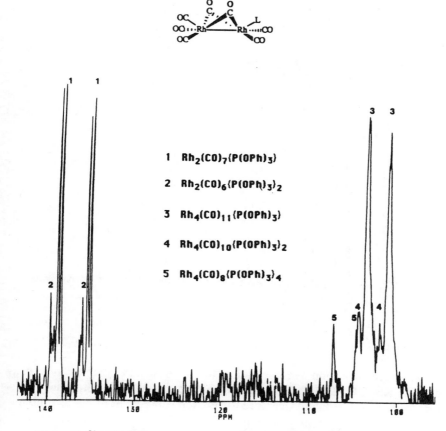

Figure 12. ^{31}P HPNMR spectrum of [Rh$_4$(CO)$_{11}$ {P(OPh)$_3$}] under 400 bar CO.

4.4. REACTION OF [Rh$_4$(CO)$_9${P(OPh)$_3$}$_3$] WITH CO

Equation (14) predicts fragmentation to equal amounts of the mono- and di-substituted dimers [Rh$_2$(CO)$_7${P(OPh)$_3$}] and < [Rh$_2$(CO)$_6${P(OPh)$_3$}$_2$].

$$[Rh_4(CO)_9\{P(OPh)_3\}_3] \xrightarrow{\text{CO}} [Rh_2(CO)_7\{P(OPh)_3\}] + [Rh_2(CO)_6\{P(OPh)_3\}_2] \quad (14)$$

Surprisingly, the fragmentation reaction gives predominantly the di-substituted dimer, [Rh$_2$(CO)$_6${P(OPh)$_3$}$_2$], Figure 13.

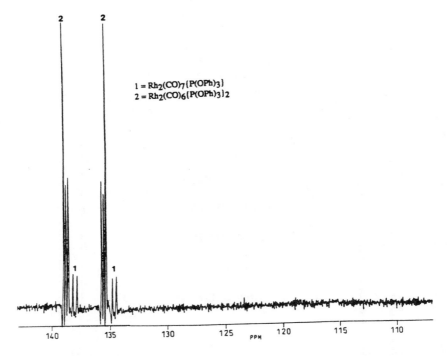

Figure 13. ^{31}P HPNMR spectrum of $[Rh_4(CO)_9 \{P(OPh)_3\}_3]$ under 400 bar CO.

On depressurisation recombination to predominantly the trisubstituted cluster $[Rh_4(CO)_9\{P(OPh)_3\}_3]$ occurs, Figure 14. We do not believe this surprising result is due to ligand redistribution subsequent to the recombination reaction since we have found such redistributions to be slow and generally give mixtures [34].

Figure 14 also shows the evolution with time of a solution of $[Rh_4(CO)_9\{P(OPh)_3\}_3]$ which has been fragmented, depressurised and allowed to recombine. An interesting feature here is the appearance of a new set of resonances at *ca.* 128 ppm. These resonances initially gain intensity and then decline as the recombination proceeds towards completion. Although we have been unable to obtain a high resolution spectrum of this new species (it is either too fluxional or too reactive) we suggest that it is a decarbonylated dimer, either $[Rh_2(CO)_5\{P(OPh)_3\}_2]$ or $[Rh_2(CO)_6\{P(OPh)_3\}]$. Horvath [29] has reported similar Co-Rh dimers, $[CoRh(CO)_6L]$.

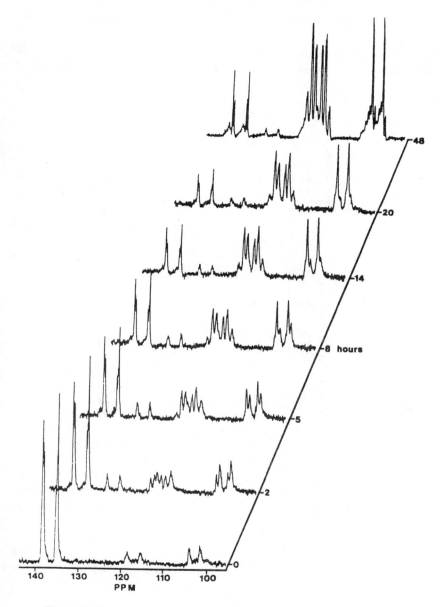

Figure 14. Recombination on depressurisation of the solution from Fig. 13.

4.5. FLUXIONALITY IN [Rh$_2$(CO)$_6${P(OPh)$_3$}$_2$] AND [Rh$_2$(CO)$_7${P(OPh)$_3$}]

NMR spectroscopy allows the study of fluxional processes occurring on the micro-second to seconds timescale. As detailed earlier most often this is

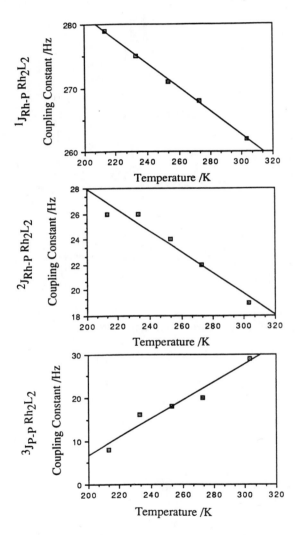

Figure 15. Variation in coupling constants with temperature for [Rh$_2$(CO)$_6$ {P(OPh)$_3$}$_2$].

done by monitoring line-shapes with changing temperature or by saturation transfer. However where a ligand is exchanging between two sites having different couplings to the metal nucleus and bond scission does not occur in the exchange process we can detect the fluxional process since the observed coupling will be the weighted average of the couplings between the ligand and the metal in the two sites.

The ^{31}P NMR spectra of $[Rh_2(CO)_7\{P(OPh)_3\}]$ and $[Rh_2(CO)_6\{P(OPh)_3\}_2]$ are temperature dependent. Figure 11 shows the ^{31}P NMR spectrum of $[Rh_2(CO)_6\{P(OPh)_3\}_2]$ recorded at 213 K and at 260 K together with the simulated spectra. The variations in coupling constants over the temperature range 213 to 300 K for the two complexes are shown graphically in Figures 15 and 16.

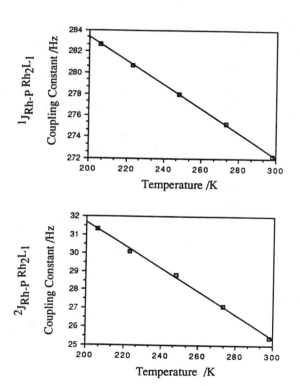

Figure 16. Variation in coupling constants with temperature for $[Rh_2(CO)_7\{P(OPh)_3\}]$.

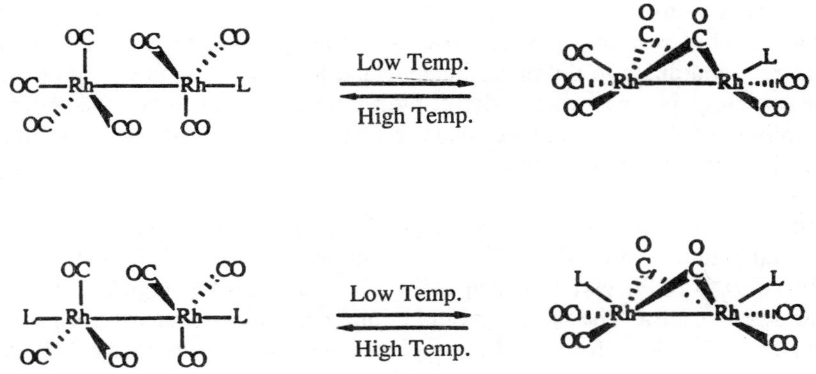

D. Brown et al.

Figure 17. Fluxional process in $[Rh_2(CO)_{8-x}L_x]$, $(x = 1,2)$ derivatives.

The major change is the increase in $^3J(P\text{–}P)$ from 10 to 28 Hz with increasing temperature, Table IV. We interpret this in terms of a fluxional process such as that shown in Figure 17.

Such a process brings the two phosphorus atoms into a transoid orientation across the Rh–Rh bond and can thus account for the increase in $^3J(P\text{–}P)$. This process is entirely analogous to that proposed to account for the equivalencing of CO groups in $[M_2(CO)_8]$, $(M = Co, Rh)$. Further support for this proposal comes from the ^{13}C NMR spectrum of $[Rh_2(CO)_6\{P(OPh)_3\}_2]$, Figure 18. After cooling to 213 K to prevent back reaction and depressurisation two resonances are observed of roughly equal intensity at 201 ppm and 196.5 ppm, $^1J(Rh\text{–}C)$ 29.4 Hz. The first of these is a broadened multiplet which we assign to $[Rh_2(CO)_6\{P(OPh)_3\}_2]$ the breadth of the resonance being due to coupling to both ^{31}P and ^{103}Rh. The resonance at higher field we assign to $[Rh_2(CO)_8]$, the triplet structure being due to coupling to ^{103}Rh. Both species undergo a fluxional process that averages the carbonyl sites even at 213 K, Figure 18, and would account for the temperature dependence of $^3J(P\text{–}P)$.

The results of our HPIR and HPNMR studies on this catalytic system are summarised in Scheme 7 below.

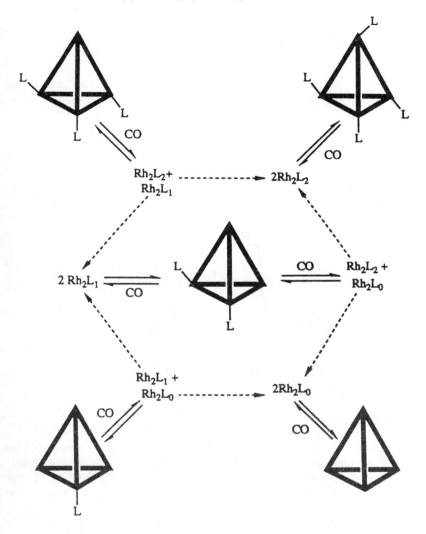

$L = P(OPh)_3$

- - - - - - ➤ = additional routes for dimer recombination.

Scheme 7. Fragmentation and recombination reactions accessible to $[Rh_4(CO)_{12-x}L_x]$ complexes determined by HPIR and HPNMR spectroscopies.

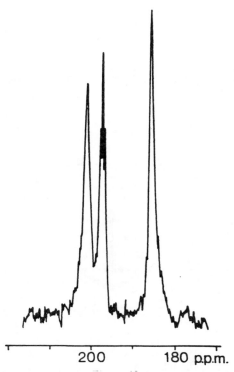

Figure 18. ^{13}C HPNMR spectrum of $[Rh_4(CO)_{10}\{P(OPh)_3\}_2]$ after fragmentation as described in the text.

In contrast to the HPIR results, the ^{31}P NMR spectra clearly reveal that up to 50% of the phosphorus may be present as Rh_4 species under the conditions of temperature and pressure used in these studies depending on the degree of phosphite substitution in the starting rhodium cluster. The higher the degree of phosphite substitution, the greater the proportion of fragmented cluster. This disparity between HPIR and HPNMR is readily explained:

(i) the intensity of NMR absorptions is proportional to the number of nuclei giving rise to the absorption whereas no such simple relation governs the intensity of IR absorptions;

(ii) the greater resolving power of NMR due to narrower linewidths and greater dispersion than in IR allows easier detection of the separate components in the mixture.

Essentially all fragmentation and recombination equilibria shown in the scheme are accessible to the catalytic system regardless of the point of entry. The major components present in a given catalytic system were found to depend on:

(i) rhodium to phosphite ratio;

(ii) rhodium/phosphite to carbon monoxide ratio;

(iii) temperature;

(iv) concentration.

It is therefore important to mimic the operating conditions of temperature, pressure and concentration as closely as possible in any study of homogeneous catalysis.

5. Conclusions

HPNMR and HPIR are complementary techniques. HPIR is particularly valuable in detecting species present in low concentration and carbonyl complexes that do not also contain phosporus ligands when C-13 labelling would be required for an NMR study. HPNMR gives detailed structural information *via* chemical shifts and couplings, is quantitative and can be used to study fluxional processes.

Recently introduced NMR techniques such as saturation transfer via DANTE selective excitation, the use of *para*-H_2 and the measurement of spectra under high pressure of a reactive gas have increased our understanding of homogeneously catalysed reactions. The use of such techniques is likely to increase in the coming years. An awareness of such techniques and an understanding of their potential is therefore vital to the chemist wishing to study homogeneous catalysis.

Acknowledgements

We would like to thank SERC for financial support (to D. Brown) and for a cooperative grant with I.C.I. We also thank Dr. R. Whyman for fruitful discussions and for recording the HPIR spectra.

University of Liverpool
Department of Chemistry
Donnan and Robert Robinson Laboratories
Liverpool L69 3BX (England)

References

1. Pruett, R.L. *Ann. N. Y. Acad. Sci.* **1977**, *295*, 239.
2. Knowles, W.S. *Acc. Chem. Res.* **1983**, *16*, 106.
3. Roth, J.F.; Craddock, J.H.; Hersham, A. and Paulik, F.E. *Chem. Tech.* **1971**, *1*, 600.
4. Coover, H.W. and Hart, R.C. *Chem. Eng. Progr.* **1982**, 72.
5. Grassi, M.; Mann, B.E.; Pickup, B.T. and Spencer, C.M. *J. Mag. Res.* **1986**, *69*, 92, and references therein.
6. Morris, G.A. and Freeman, R. *J. Mag. Res.* **1978**, *29*, 433.
7. Allevi, C.; Heaton, B.T.; Seregni, C.; Strona, L.; Goodfellow, R.J.; Chini, P. and Martinengo, S. *J. Chem. Soc. Dalton Trans.* **1986**, 1375 and references therein.
8. Allevi, C.; Bordoni, S.; Clavering, C.P.; Heaton, B.T. and Iggo, J.A. *Organometallics* **1989**, *8*, 385.
9. Elsevier, C.J.; Ernsting, J.M. and de Lange, W.G.J. *J. Chem. Soc. Chem. Comm.* **1989**, 585.
10. Osborne, J.A.; Jardine, F.H.; Wilkinson, G. and Young, J.F. *J. Chem. Soc. A* **1966**, 1711.
11. Halpern, J. *Inorg. Chim. Acta* **1981**, *50*, 11.
12. Brown, J.M.; Evans, P.L. and Lucy, A.R. *J. Chem. Soc. Perkin Trans. II* **1987**, 1589.
13. Bowers, R. and Weitekamp, D.P. *J. Am. Chem. Soc.* **1987**, *109*, 5541.
14. Tolman, C.A. and Faller, J.W. *Metal-Phosphine Complexes in Homogeneous Catalysis*; L.H. Pignolet, Plenum Press, New York, 1984; pp 13 and references therein.
15. Brown, J.M.; Canning, L.R.; Downs, A.J. and Forster, A.M. *J. Organometal. Chem.* **1983**, *255*, 103.
16. Kagan, H.B. *Comp. Organomet. Chem.*, 8; Wilkinson, G. and Stone, F.G.A., Pergamon, Oxford, 1982; 463.
17. Landis, C.R. and Halpern, J. *J. Am. Chem. Soc.* **1987**, *109*, 1746.
18. Brown, J.M.; Chaloner, P.A. and Morris, G.A. *J. Chem. Soc. Perkin II* **1987**, 1583.
19. Whyman, R. *Laboratory Methods in Infrared Spectroscopy*; R. G. J. Miller and B.C. Stace, Heyden, London, 1972; pp 149.
20. Whyman, R. *Laboratory Methods in Vibrational Spectroscopy*; Willis, H.A.;Van der Maas and Miller, R. G., J. Wiley, 1987.
21. Jonas, J.; Hasha, D.L.; Lamb, W.J.; Hoffman, G.A. and Eguchi, T. *J. Mag. Res.* **1981**, *42*, 169, and references therein.
22. Heaton, B.T.; Strona, L.; Jonas, J.; Eguchi, T. and Hoffman, G.A. *J. Chem. Soc. Dalton Trans.* **1982**, 1159.
23. Iggo, J.A. unpublished results.
24. Fukushima, E. and Roeder, S.B.W. *Experimental Pulse NMR: A Nuts and Bolts Approach*; Addison-Wesley Publishing Company, Reading, Massachusetts, U.S.A., 1981.
25. Van der Velde, D.G. and Jonas, J. *J. Mag. Res.* **1987**, *71*, 480.
26. Roe, D.C. *J. Mag. Res.* **1985**, *63*, 388; Horvath, I.T.; Kastrup, R.V.; Oswald, A.A. and Mozeleski, E.J. *Catal. Letts.* **1989**, *2*, 85.
27. Whyman, R. *J. Chem. Soc. Dalton Trans.* **1972**, 1375.
28. Vidal, J.L. and Walker, W.E. *Inorg. Chem.* **1981**, *20*, 249.
29. Horvath, I.T.; Bor, G.; Garland, M. and Pino, P. *Organometallics* **1986**, *5*, 1441.
30. Kiso, Y.; Tanaka, M.; Hayashi, T. and Saeki, K. *J. Organometal. Chem.* **1987**, *322*, C32.

31. Pottier, Y.; Mortreux, A. and Petit, F. *J. Organometal. Chem.* **1989**, *370*, 333.
32. Capron-Cotigny, G. and Poilblanc, R. *Bull. Soc. Chim. France* **1967**, *4*, 1440.
33. PANIC is a spectral simulation programme supplied by Bruker Spectrospin Limited.
34. Brown, D.T.; Heaton, B.T. and Iggo, J.A. unpublished results.
35. Allevi, C.; Golding, M.; Heaton B. T.; Ghilardi, C.A.; Midollini, S. and Orlandi, A. *J. Organometal. Chem.* **1987**, *326*, C19.
36. Heaton, B.T.; Longhetti, L.; Garleschelli, L. and Sartorelli, U. *J. Organometal. Chem.* **1980**, *192*, 431.
37. Tamura, M.; Ishino, M.; Deguchi, T. and Nakamura, S. *J. Organometal. Chem.* **1986**, *312*, C75.
38. Rosas, N. ; Marquez, C.; Hernandez, H. and Gomez, R. *J. Mol. Catal.* **1988**, *48*, 59 and references therein.
39. Csontos, G. ; Heil, B.; Markó, L. and Chini, P. *Hung. J. Ind. Chem.* **1973**, *1*, 53; Csontos, G.; Heil, B. and Markó, L. *Ann. N.Y. Acad. Sci.* **1974**, *239*, 47.

INDEX

Catalysis by Metal Complexes

Series Editors:
R. Ugo, *University of Milan, Milan, Italy*
B. R. James, *University of British Columbia, Vancouver, Canada*

1.* F. J. McQuillin: *Homogeneous Hydrogenation in Organic Chemistry.* 1976
ISBN 90-277-0646-8

2. P. M. Henry: *Palladium Catalyzed Oxidation of Hydrocarbons.* 1980
ISBN 90-277-0986-6

3. R. A. Sheldon: *Chemicals from Synthesis Gas.* Catalytic Reactions of CO and H_2. 1983
ISBN 90-277-1489-4

4. W. Keim (ed.): *Catalysis in C_1 Chemistry.* 1983
ISBN 90-277-1527-0

5. A. E. Shilov: *Activation of Saturated Hydrocarbons by Transition Metal Complexes.* 1984
ISBN 90-277-1628-5

6. F. R. Hartley: *Supported Metal Complexes.* A New Generation of Catalysts. 1985
ISBN 90-277-1855-5

7. Y. Iwasawa (ed.): *Tailored Metal Catalysts.* 1986
ISBN 90-277-1866-0

8. R. S. Dickson: *Homogeneous Catalysis with Compounds of Rhodium and Iridium.* 1985
ISBN 90-277-1880-6

9. To be announced later.

10. A. Mortreux and F. Petit (eds.): *Industrial Applications of Homogeneous Catalysis.* 1988
ISBN 90-277-2520-9

11. N. Farrell: *Transition Metal Complexes as Drugs and Chemotherapeutic Agents.* 1989
ISBN 90-277-2828-3

12. A.F. Noels, M. Graziani and A.J. Hubert (eds.): *Metal Promoted Selectivity in Organic Synthesis.* 1991
ISBN 0-7923-1184-1

Kluwer Academic Publishers – Dordrecht / Boston / London